AF696255

de Gruyter Studies in Mathematics 31

Editors: Carlos Kenig · Andrew Ranicki · Michael Röckner

de Gruyter Studies in Mathematics

1 Riemannian Geometry, 2nd rev. ed., *Wilhelm P. A. Klingenberg*
2 Semimartingales, *Michel Métivier*
3 Holomorphic Functions of Several Variables, *Ludger Kaup and Burchard Kaup*
4 Spaces of Measures, *Corneliu Constantinescu*
5 Knots, 2nd rev. and ext. ed., *Gerhard Burde and Heiner Zieschang*
6 Ergodic Theorems, *Ulrich Krengel*
7 Mathematical Theory of Statistics, *Helmut Strasser*
8 Transformation Groups, *Tammo tom Dieck*
9 Gibbs Measures and Phase Transitions, *Hans-Otto Georgii*
10 Analyticity in Infinite Dimensional Spaces, *Michel Hervé*
11 Elementary Geometry in Hyperbolic Space, *Werner Fenchel*
12 Transcendental Numbers, *Andrei B. Shidlovskii*
13 Ordinary Differential Equations, *Herbert Amann*
14 Dirichlet Forms and Analysis on Wiener Space, *Nicolas Bouleau and Francis Hirsch*
15 Nevanlinna Theory and Complex Differential Equations, *Ilpo Laine*
16 Rational Iteration, *Norbert Steinmetz*
17 Korovkin-type Approximation Theory and its Applications, *Francesco Altomare and Michele Campiti*
18 Quantum Invariants of Knots and 3-Manifolds, *Vladimir G. Turaev*
19 Dirichlet Forms and Symmetric Markov Processes, *Masatoshi Fukushima, Yoichi Oshima and Masayoshi Takeda*
20 Harmonic Analysis of Probability Measures on Hypergroups, *Walter R. Bloom and Herbert Heyer*
21 Potential Theory on Infinite-Dimensional Abelian Groups, *Alexander Bendikov*
22 Methods of Noncommutative Analysis, *Vladimir E. Nazaikinskii, Victor E. Shatalov and Boris Yu. Sternin*
23 Probability Theory, *Heinz Bauer*
24 Variational Methods for Potential Operator Equations, *Jan Chabrowski*
25 The Structure of Compact Groups, *Karl H. Hofmann and Sidney A. Morris*
26 Measure and Integration Theory, *Heinz Bauer*
27 Stochastic Finance, *Hans Föllmer and Alexander Schied*
28 Painlevé Differential Equations in the Complex Plane, *Valerii I. Gromak, Ilpo Laine and Shun Shimomura*
29 Discontinuous Groups of Isometries in the Hyperbolic Plane, *Werner Fenchel and Jakob Nielsen*
30 The Reidemeister Torsion of 3-Manifolds, *Liviu I. Nicolaescu*

Susanne Schmitt · Horst G. Zimmer

Elliptic Curves

A Computational Approach

With an Appendix by Attila Pethö

Walter de Gruyter
Berlin · New York

Authors

Susanne Schmitt
Max-Planck-Institut für Informatik (MPII)
Algorithms and Complexity Group (AG1)
Stuhlsatzenhausweg 85
66123 Saarbrücken
Germany
E-Mail: sschmitt@mpi-sb.mpg.de

Horst Günter Zimmer
Fachbereich Mathematik (Bau 27.1)
Universität des Saarlandes
Postfach 15 11 50
66041 Saarbrücken
Germany
E-Mail: zimmer@math.uni-sb.de

Series Editors

Carlos E. Kenig
Department of Mathematics
University of Chicago
5734 University Ave
Chicago, IL 60637
USA

Andrew Ranicki
Department of Mathematics
University of Edinburgh
Mayfield Road
Edinburgh EH9 3JZ
Scotland

Michael Röckner
Fakultät für Mathematik
Universität Bielefeld
Universitätsstraße 25
33615 Bielefeld
Germany

Mathematics Subject Classification 2000: 11-01, 14-01; 11G05, 11G07, 11-04, 11G20, 11G40, 11G50, 14-04, 14H52

Keywords: elliptic curves, Weierstrass function, height, torsion-group, rank, Mordell–Weil group, algorithms, cryptography

♾ Printed on acid-free paper which falls within the guidelines of the ANSI to ensure permanence and durability.

Library of Congress – Cataloging-in-Publication Data

Schmitt, Susanne.
Elliptic curves : a computational approach / Susanne Schmitt, Horst G. Zimmer ; with an appendix by Attila Pethö.
p. cm. – (De Gruyter studies in mathematics ; 31)
Includes bibliographical references and index.
ISBN 3-11-016808-1 (acid-free paper)
1. Curves, Elliptic. I. Zimmer, Horst G. II. Title. III. Series.
QA567.2.E44S35 2003
516.3'52–dc21 2002041531

ISBN 3-11-016808-1

Bibliographic information published by Die Deutsche Bibliothek

Die Deutsche Bibliothek lists this publication in the Deutsche Nationalbibliografie; detailed bibliographic data is available in the Internet at <http://dnb.ddb.de>.

© Copyright 2003 by Walter de Gruyter GmbH & Co. KG, 10785 Berlin, Germany.
All rights reserved, including those of translation into foreign languages. No part of this book may be reproduced in any form or by any means, electronic or mechanical, including photocopy, recording, or any information storage and retrieval system, without permission in writing from the publisher.
Printed in Germany.

Cover design: Rudolf Hübler, Berlin.
Typeset using the authors' TEX files: I. Zimmermann, Freiburg.
Printing and binding: Hubert & Co. GmbH & Co. KG, Göttingen.

Preface

The purpose of the present text is to give an elementary introduction to the arithmetic of elliptic curves over number fields from a computational point of view. The text is therefore equipped with the corresponding algorithms. Many examples and, of course, some exercises are added. In all those regards we make extensive use of the availability of computers.

The arithmetic of elliptic curves can become almost arbitrarily sophisticated. That is why an elementary introduction requires a certain restriction on the selection of topics treated. However, the most important topics are dealt with in the book. They include the determination of torsion groups, computations concerning the Mordell–Weil group, especially concerning the rank and a basis of that group, height calculations, and the determination of integral and, more generally, S-integral points. To avoid overlapping we occasionally, instead of giving proofs, cite the books [204], [207] of Silverman. For some more exercises, we also refer to the books of Silverman which contain a lot of exercises related to the present text as well. We should also mention that there are some survey articles on this topic, for instance those of Frey [69], Stroeker [216], Zagier [245] and the second author [254], [255]

Elliptic curves admit many applications, in pure mathematics and in computer science. They play an important role in the proof by Faltings of Mordell's conjecture and in the proof by Wiles of Fermat's last theorem on the one hand, and they are used e.g. for factoring integers, deriving properties of prime numbers, and in cryptography on the other (see Jacobson, Menezes, Stein [104]). Thus cryptography enters the picture here too.

One aim of the book is also to inform about results obtained in the research group of the second author. Specifically, the research group was developing the computer algebra system SIMATH. This system will be further developed from the year 2002 on by Ken Nakamula and his group in Tokyo and by others.

The SIMATH package focuses on elliptic curves. Therefore it plays an important part in the book: The algorithms are implemented in SIMATH and the examples are produced by SIMATH. Of course, among other things, SIMATH contains all the elementary algorithms used in the book. More about SIMATH can be found on `http://diana.math.uni-sb.de/~simath/`.

Elliptic curves are a special case of diophantine equations. It is therefore in order to briefly consider in an appendix how to solve general diophantine equations. The LLL-algorithm plays an essential role here. In addition, lower bounds for linear forms in logarithms are important. Finally, the p-adic methods used are outlined in Appendix A.

We hope that the text will be read by mathematicians as well as computer scientists, and scientists from industry. In view of the applications in cryptography, it is especially important for computer scientists to know some elements of the arithmetic of elliptic curves.

The bibliography does not comprise *all* the books or papers relevant to the field of elliptic curves because it was nearly impossible in view of its extend.

The authors wish to express their heartfelt thanks to F. Lemmermeyer from the University of San Marcos, and M. Kida from the University of Electro-Communication Chofu, Tokyo, for their detailed and helpful reading of the manuscript. F. Lemmermeyer and M. Kida also made a number of suggestions for improvements and extensions. Many of them could be incorporated in the text.

A. Pethő has not only written Appendix A but also essentially contributed to the manuscript. Therefore, thanks are due to him too.

Last but not least we thank Dr. M. Karbe from the de Gruyter Verlag for his permanent engagement for the manuscript and for his patience with the authors.

Saarbrücken, September 2003

Susanne Schmitt
Horst-Günter Zimmer

Contents

Chapter 1
Elliptic curves

This chapter serves as an introduction to the theory of elliptic curves over arbitrary fields. We define several normal forms, the addition law and the multiplication formulas. Then we present first applications of elliptic curves: factorization and primality testing. In the last section we define isogenies and endomorphisms.

Throughout this book we denote the ring of rational integers by $\mathbb{Z}$, the set of prime numbers by $\mathbb{P}$, the positive integers by $\mathbb{N}$, whereas $\mathbb{N}_0$ is the set of non negative integers. The finite field with q elements is $\mathbb{F}_q$. We denote the field of rational numbers by $\mathbb{Q}$, the field of real numbers by $\mathbb{R}$, and the field of complex numbers by $\mathbb{C}$.

Further, in this chapter, $\mathbb{K}$ is an arbitrary field with algebraic closure $\overline{\mathbb{K}}$ and characteristic $\operatorname{char}(\mathbb{K})$.

1.1 Normal forms

Elliptic curves can be given in several normal forms. In this section we introduce Weierstraß normal forms and the Legendre normal form. We also define some constants related to the equations and look at birational transformations.

We first recall some basic facts from algebraic geometry (see for example Fulton [73], Mumford [151], Shafarevich [199]).

Definition 1.1. a) The *affine n-space* is $\mathbb{A}^n := \overline{\mathbb{K}}^n$. The affine n-space over $\mathbb{K}$ is $\mathbb{A}^n(\mathbb{K}) := \mathbb{K}^n$.

b) Two elements $(x_0, \ldots, x_n), (x_0', \ldots, x_n') \in \mathbb{A}^{n+1}(\mathbb{K})$ are *equivalent* if there exists a non-zero $\lambda \in \mathbb{K}$ with

$$x_i = \lambda x_i' \quad \text{for } i = 0, \ldots, n.$$

The equivalence class of $(x_0, \ldots, x_n)$ for this equivalence relation is written as $[x_0 : \ldots : x_n]$.

c) The *projective n-space* is

$$\mathbb{P}^n := \{[x_0 : \ldots : x_n] \in \overline{\mathbb{K}}^{n+1} : \text{ not all } x_i = 0\},$$

the projective n-space over $\mathbb{K}$ is

$$\mathbb{P}^n(\mathbb{K}) := \{[x_0 : \ldots : x_n] \in \mathbb{K}^{n+1} : \text{ not all } x_i = 0\}.$$

d) A polynomial $F \in \mathbb{K}[X, Y, Z]$ is called *homogeneous of degree d* if

$$F(\lambda X, \lambda Y, \lambda Z) = \lambda^d F(X, Y, Z) \quad \text{for all } \lambda \in \mathbb{K}.$$

e) A polynomial $f(X, Y) \in \mathbb{K}[X, Y]$ of total degree d can be *homogenised* by defining

$$F(X, Y, Z) := Z^d f\left(\frac{X}{Z}, \frac{Y}{Z}\right) \in \mathbb{K}[X, Y, Z].$$

A homogeneous polynomial $F(X, Y, Z) \in \mathbb{K}[X, Y, Z]$ can be *dehomogenised* by defining

$$f(X, Y) := F(X, Y, 1) \in \mathbb{K}[X, Y].$$

f) A *plane projective algebraic curve* over $\mathbb{K}$ is the set of roots in $\overline{\mathbb{K}}$ of a non-constant homogeneous polynomial $F(X, Y, Z) \in \mathbb{K}[X, Y, Z]$,

$$C := C(F) = \{[x : y : z] \in \mathbb{P}^2 : F(x, y, z) = 0\}.$$

We define

$$C(\mathbb{K}) := C(F)(\mathbb{K}) = \{[x : y : z] \in \mathbb{P}^2(\mathbb{K}) : F(x, y, z) = 0\}$$

the set of $\mathbb{K}$-rational points of C. A *point at infinity* of this curve is a point $P = [x : y : z] \in C$ with $z = 0$.

g) A *plane affine algebraic curve* over $\mathbb{K}$ is the set of roots in $\overline{\mathbb{K}}$ of a non-constant polynomial $f(X, Y) \in \mathbb{K}[X, Y]$,

$$C := C(f) = \{(x, y) \in \mathbb{A}^2 : f(x, y) = 0\}.$$

We define

$$C(\mathbb{K}) := C(f)(\mathbb{K}) = \{(x, y) \in \mathbb{A}^2(\mathbb{K}) : f(x, y) = 0\},$$

the set of $\mathbb{K}$-rational points of C.

h) Let $C = C(f)$ be a plane affine algebraic curve over the field $\mathbb{K}$. The *function field of* $C|\mathbb{K}$ is the quotient field of $\mathbb{K}[X, Y]/(f)$. It is denoted by $\mathbb{K}(C)$.

Let $C = C(F)$ be a plane projective algebraic curve defined by the homogeneous polynomial $F(X, Y, Z)$. Let f be the dehomogenised polynomial $f(X, Y) = F(X, Y, 1)$. Then the plane affine algebraic curve $C(f)$ together with the points at infinity correspond to the projective curve $C(F)$. Using this correspondence we consider in this text the points at infinity as additional rational points on the affine curve.

As examples for plane algebraic curves, we define Weierstraß equations.

Definition 1.2. An equation of the form

$$Y^2 + a_1XY + a_3Y = X^3 + a_2X^2 + a_4X + a_6$$

with $a_1, a_2, a_3, a_4, a_6 \in \mathbb{K}$ is a *long Weierstraß normal form*.

The projective long Weierstraß normal form is given by

$$Y^2Z + a_1XYZ + a_3YZ^2 = X^3 + a_2X^2Z + a_4XZ^2 + a_6Z^3.$$

Here we have one point at infinity: $\mathcal{O} = [0 : 1 : 0]$. In affine representation, this is the point $\mathcal{O} = (\infty, \infty)$.

As examples for curves in Weierstraß form we consider the following three curves

$$\begin{aligned} C_1 : Y^2 &= X^3, \\ C_2 : Y^2 &= X^3 + X^2, \\ C_3 : Y^2 &= X^3 + X. \end{aligned}$$

The corresponding projective curves are

$$\begin{aligned} C_1 : Y^2Z &= X^3, \\ C_2 : Y^2Z &= X^3 + X^2Z, \\ C_3 : Y^2Z &= X^3 + XZ^2. \end{aligned}$$

All three curves have the two $\mathbb{K}$-rational points $P = (0, 0)$ and $\mathcal{O}$, which, in projective representation, are $P = [0 : 0 : 1]$ and $\mathcal{O} = [0 : 1 : 0]$.

Definition 1.3. We consider an equation in long Weierstraß normal form with coefficients $a_1, a_2, a_3, a_4, a_6 \in \mathbb{K}$. The *Tate values* are

$$\begin{aligned} b_2 &:= a_1^2 + 4a_2, \\ b_4 &:= 2a_4 + a_1a_3, \\ b_6 &:= a_3^2 + 4a_6, \\ b_8 &:= a_1^2a_6 + 4a_2a_6 - a_1a_3a_4 + a_2a_3^2 - a_4^2, \\ c_4 &:= b_2^2 - 24b_4, \\ c_6 &:= -b_2^3 + 36b_2b_4 - 216b_6. \end{aligned}$$

Furthermore, the *discriminant* is

$$\Delta := -b_2^2b_8 - 8b_4^3 - 27b_6^2 + 9b_2b_4b_6$$

and the *j-invariant*

$$j := \frac{c_4^3}{\Delta}.$$

These constants satisfy the relations

$$4b_8 = b_2b_6 - b_4^2 \quad \text{and} \quad 12^3\Delta = c_4^3 - c_6^2.$$

Definition 1.4. Let a plane algebraic curve C be defined by the polynomial equation $f(X, Y) = 0$. Then $P = (x_0, y_0) \in C$ is a *singular point* of C if and only if

$$\frac{\partial f}{\partial X}(x_0, y_0) = 0 \quad \text{and} \quad \frac{\partial f}{\partial Y}(x_0, y_0) = 0.$$

The singular point is a *double point*, if only the first partial differentials vanish. A double point is a *node*, if the point has two different tangents, a *cusp* if the two tangents coincide. A curve without singular points is called *nonsingular*.

Proposition 1.5. *Curves given by an equation in long Weierstraß normal form have the following classification:*

(i) *The curve is nonsingular* $\Leftrightarrow \Delta \neq 0$. *Otherwise the curve is singular with exactly one singular point.*

(ii) *The curve has a node* $\Leftrightarrow \Delta = 0$ *and* $c_4 \neq 0$.

(iii) *The curve has a cusp* $\Leftrightarrow \Delta = 0$ *and* $c_4 = 0$.

Proof. Silverman [204], Chapter III, Proposition 1.4. See also Exercise 4. □

When we look at the three examples above, we have for their discriminants

$$\Delta_{C_1} = 0, \quad \Delta_{C_2} = 0, \quad \Delta_{C_3} = -64.$$

We further have

$$c_{4,C_1} = 0, \quad c_{4,C_2} = 16, \quad c_{4,C_3} = -48.$$

If $\operatorname{char}(\mathbb{K}) = 2$, then all three curves are singular and have a cusp. If $\operatorname{char}(\mathbb{K}) \neq 2$, the curve C_1 has a cusp, the curve C_2 has a node and the curve C_3 is nonsingular. In all singular cases, the singular point is $P = (0, 0)$. This can be seen by looking at the partial derivatives, which we will do for the curve C_1. We have

$$C_1 : f(X, Y) = Y^2 - X^3 = 0,$$

and the derivatives are

$$\frac{\partial f}{\partial X} = -3X^2, \quad \frac{\partial f}{\partial Y} = 2Y.$$

In any characteristic, the three equations

$$\begin{aligned} Y^2 - X^3 &= 0 \\ -3X^2 &= 0 \\ 2Y &= 0 \end{aligned}$$

have the only solution $X = Y = 0$.

An *elliptic curve* over $\mathbb{K}$ is a nonsingular curve of genus 1 over $\mathbb{K}$ together with one $\mathbb{K}$-rational point. (For the definition of the genus see Mumford [151] or Shafarevich [199].) The equation for such a curve can (at least in theory) be transformed to a long Weierstraß normal form. The specified $\mathbb{K}$-rational point is then transformed to the point at infinity. Such transformations are given for example in Cassels [30], Chapter 8, and in Cohen [34], Chapter 7. For the sake of simplicity, we only consider elliptic curves in long Weierstraß normal form. So we have the definition:

Definition 1.6. An *elliptic curve* over $\mathbb{K}$ is a curve given in long Weierstraß normal form over $\mathbb{K}$ with discriminant $\neq 0$ (and the specified point at infinity).

We give two examples for elliptic curves which are *not* given in Weierstraß normal form. These are the Fermat curves

$$F_1 : U^3 + V^3 = 1 \quad \text{and} \quad F_2 : U^4 + V^4 = 1.$$

The curve F_1 has one point at infinity $\mathcal{O}_{F_1}$, which can be seen by using the projective representation:

$$F_1 : U^3 + V^3 = W^3, \quad \mathcal{O}_{F_1} = [1 : -1 : 0].$$

The transformation

$$U = \frac{6}{X} + \frac{Y}{6X}, \quad V = \frac{6}{X} - \frac{Y}{6X}$$

yields the Weierstraß equation

$$E_{F_1} : Y^2 = X^3 - 432,$$

where $\mathcal{O}_{F_1}$ transforms to $\mathcal{O} = [0 : 1 : 0]$. The second curve F_2 has no point at infinity over $\mathbb{Q}$. The transformation

$$X = \frac{4}{1-V} - 2, \quad Y = \frac{4U^2}{(1-V)^2}$$

yields the Weierstraß equation

$$E_{F_2} : Y^2 = X^3 + 4X,$$

where the point $(0, 1)$ on F_2 maps to $\mathcal{O}$.

Note that F_2 has the point at infinity $[1 : \sqrt[4]{-1} : 0]$ over the complex numbers. This point transforms to the point $(-2, -4i)$ of $E_{F_2}(\mathbb{C})$.

When we look at variable transformations between the elliptic curves

$$E : Y^2 + a_1XY + a_3Y = X^3 + a_2X^2 + a_4X + a_6$$

and

$$E' : (Y')^2 + a_1'X'Y' + a_3'Y' = (X')^3 + a_2'(X')^2 + a_4'X' + a_6',$$

(both defined over $\mathbb{K}$) we want such transformations which map a Weierstraß normal form into another one. The only variable transformation which does this is of the form

$$X = u^2 X' + r, \quad Y = u^3 Y' + u^2 s X' + t,$$

where $u, r, s, t \in \mathbb{K}$ with $u \neq 0$. The inverse transformation is

$$X' = \frac{1}{u^2}(X - r), \quad Y' = \frac{1}{u^3}(Y - sX + sr - t).$$

Such transformations are *birational*. We then have

$$\begin{aligned}
ua'_1 &= a_1 + 2s, \\
u^2 a'_2 &= a_2 - sa_1 + 3r - s^2, \\
u^3 a'_3 &= a_3 + ra_1 + 2t, \\
u^4 a'_4 &= a_4 - sa_3 + 2ra_2 - (t + rs)a_1 + 3r^2 - 2st, \\
u^6 a'_6 &= a_6 + ra_4 + r^2 a_2 + r^3 - ta_3 - t^2 - rta_1, \\
u^2 b'_2 &= b_2 + 12r, \\
u^4 b'_4 &= b_4 + rb_2 + 6r^2, \\
u^6 b'_6 &= b_6 + 2rb_4 + r^2 b_2 + 4r^3, \\
u^8 b'_8 &= b_8 + 3rb_6 + 3r^2 b_4 + r^3 b_2 + 3r^4, \\
u^4 c'_4 &= c_4, \\
u^6 c'_6 &= c_6, \\
u^{12} \Delta' &= \Delta, \\
j' &= j.
\end{aligned}$$

Two equations in Weierstraß normal form are *isomorphic* if there is such a birational transformation between them.

Theorem 1.7. *Let $E|\mathbb{K}$ be an elliptic curve,* $\operatorname{char}(\mathbb{K}) \neq 2, 3$. *Then there exists a birational transformation $\phi : E \to E'$ to a curve $E'|\mathbb{K}$ of the form*

$$E' : Y^2 = X^3 + AX + B$$

with $A, B \in \mathbb{K}$. We say that the curve has a representation in short Weierstraß normal form.

Proof. We sketch the proof. If $\operatorname{char}(\mathbb{K}) \neq 2$ replacing Y by $\frac{1}{2}(Y - a_1 X - a_3)$ leads to

$$E'' : Y^2 = 4X^3 + b_2 X^2 + 2b_4 X + b_6.$$

Assuming further $\operatorname{char}(\mathbb{K}) \neq 2, 3$ and replacing (X, Y) by $\left(\frac{X - 3b_2}{36}, \frac{2Y}{216}\right)$, we have

$$E' : Y^2 = X^3 - 27c_4 X - 54c_6.$$

□

For the short Weierstraß normal form, the discriminant and the j-invariant are

$$\Delta = -16(4A^3 + 27B^2), \quad j = \frac{-12^3(4A)^3}{\Delta}.$$

The only birational transformations which leave short Weierstraß normal form fixed are of the form

$$X = u^2 X', \quad Y = u^3 Y'$$

for $u \in \mathbb{K}^*$. We then have

$$u^4 A' = A, \quad u^6 B' = B, \quad u^{12}\Delta' = \Delta.$$

In general, the following holds:

Proposition 1.8. *Two elliptic curves in Weierstraß normal form are isomorphic over* $\overline{\mathbb{K}} \Leftrightarrow$ *they have the same* j*-invariant.*

Proof. We see from the formulas on page 6 that two isomorphic elliptic curves in Weierstraß normal form have the same j-invariant. For the converse direction of the proof, assume $\mathrm{char}(\mathbb{K}) \neq 2, 3$ (for $\mathrm{char}(\mathbb{K}) = 2$ or $\mathrm{char}(\mathbb{K}) = 3$ see Silverman [204], A.1.2.b)). Then the curves can be given in short Weierstraß normal form:

$$E : Y^2 = X^3 + AX + B, \quad E' : (Y')^2 = (X')^3 + A'X' + B'.$$

We want to find an isomorphism of the form

$$X = u^2 X', \quad Y = u^3 Y'.$$

We have

$$\begin{aligned} j &= j' \\ \Leftrightarrow \quad (4A)^3(4(A')^3 + 27(B')^2) &= (4A')^3(4A^3 + 27B^2) \\ \Leftrightarrow \quad A^3(B')^2 &= (A')^3 B^2. \end{aligned}$$

If $A = 0$ then $B \neq 0$, hence $A' = 0$ and $B' \neq 0$. In this case we can take $u = \left(\frac{B}{B'}\right)^{1/6}$.

If $B = 0$ then $A \neq 0$, hence $B' = 0$ and $A' \neq 0$. In this case we can take $u = \left(\frac{A}{A'}\right)^{1/4}$.

If $AB \neq 0$ then $A'B' \neq 0$. Indeed, if $A'B' = 0$, then $A' = 0$ and $B' = 0$, hence $\Delta' = 0$, which is excluded. We have

$$(A')^3 B^2 = A^3 (B')^2 \Leftrightarrow \frac{B^2}{(B')^2} = \frac{A^3}{(A')^3} \Leftrightarrow \left(\frac{B}{B'}\right)^{1/6} = \left(\frac{A}{A'}\right)^{1/4}.$$

In this case we can take $u = \left(\frac{A}{A'}\right)^{1/4} = \left(\frac{B}{B'}\right)^{1/6}$. □

Proposition 1.9. *For each $j_0 \in \mathbb{K}$ there exists an elliptic curve defined over $\mathbb{K}$ with j-invariant j_0.*

Proof. (See also the article of Deuring [47].)

char($\mathbb{K}$)	j_0	Elliptic curve
$\neq 2, 3$	0	$Y^2 = X^3 + 1$
	12^3	$Y^2 = X^3 + X$
	$\neq 0, 12^3$	$Y^2 = X^3 + 3\kappa X + 2\kappa$ with $\kappa = \frac{j_0}{12^3 - j_0}$
2	0	$Y^2 + Y = X^3$
	$\neq 0$	$Y^2 + XY = X^3 + X^2 + j_0^{-1}$
3	0	$Y^2 = X^3 + X$
	$\neq 0$	$Y^2 = X^3 + X^2 - j_0^{-1}$

□

Definition 1.10. Let $\mathbb{K}$ be a field with char($\mathbb{K}$) $\neq 2$. An elliptic curve $E|\mathbb{K}$ is given in *Legendre normal form*, if it is given by an equation of the form

$$E = E_\lambda : Y^2 = X(X-1)(X-\lambda)$$

with $\lambda \in \mathbb{K}\backslash\{0, 1\}$.

Legendre normal forms have the following properties:

Proposition 1.11. *Let $\mathbb{K}$ be a field with* char($\mathbb{K}$) $\neq 2$.

a) *Let $E_\lambda|\mathbb{K}$ be an elliptic curve in Legendre normal form. The j-invariant $j(\lambda)$ of E_λ is*

$$j(\lambda) = 2^8 \frac{(\lambda^2 - \lambda + 1)^3}{\lambda^2(\lambda - 1)^2}.$$

b) *Two elliptic curves E_λ, $E_{\lambda'}|\mathbb{K}$ in Legendre normal form are equivalent if and only if*

$$\lambda' \in \left\{\lambda, 1-\lambda, \frac{1}{\lambda}, \frac{1}{1-\lambda}, \frac{\lambda-1}{\lambda}, \frac{\lambda}{\lambda-1}\right\}.$$

Proof. Silverman [204] Chapter III, Proposition 1.7 □

Hasse (see the appendix of the book [177] of Roquette) introduced an invariant γ which together with the invariant j characterizes an elliptic curve E over a field $\mathbb{K}$ up to $\mathbb{K}$-isomorphisms. This invariant γ is defined in the following way. If, in the short Weierstraß normal form

$$E: Y^2 = X^3 + AX + B \quad (A, B \in \mathbb{K}) \tag{1.1}$$

(so that $\mathrm{char}(\mathbb{K}) \neq 2, 3$) both coefficients A and B are different from 0, so that $j \neq 0, 12^3$, we define

$$\gamma \equiv -\frac{1}{2} \cdot \frac{A}{B} \pmod{\mathbb{K}^{*2}}.$$

The curve E is then isomorphic over $\mathbb{K}$ to the elliptic curve

$$E': Y^2 = X^3 - \frac{1}{(4\gamma)^2} \cdot \frac{27j}{j - 12^3} X + \frac{2}{(4\gamma)^3} \cdot \frac{27j}{j - 12^3} \tag{1.2}$$

over $\mathbb{K}$.

If $\mathrm{char}(\mathbb{K}) \neq 2$, we take the normal form

$$\bar{E}: Y^2 = 4X^3 + b_2X^2 + 2b_4X + b_6 \quad (b_2, b_4, b_6 \in \mathbb{K})$$

(see the proof of Theorem 1.7) with absolute invariant

$$j = \frac{(b_2^2 - 24b_4)^3}{\Delta}$$

which reduces to

$$\bar{E}: Y^2 = X^3 + b_2X^2 - b_4X + b_6 \quad (b_2, b_4, b_6 \in \mathbb{K})$$

with absolute invariant

$$j = \frac{b_2^6}{\Delta}$$

if $\mathrm{char}(\mathbb{K}) = 3$. We then define for $b_2 \neq 0$, that is, $j \neq 0$,

$$\gamma \equiv b_2 \pmod{\mathbb{K}^{*2}}.$$

By the assumption $b_2 \neq 0 \Leftrightarrow \gamma \neq 0$ and by the transformation

$$X \mapsto X' = X + \frac{b_4}{b_2}, \qquad Y \mapsto Y' = Y$$

we obtain from $\bar{E}$ the elliptic curve (in new notation)

$$E: Y^2 = X^3 + b_2X^2 + b_6' \quad (b_2, b_6' \in \mathbb{K}) \tag{1.3}$$

with coefficient $b_6' := b_6 - \frac{b_4^3}{b_2^3} - \frac{b_4^2}{b_2}$. This curve is isomorphic to the elliptic curve

$$E': Y^2 = X^3 + \gamma X^2 - \frac{\gamma^3}{j} \quad (\gamma, j \in \mathbb{K}). \tag{1.4}$$

In the above table (Proposition 1.9) we have then $\gamma = 1$.

If $\mathrm{char}(\mathbb{K}) = 2$, we proceed in a different manner. We start with the long Weierstraß form

$$\bar{E} : Y^2 + (a_1 X + a_3)Y = X^3 + a_2 X^2 + a_4 X + a_6 \quad (a_1, a_2, a_3, a_4, a_6 \in \mathbb{K})$$

and have as before

$$j = \frac{c_4^3}{\Delta}.$$

If $\mathrm{char}(\mathbb{K}) = 2$, the absolute invariant reduces to

$$j = \frac{a_1^{12}}{\Delta}.$$

Now we utilize the assumption $j \neq 0 \Leftrightarrow a_1 \neq 0$. The transformation

$$X \mapsto X' = \frac{a_1 X + a_3}{a_1^3}, \qquad Y \mapsto Y' = \frac{Y}{a_1^3}$$

then leads to the elliptic curve (in new notation)

$$\tilde{E} : Y^2 + XY = X^3 + a_2 X^2 + a_4 X + a_6 \quad (a_2, a_4, a_6 \in \mathbb{K})$$

with absolute invariant

$$j = \frac{1}{\Delta}.$$

Here, the discriminant is

$$\Delta = a_4^2 + a_6.$$

The further transformation

$$X \mapsto X' = X, \quad Y \mapsto Y' = Y + a_4$$

yields the elliptic curve (again in new notation)

$$E : Y^2 + XY = X^3 + a_2 X^2 + a_6' \quad (a_2, a_6' \in \mathbb{K}) \tag{1.5}$$

with coefficient

$$a_6' := a_6 + a_4^2.$$

In this case we define the invariant

$$\gamma \equiv a_2 \ (\mathrm{mod}\ \pi(\mathbb{K}^+)),$$

where π is the function

$$\pi(Z) := Z(Z+1)$$

and $\mathbb{K}^+$ denotes the additive group of the field $\mathbb{K}$. Then E is birationally isomorphic to the elliptic curve

$$E' : Y^2 + XY = X^3 + \gamma X^2 + \frac{1}{j} \quad (\gamma, j \in \mathbb{K}). \tag{1.6}$$

In the table (Proposition 1.9) again the relation $\gamma = 1$ holds.

We thus have established

Proposition 1.12. *The elliptic curve $E|\mathbb{K}$ in Weierstraß normal form* (1.1) *if* $\mathrm{char}(\mathbb{K}) \neq 2, 3$, (1.3) *if* $\mathrm{char}(\mathbb{K}) = 3$, *and* (1.5) *if* $\mathrm{char}(\mathbb{K}) = 2$, *each with $j \neq 0$, is isomorphic over $\mathbb{K}$ to $E'|\mathbb{K}$. In particular, these curves have the same invariants j and γ.*

We calculate γ for the Legendre normal form

$$\bar{E} : Y^2 = X(X-1)(X-\lambda) \quad (\lambda \in \mathbb{K}),$$

where $j \neq 0$ and $\mathrm{char}(\mathbb{K}) \neq 2, 3$. (If $\mathrm{char}(\mathbb{K}) = 3$, $\gamma \equiv -(\lambda + 1) \pmod{\mathbb{K}^{*2}}$.) The curve has Weierstraß normal form

$$E : Y^2 = X^3 - \frac{1}{3}(\lambda^2 - \lambda + 1)X - \frac{2}{3^3}\left(\lambda^3 - \frac{3}{2}\lambda^2 - \frac{3}{2}\lambda + 1\right).$$

Hence the invariant of the Legendre normal form is

$$\gamma \equiv -\frac{1}{2} \cdot \frac{\lambda^2 - \lambda + 1}{(2\lambda - 1)(\lambda + 1)(\lambda - 2)} \quad (\mathrm{mod}\ \mathbb{K}^{*2}).$$

Here $(2\lambda - 1)(\lambda + 1)(\lambda - 2) \neq 0$. The assumption $j \neq 0$ implies that $\lambda \neq -\rho$, where $\rho \in \mathbb{C}$ is a certain non-trivial third root of unity.

1.2 The addition law

The most important fact about elliptic curves is that the points on the curve form an (additive) abelian group. In this section we give the addition law of this group.

Let $E|\mathbb{K}$ be an elliptic curve given in (long) Weierstraß normal form over any field $\mathbb{K}$. The set of rational points of E over $\mathbb{K}$ is

$$E(\mathbb{K}) = \{(x, y) \in E : x, y \in \mathbb{K}\} \cup \{\mathcal{O}\},$$

where $\mathcal{O}$ is the point at infinity.

Elliptic curves can have finitely or infinitely many rational points. We use the following applications of Bézout's theorem:

Theorem 1.13. a) *A line intersects an elliptic curve in exactly 3 points (counting multiplicities).*

b) *Let C and C' be two cubic curves over an infinite field $\mathbb{K}$ intersecting in exactly nine points in $\mathbb{P}(\overline{\mathbb{K}})$. If C'' is a plane cubic curve over $\mathbb{K}$ going through eight of the intersection points, then it goes through the ninth.*

Proof. See for example Fulton [73] or Gibson [82]. □

Definition 1.14. Let $E|\mathbb{K}$ be an elliptic curve and $P_1, P_2 \in E(\mathbb{K})$ be two (not necessary distinct) points. The line through P_1 and P_2 (i.e. the secant) intersects the elliptic curve in a third point P_3'. We consider the line through P_3' and $\mathcal{O}$. This line intersects the curve in a third point P_3. We define

$$P_1 + P_2 := P_3.$$

(If $P_1 = P_2$, one has to take the tangent line at E in P_1.)

Theorem 1.15. *Let $E|\mathbb{K}$ be an elliptic curve over $\mathbb{K}$. The set $E(\mathbb{K})$ of rational points is an additive abelian group under the addition defined above, with the point at infinity $\mathcal{O}$ as identity element.*

If $\mathbb{K}$ is a number field, the group $E(\mathbb{K})$ is called the *Mordell–Weil group* of E over $\mathbb{K}$.

Proof. The following properties of the addition are easy to obtain:

- For $P_1, P_2 \in E(\mathbb{K})$ we have $P_1 + P_2 \in E(\mathbb{K})$: In Theorem 1.16 we shall give formulas for the addition on elliptic curves in long Weierstraß normal form. These formulas prove the assertion.
- Identity element: $\mathcal{O}$
- Commutativity: $P_1 + P_2 = P_2 + P_1$
- Inverse: Let P' be the third point of intersection of the line through P and $\mathcal{O}$ with the curve. Then $P + P' = \mathcal{O}$, therefore $P' = -P$.

It remains to show that the addition is associative. Let $P_1, P_2, P_3 \in E(\mathbb{K})$. We have to show that (see e.g. Chahal [31])

$$(P_1 + P_2) + P_3 = P_1 + (P_2 + P_3)$$

$$\Leftrightarrow \; -((P_1 + P_2) + P_3) = -(P_1 + (P_2 + P_3)).$$

We define the following lines (secants or tangents if two of the points coincide):
L_1: Line through P_1 and P_2. This line intersects the curve in the third point $-(P_1+P_2)$.

L_2: Line through P_3 and $(P_1 + P_2)$. This line intersects the curve in the third point $-((P_1 + P_2) + P_3)$.
L_3: Line through $(P_2 + P_3)$ and $\mathcal{O}$. This line intersects the curve in the third point $-(P_2 + P_3)$.
L_1': Line through P_2 and P_3. This line intersects the curve in the third point $-(P_2+P_3)$.
L_2': Line through P_1 and $(P_2 + P_3)$. This line intersects the curve in the third point $-(P_1 + (P_2 + P_3))$.
L_3': Line through $(P_1 + P_2)$ and $\mathcal{O}$. This line intersects the curve in the third point $-(P_1 + P_2)$.

Then we define the cubic curves

$$C := L_1 \cup L_2 \cup L_3, \quad C' := L_1' \cup L_2' \cup L_3'.$$

The curves C and E have no common components (because C is a union of 3 lines). An application of the theorem of Bézout states that such curves have exactly 9 common points. For the curves C and E these are the points

$$\mathcal{O}, P_1, P_2, P_3, (P_1 + P_2), -(P_1 + P_2), (P_2 + P_3), -(P_2 + P_3), -((P_1 + P_2) + P_3).$$

The curve C' intersects at the first 8 of the common points of C and E. Therefore C' intersects also at the 9-th common point. On the other hand C' has exactly 9 common points with E:

$$\mathcal{O}, P_1, P_2, P_3, (P_1 + P_2), -(P_1 + P_2), (P_2 + P_3), -(P_2 + P_3), -(P_1 + (P_2 + P_3)).$$

Hence

$$-((P_1 + P_2) + P_3) = -(P_1 + (P_2 + P_3)).$$ □

Theorem 1.16 (Addition theorem). *Let $E|\mathbb{K}$ be an elliptic curve over the field $\mathbb{K}$ in long Weierstraß normal form and let $P_1 = (x_1, y_1)$, $P_2 = (x_2, y_2) \in E(\mathbb{K})$. Then*

a) $-P_1 = (x_1, -y_1 - a_1x_1 - a_3)$.

b) *If $x_1 = x_2$ and $y_2 + y_1 + a_1x_1 + a_3 = 0$, i.e. if $P_1 = -P_2$, then*

$$P_1 + P_2 = \mathcal{O}.$$

c) *Let $P_1 \neq -P_2$. Define*

$$\lambda = \frac{y_2 - y_1}{x_2 - x_1},$$
$$\nu = \frac{y_1x_2 - y_2x_1}{x_2 - x_1} = y_1 - \lambda x_1,$$

if $x_1 \neq x_2$, and

$$\lambda = \frac{3x_1^2 + 2a_2x_1 + a_4 - a_1y_1}{2y_1 + a_1x_1 + a_3},$$
$$\nu = \frac{-x_1^3 + a_4x_1 + 2a_6 - a_3y_1}{2y_1 + a_1x_1 + a_3} = y_1 - \lambda x_1,$$

if $x_1 = x_2$.
Then $P_1 + P_2 = P_3 = (x_3, y_3)$ *with coordinates*

$$\begin{aligned} x_3 &= \lambda^2 + a_1\lambda - a_2 - x_1 - x_2, \\ y_3 &= -(\lambda + a_1)x_3 - \nu - a_3 = \lambda(x_1 - x_3) - y_1 - a_1x_3 - a_3. \end{aligned}$$

Proof. We write

$$E : f(X, Y) = Y^2 + a_1XY + a_3Y - X^3 - a_2X^2 - a_4X - a_6 = 0.$$

a) The line L through $P_1 = (x_1, y_1) \in E(\mathbb{K})$ and $\mathcal{O}$ is

$$L : X = x_1.$$

We compute the intersection point $P_1' = (x_1, y_1')$ of L and E:

$$\begin{aligned} f(x_1, Y) &= Y^2 + (a_1x_1 + a_3)Y - (x_1^3 + a_2x_1^2 + a_4x_1 + a_6) \\ &= (Y - y_1)(Y - y_1') \\ &= Y^2 + (-y_1 - y_1')Y + y_1y_1'. \end{aligned}$$

Comparing coefficients, we see that

$$y_1' = -y_1 - a_1x_1 - a_3.$$

The third intersection point of L with E is therefore

$$P_1' = (x_1, -y_1 - a_1x_1 - a_3).$$

With Theorem 1.15, this point P_1' is equal to $-P_1$.

b) follows from a).

c) We assume that $P_1 \neq -P_2$. First let $x_1 = x_2$, that means $P_1 = P_2$. The tangent in P_1 at E is

$$L : f_X(x_1, y_1)(X - x_1) + f_Y(x_1, y_1)(Y - y_1) = 0$$

with the partial derivatives

$$f_X(x_1, y_1) = -(3x_1^2 + 2a_2x_1 + a_4 - a_1y_1), \quad f_Y(x_1, y_1) = 2y_1 + a_1x_1 + a_3.$$

The assumption of $P_1 \neq -P_2$ implies $f_Y(x_1, y_1) \neq 0$. Therefore we write

$$\begin{aligned} L : Y &= \frac{-f_X(x_1, y_1)}{f_Y(x_1, y_1)}(X - x_1) + y_1 \\ &= \frac{-f_X(x_1, y_1)}{f_Y(x_1, y_1)}X + \frac{x_1 f_X(x_1, y_1) + y_1 f_Y(x_1, y_1)}{f_Y(x_1, y_1)} \\ &= \lambda X + \nu \end{aligned}$$

with λ and ν as in the theorem.

In the other case, i.e. if $x_1 \neq x_2$ and thus also $P_1 \neq P_2$, the line through P_1 and P_2 is

$$L : \frac{Y - y_1}{X - x_1} = \frac{y_2 - y_1}{x_2 - x_1}.$$

Therefore we have

$$L : Y = \frac{y_2 - y_1}{x_2 - x_1} X + \frac{y_2 - y_1}{x_2 - x_1}(-x_1) + y_1 = \lambda X + \nu$$

with λ and ν as in the theorem.

In both cases the line through P_1 and P_2 is given by the equation

$$L : Y = \lambda X + \nu.$$

The third intersection point of the line L with E is a point $P_3' = (x_3', y_3')$ (by assumption $P_3' \neq \mathcal{O}$). We now compute this intersection point:

$$\begin{aligned} & f(X, \lambda X + \nu) \\ &\quad = (\lambda X + \nu)^2 + a_1 X(\lambda X + \nu) + a_3(\lambda X + \nu) - X^3 - a_2 X^2 - a_4 X - a_6 \\ &\quad = -X^3 + (\lambda^2 + a_1\lambda - a_2)X^2 + (2\lambda\nu + a_1\nu + a_3\lambda - a_4)X \\ &\qquad + (\nu^2 + a_3\nu - a_6) \\ &\quad = -(X - x_1)(X - x_2)(X - x_3') \\ &\quad = -X^3 + (x_1 + x_2 + x_3')X^2 + (-x_1x_2 - x_1x_3' - x_2x_3')X + x_1x_2x_3'. \end{aligned}$$

Comparing coefficients, we see that

$$x_3' = \lambda^2 + a_1\lambda - a_2 - x_1 - x_2.$$

Since P_3' is a point of L, one has

$$y_3' = \lambda x_3' + \nu.$$

The point $P_3 = (x_3, y_3) = P_1 + P_2$ is $-P_3'$. According to a) this point has the coordinates

$$\begin{aligned} x_3 &= x_3' = \lambda^2 + a_1\lambda - a_2 - x_1 - x_2, \\ y_3 &= -y_3' - a_1 x_3' - a_3 = -(\lambda + a_1)x_3 - \nu - a_3. \end{aligned}$$ □

For $m \in \mathbb{Z}$ we write

$$mP \text{ for } \begin{cases} \sum_{j=1}^{m} P & \text{if } m > 0, \\ \mathcal{O} & \text{if } m = 0, \\ \sum_{j=1}^{-m}(-P) & \text{if } m < 0. \end{cases}$$

As a first example we consider the elliptic curve

$$E : Y^2 = X^3 + 1$$

over $\mathbb{Q}$. This curve has a $\mathbb{Q}$-rational point $P = (2, -3)$. We compute

$$2P = (0, -1), \quad 3P = (-1, 0), \quad 4P = (0, 1), \quad 5P = (2, 3), \quad 6P = \mathcal{O}.$$

We thus see that $5P = -P$. The point P is a point of order 6.

As a second example consider the elliptic curve

$$E : Y^2 = X^3 - 10X$$

over $\mathbb{Q}$ and the points $P = (-1, 3)$, $Q = (0, 0) \in E(\mathbb{Q})$. We have

$$P + Q = (10, 30).$$

The point Q is a point of order 2:

$$2Q = \mathcal{O}.$$

The point P is of infinite order, which we shall prove in Chapter 6, Section 6.6:

$$2P = (121/36, 451/216), \quad 3P = (-57121/24649, -12675843/3869893),$$

$$4P = (761815201/29289744, -20870873704079/158516094528), \ldots.$$

Definition 1.17. Let $E|\mathbb{K}$ be an elliptic curve and $n \in \mathbb{N}$. The set

$$E[n] := \{P \in E : nP = \mathcal{O}\}$$

is called the set of *n-division points* of E. The $\mathbb{K}$-rational n-division points of E are

$$E(\mathbb{K})[n] := \{P \in E(\mathbb{K}) : nP = \mathcal{O}\}.$$

Hence $E[n] = E(\overline{\mathbb{K}})[n]$ for the algebraic closure $\overline{\mathbb{K}}$ of $\mathbb{K}$

In applications of elliptic curves one often has to add points by the addition law. We now describe different methods for the representation of points on elliptic curves. These methods lead to different addition formulas. In applications one uses those formulas which need the smallest amount of computing time.

In what follows we denote the computing times for taking the inverse in the field $\mathbb{K}$ by I, for multiplication in $\mathbb{K}$ by M, and for squaring in $\mathbb{K}$ by S. We neglect, as usual, addition, subtraction, and multiplication by a small field constant. The computing time for an addition of two different points is denoted by $t(P_1 + P_2)$, the computing time for doubling by $t(2P)$.

The most general representation of points on elliptic curves is the *affine representation on the long Weierstraß normal form*. The addition formulas for this representation are given above. This representation can be used for any field of an arbitrary characteristic.

If one has characteristic > 3 or 0, the equation can be transformed into short Weierstraß normal form

$$E : Y^2 = X^3 + a_4 X + a_6.$$

We consider the *affine representation on short Weierstraß normal form.* Let

$$P_1 = (x_1, y_1), P_2 = (x_2, y_2) \in E, \quad P_1 \neq -P_2.$$

Then, by Theorem 1.16, $P_1 + P_2 = (x_3, y_3)$ is given by

$$\begin{aligned} x_3 &= \lambda^2 - x_1 - x_2, \\ y_3 &= \lambda(x_1 - x_3) - y_1, \end{aligned}$$

with

$$\lambda = \begin{cases} \dfrac{y_2 - y_1}{x_2 - x_1} & \text{if } P_1 \neq P_2, \\ \dfrac{3x_1^2 + a_4}{2y_1} & \text{if } P_1 = P_2. \end{cases}$$

With this representation, we have

$$t(P_1 + P_2) = I + 2M + S, \quad t(2P_1) = I + 2M + 2S.$$

When looking at elliptic curves in Weierstraß normal form, we also encounter projective coordinates. For sake of simplicity we only consider *projective representations on short Weierstraß normal forms.* Let

$$E : Y^2 Z = X^3 + a_4 X Z^2 + a_6 Z^3$$

be the projective elliptic curve and

$$P_1 = [x_1 : y_1 : z_1], P_2 = [x_2 : y_2 : z_2] \in E, \quad P_1 \neq -P_2.$$

Then $P_1 + P_2 = [x_3 : y_3 : z_3]$ with

$$\begin{aligned} x_3 &= bc, \\ y_3 &= a(b^2 x_1 z_2 - c) - b^3 y_1 z_2, \\ z_3 &= b^3 z_1 z_2, \end{aligned}$$

where

$$\begin{aligned} a &:= y_2 z_1 - y_1 z_2, \\ b &:= x_2 z_1 - x_1 z_2, \\ c &:= a^2 z_1 z_2 - b^3 - 2b^2 x_1 z_2, \end{aligned}$$

if $P_1 \neq P_2$ and

$$\begin{aligned} x_3 &= 2db, \\ y_3 &= a(4c-d) - 8y_1^2 b^2, \\ z_3 &= 8b^3, \end{aligned}$$

where

$$\begin{aligned} a &:= a_4 z_1^2 + 3x_1^2, \\ b &:= y_1 z_1, \\ c &:= x_1 y_1 b, \\ d &:= a^2 - 8c, \end{aligned}$$

if $P_1 = P_2$. Here we obtain

$$t(P_1 + P_2) = 12M + 2S, \quad t(2P_1) = 6M + 5S.$$

Another possible representation is the representation in Jacobian coordinates. Therefore we set in the affine short Weierstraß normal form

$$\frac{X}{Z^2}, \quad \frac{Y}{Z^3}$$

and get the *short Weierstraß normal form in Jacobian coordinates*

$$E_J : Y^2 = X^3 + a_4 X Z^4 + a_6 Z^6.$$

The points

$$P_{1,J} = (x_1, y_1, z_1)_J, \quad P_{2,J} = (x_2, y_2, z_2)_J \in E_J$$

represent the affine points

$$P_1 = \left(\frac{x_1}{z_1^2}, \frac{y_1}{z_1^3}\right), \quad P_2 = \left(\frac{x_2}{z_2^2}, \frac{y_2}{z_2^3}\right).$$

We have $P_{1,J} + P_{2,J} = (x_3, y_3, z_3)_J$ with

$$\begin{aligned} x_3 &= -c^3 - 2a_1 c^2 + d^2, \\ y_3 &= -b_1 c^3 + d(a_1 c^2 - x_3), \\ z_3 &= z_1 z_2 c, \end{aligned}$$

where

$$\begin{aligned} a_1 &:= x_1 z_2^2, \\ a_2 &:= x_2 z_1^2, \\ b_1 &:= y_1 z_2^3, \\ b_2 &:= y_2 z_1^3, \\ c &:= a_2 - a_1, \\ d &:= b_2 - b_1, \end{aligned}$$

if $P_1 = \pm P_2$ and

$$\begin{aligned} x_3 &= -2a + b^2, \\ y_3 &= -8y_1^4 + b(a - x_3), \\ z_3 &= 2y_1 z_1, \end{aligned}$$

where

$$\begin{aligned} a &:= 4x_1 y_1^2, \\ b &:= 3x_1^2 + a_4 z_1^4, \end{aligned}$$

if $P_1 = P_2$.

Using Jacobian coordinates we obtain in this case

$$t(P_1 + P_2) = 12M + 4S, \quad t(2P_1) = 3M + 6S.$$

For more possible representations we refer to the paper of Cohen, Miyaji, and Ono [35], in which the authors discuss the representations defined above and also the use of mixed coordinates.

1.3 Multiplication formulas

Apart from the addition formulas there are also multiplication formulas for elliptic curves. They seem to be first given for elliptic curves in short Weierstraß normal form (see for example Cassels [25]). We define them in their most general form as in the reports of Zimmer [254], [255].

Definition 1.18. Let $E|\mathbb{K}$ be an elliptic curve E in long Weierstraß normal form over the field $\mathbb{K}$ (with $\operatorname{char}(\mathbb{K}) \neq 2$). Define the *multiplication polynomials*

$$\begin{aligned} \phi_1(X, Y) &:= X, \\ \phi_2(X, Y) &:= X^4 - b_4 X^2 - 2b_6 X - b_8, \\ \psi_0(X, Y) &:= 0, \\ \psi_1(X, Y) &:= 1, \\ \psi_2(X, Y) &:= 2Y + a_1 X + a_3, \\ \psi_3(X, Y) &:= 3X^4 + b_2 X^3 + 3b_4 X^2 + 3b_6 X + b_8, \\ \psi_4(X, Y) &:= \psi_2(X, Y)[2X^6 + b_2 X^5 + 5b_4 X^4 + 10b_6 X^3 \\ &\qquad + 10b_8 X^2 + (b_2 b_8 - b_4 b_6)X + (b_4 b_8 - b_6^2)], \\ \Omega_1(X, Y) &:= Y, \\ 2\Omega_1'(X, Y) &:= 2\Omega_1(X, Y) + a_1 X + a_3, \end{aligned}$$

and for $m \geq 2$:

$$\begin{aligned}
\phi_m(X, Y) &:= X\psi_m^2(X, Y) - \psi_{m-1}(X, Y)\psi_{m+1}(X, Y),\\
\psi_{2m+1}(X, Y) &:= \psi_{m+2}(X, Y)\psi_m(X, Y)^3 - \psi_{m-1}(X, Y)\psi_{m+1}(X, Y)^3,\\
\psi_{2m}(X, Y) &:= 2\psi_m(X, Y)\Omega'_m(X, Y),\\
2\Omega_m(X, Y) &:= 2\Omega'_m(X, Y) - \psi_m(X, Y)[a_1\phi_m(X, Y) + a_3\psi_m(X, Y)^2],\\
2\psi_2(X, Y)\Omega'_m(X, Y) &:= \psi_{m+2}(X, Y)\psi_{m-1}(X, Y)^2 - \psi_{m-2}(X, Y)\psi_{m+1}(X, Y)^2.
\end{aligned}$$

The polynomials ψ_m are also called *m-division polynomials*. In the case of $\mathrm{char}(\mathbb{K}) = 2$, we define the polynomials ϕ_m and ψ_m as above, but do not consider the polynomials Ω_m and Ω'_m.

Theorem 1.19. *Let $E|\mathbb{K}$ be an elliptic curve in long Weierstraß normal form over the field $\mathbb{K}$ ($\mathrm{char}(\mathbb{K}) \neq 2$), $P = (x, y) \in E(\mathbb{K})$, and $m \in \mathbb{N}$. Then*

$$mP = \left(\frac{\phi_m(x, y)}{\psi_m(x, y)^2}, \frac{\Omega_m(x, y)}{\psi_m(x, y)^3}\right).$$

Proof. This is proved by using the Weierstraß $\wp$ functions, see e.g. Folz [64]. We shall also give a proof in connection with Theorem 2.26. □

The multiplication polynomials have the following properties:

Proposition 1.20. *Let $E|\mathbb{K}$ be an elliptic curve in long Weierstraß normal form over the field $\mathbb{K}$, $m \in \mathbb{N}$. Assign to*

$$\begin{Bmatrix} X \\ Y \\ a_i \\ b_j \end{Bmatrix} \quad \textit{the weight} \quad \begin{Bmatrix} 2 \\ 3 \\ i \\ j \end{Bmatrix}.$$

a) *The polynomial ϕ_m is a polynomial in X over $\mathbb{Z}[b_2, b_4, b_6, b_8]$ of degree m^2 with leading coefficient 1, where every monomial has weight $2m^2$.*

b) *Let m be $\begin{Bmatrix} \text{even} \\ \text{odd} \end{Bmatrix}$. The expression $\begin{Bmatrix} \psi_2^{-1}\psi_m \\ \psi_m \end{Bmatrix}$ is a polynomial in X over $\mathbb{Z}[b_2, b_4, b_6, b_8]$ of degree $\begin{Bmatrix} \frac{m^2-4}{2} \\ \frac{m^2-1}{2} \end{Bmatrix}$ with leading coefficient $\begin{Bmatrix} \frac{m}{2} \\ m \end{Bmatrix}$, where every monomial has weight $\begin{Bmatrix} m^2-4 \\ m^2-1 \end{Bmatrix}$.*

c) *If* $\operatorname{char}(\mathbb{K}) \neq 2$ *let* m *be* $\left\{\begin{array}{c}\text{even}\\ \text{odd}\end{array}\right\}$. *The expression* $\left\{\begin{array}{c}2\Omega'_m\\ 2\psi_2^{-1}\Omega'_m\end{array}\right\}$ *is a polynomial in* X *over* $\mathbb{Z}[b_2, b_4, b_6, b_8]$ *of degree* $\left\{\begin{array}{c}\frac{3m^2}{2}\\ \frac{3(m^2-1)}{2}\end{array}\right\}$ *with leading coefficient* $\left\{\begin{array}{c}2\\ 1\end{array}\right\}$, *where every monomial has weight* $\left\{\begin{array}{c}3m^2\\ 3(m^2-1)\end{array}\right\}$.

Proof. This is proved by induction for $m \in \mathbb{N}$ and reduction modulo the long Weierstraß equation (see Exercise 1.6.12). □

By Proposition 1.20 we write in the following $\phi_m(X)$ instead of $\phi_m(X, Y)$ and $\psi_m^2(X)$ instead of $\psi_m^2(X, Y)$ for $m \in \mathbb{N}$.

Proposition 1.21. *Let* $E|\mathbb{K}$ *be an elliptic curve in long Weierstraß normal form over the field* $\mathbb{K}$, $m \in \mathbb{N}$. *Assign to* b_j *the weight* j. *Then*

$$\psi_m^2(X) = \sum_{i=0}^{m^2-1} f_i X^{m^2-(1+i)}$$

is a polynomial in X *of degree* m^2-1 *with leading coefficient* $f_0 = m^2$. *The coefficients* f_i *for* $i = 0, \ldots, m^2-1$, *are polynomials in* b_2, b_4, b_6, *and* b_8 *with integral coefficients and weight* $2i$.

Proof. This follows from Proposition 1.20 and the fact that

$$\psi_2^2(X) = 4X^3 + b_2X^2 + 2b_4X + b_6$$

is a polynomial in X of degree 3, where the i-th coefficient is a polynomial f_i in b_2, b_4, b_6 with weight $2i$. □

Proposition 1.22. *Let* $E|\mathbb{K}$ *be an elliptic curve in long Weierstraß normal form over the field* $\mathbb{K}$, $p \in \mathbb{P}$ *a prime number, and* $r \in \mathbb{N}$. *Then*

$$\psi_{p^r}^2(X)/\psi_{p^{r-1}}^2(X) = \sum_{i=0}^{(p^2-1)p^{2(r-1)}} h_i X^{(p^2-1)p^{2(r-1)}-i}$$

is a polynomial in X *of degree* $(p^2-1)p^{2(r-1)}$ *with leading coefficient* $h_0 = p^2$. *Further* h_i *for* $i = 1, \ldots, (p^2-1)p^{2(r-1)}$ *is a polynomial in* b_2, b_4, b_6, *and* b_8 *with integral coefficients and weight* $2i$. *For* $i = 0, \ldots, \frac{p-1}{2}p^{r-1} - 1$ *the coefficients* h_i *are divisible by* p^2.

Proof. See Folz [64]. (See also Cassels [25] and Zimmer [251].) □

We need the first of the following two formulas to show that the polynomials ϕ_m and ψ_m^2 are relatively prime. The second formula is needed in Chapter 5 in the proof of Theorem 5.33.

Proposition 1.23. *Let $E|\mathbb{K}$ be an elliptic curve in long Weierstraß normal form over the field $\mathbb{K}$.*

$$\begin{aligned}\Delta &= \{-48X^2 - 8b_2X + b_2^2 - 32b_4\}\phi_2(X)\\ &\quad + \{12X^3 - b_2X^2 - 10b_4X + b_2b_4 - 27b_6\}\psi_2^2(X), \end{aligned}\tag{1.7}$$

$$\begin{aligned}\Delta X^7 &= \{\Delta X^3 + 4(b_2b_4b_8 - b_6b_8 - b_2b_6^2)X^2 + (12b_4^2b_8 - b_2b_6b_8 - 11b_4b_6^2)X\\ &\quad + 6b_6(b_4b_8 - b_6^2)\}\phi_2(X) + \{-(b_2b_4b_8 - b_6b_8 - b_2b_6^2)X^3\\ &\quad + (5b_4^2b_8 - b_2b_6b_8 - 4b_4b_6^2)X^2 + (13b_4b_6b_8 - b_2b_8^2 - 12b_6^3)X\\ &\quad + 6b_8(b_4b_8 - b_6^2)\}\psi_2^2(X).\end{aligned}\tag{1.8}$$

Proof. These relations can be directly computed. □

Proposition 1.24. *Let $E|\mathbb{K}$ be an elliptic curve in long Weierstraß normal form over the field $\mathbb{K}$ and $m \in \mathbb{N}$. Then the polynomials ϕ_m and ψ_m^2 are relatively prime.*

Proof. Suppose m is the smallest natural number such that $\phi_m(X)$ and $\psi_m^2(X)$ have a common nonconstant irreducible factor $\theta(X)$. Then $m > 1$.

1. Case $m = 2k$. Consider Equation (1.7) with $\frac{\phi_k(X)}{\psi_k^2(X)}$:

$$\begin{aligned}\Delta = &\left(-48\frac{\phi_k^2(X)}{\psi_k^4(X)} - 8b_2\frac{\phi_k(X)}{\psi_k^2(X)} + b_2^2 - 32b_4\right)\phi_2\left(\frac{\phi_k(X)}{\psi_k^2(X)}\right)\\ &+\left(12\frac{\phi_k^3(X)}{\psi_k^6(X)} - b_2\frac{\phi_k^2(X)}{\psi_k^4(X)} - 10b_4\frac{\phi_k(X)}{\psi_k^2(X)} + b_2b_4 - 27b_6\right)\psi_2^2\left(\frac{\phi_k(X)}{\psi_k^2(X)}\right)\end{aligned}$$

Now we need the following two formulas (see Exercise 1.6.13). For positive integers $n, k \in \mathbb{N}$:

$$\phi_{nk}(X) = \psi_k^{2n^2}(X)\phi_n\left(\frac{\phi_k(X)}{\psi_k^2(X)}\right), \tag{1.9}$$

$$\psi_{nk}^2(X) = \psi_k^{2n^2}(X)\psi_n^2\left(\frac{\phi_k(X)}{\psi_k^2(X)}\right). \tag{1.10}$$

We use these formulas with $n = 2$. Then:

$$\Delta = \left(-48\frac{\phi_k^2(X)}{\psi_k^4(X)} - 8b_2\frac{\phi_k(X)}{\psi_k^2(X)} + b_2^2 - 32b_4\right)\frac{\phi_{2k}(X)}{\psi_k^8(X)}$$
$$+\left(12\frac{\phi_k^3(X)}{\psi_k^6(X)} - b_2\frac{\phi_k^2(X)}{\psi_k^4(X)} - 10b_4\frac{\phi_k(X)}{\psi_k^2(X)} + b_2b_4 - 27b_6\right)\frac{\psi_{2k}^2(X)}{\psi_k^8(X)}.$$

Multiplying by $\psi_k^{14}(X)$ we get

$$\begin{aligned}\Delta\psi_k^{14}(X) &= \{-48\phi_k^2(X) - 8b_2\phi_k(X)\psi_k^2(X) + (b_2^2 - 32b_4)\psi_k^4(X)\}\phi_{2k}(X)\psi_k^2(X)\\ &\quad + \{12\phi_k^3(X) - b_2\phi_k^2(X)\psi_k^2(X) - 10b_4\phi_k(X)\psi_k^4(X)\\ &\quad + (b_2b_4 - 27b_6)\psi_k^6(X)\}\psi_{2k}^2(X).\end{aligned}$$

As $\theta(X)$ divides $\phi_{2k}(X)$ and $\psi_{2k}^2(X)$, it also divides $\psi_k^{14}(X)$. Then it follows that $\theta(X)$ divides $\psi_k^2(X)$, as $\theta(X)$ is irreducible.

Using again formula (1.9) with $n = 2$, we get

$$\phi_{2k}(X) = \phi_k^4(X) - b_4\phi_k^2(X)\psi_k^4(X) - 2b_6\phi_k(X)\psi_k^6(X) - b_8\psi_k^8(X).$$

It follows that $\theta(X)$ divides $\phi_k(X)$. This contradicts the fact that m is minimal with $\theta(X)$ dividing both $\phi_m(X)$ and $\psi_m^2(X)$.

2. Case $m = 2k + 1$ is odd. From the formula for ϕ_m we obtain

$$\psi_{m-1}^2(X)\psi_{m+1}^2(X) = (X\psi_m^2(X) - \phi_m(X))^2.$$

It follows that $\theta(X)$ divides $\psi_{m-1}^2(X)$ or $\psi_{m+1}^2(X)$. Assume that $\theta(X)$ divides $\psi_{m+1}^2(X)$. Then it also divides

$$\begin{aligned}\phi_{m+1}^2(X) &= (X\psi_{m+1}^2(X) - \psi_m(X)\psi_{m+2}(X))^2\\ &= X^2\psi_{m+1}^4(X) - 2X\psi_{m+1}^2(X)\psi_m(X)\psi_{m+2}(X) + \psi_m^2(X)\psi_{m+2}^2(X);\end{aligned}$$

hence $\theta(X)$ divides $\phi_{m+1}(X)$. This means that we are in the first case. We have shown then that $\theta(X)$ divides $\phi_{\frac{m+1}{2}}(X)$ and $\psi_{\frac{m+1}{2}}^2(X)$, which is a contradiction to the choice of m. □

An important application of the division polynomials is the following proposition.

Proposition 1.25. *Let $E|\mathbb{K}$ be an elliptic curve in long Weierstraß normal form over the field $\mathbb{K}$ and $P = (x_0, y_0) \in E(\mathbb{K})$. Let $m \in \mathbb{N}$ be relatively prime to* $\mathrm{char}(\mathbb{K})$. *Then*

$$mP = \mathcal{O} \Leftrightarrow \psi_m(x_0, y_0) = 0.$$

Proof. If $\psi_m(x_0, y_0) = 0$ then it follows from the Theorem 1.19 and the fact that the polynomials ϕ_m and ψ_m^2 have no common root, that $mP = \mathcal{O}$. On the other hand, if $\psi_m(x_0, y_0) \neq 0$, then

$$\frac{\phi_m(x_0, y_0)}{\psi_m(x_0, y_0)^2}, \frac{\Omega_m(x_0, y_0)}{\psi_m(x_0, y_0)^3} \in \mathbb{K},$$

therefore $mP \neq \mathcal{O}$. (For $\mathrm{char}(\mathbb{K}) = 2$ consider only the first coordinate.) □

1.4 Factorization and primality test

First applications of elliptic curves are the factorization and the primality test using elliptic curves. Both methods are based on the following proposition.

Proposition 1.26. *Let $E : Y^2 = X^3 + a_4X + a_6$ be an elliptic curve over $\mathbb{Q}$ with $a_4, a_6 \in \mathbb{Z}$, $n \in \mathbb{N}$ and* $\gcd(\Delta, n) = 1$. *Let $P_1, P_2 \in E(\mathbb{Q})$ with coordinates which have denominator prime to n, $P_1 \neq -P_2$. Then the point $P_1 + P_2$ has coordinates with denominator not prime to n if and only if there exists a prime $p \mid n$ such that $P_1 + P_2 \equiv \mathcal{O} \pmod{p}$.*

Note that we consider here $\bmod\ p$ reduced points of an elliptic curve. A point is reduced $\bmod\ p$ by reducing the coordinates:

$$(x, y) \pmod{p} \equiv (x \pmod{p}, y \pmod{p}).$$

If p divides the denominator of a coordinate, the point reduces to $\mathcal{O}$.

Proof of Proposition 1.26. First let $P_1 = (x_1, y_1)$, $P_2 = (x_2, y_2)$ and $P_1 + P_2$ have coordinates whose denominators are prime to n. Suppose that there exists a prime $p|n$ with $P_1 + P_2 \equiv \mathcal{O} \pmod{p}$. This p can not be equal to 2, as $\gcd(\Delta, n) = 1$. There are two possibilities.

If $P_1 = P_2$, consider the denominator of the coordinates of $P_1 + P_2 = 2P_1$. As $2P_1 \equiv \mathcal{O} \pmod{p}$, it follows that $y_1 \equiv 0 \pmod{p}$. On the other hand, no divisor of n divides the denominator of the coordinates of $2P_1$. As p divides y_1 it must also divide the numerator. In this case, it follows that p divides $x_1^3 + a_4x_1 + a_6$ and the derivative $3x_1^2 + a_4$. Therefore P_1 is a singular point modulo p, that means that p divides the discriminant Δ (see also (1.7)), which is not possible.

If $P_1 \neq P_2$ and $x_1 \not\equiv x_2 \pmod{p}$ then it follows easily from the addition formulas that

$$P_1 + P_2 \not\equiv \mathcal{O} \pmod{p}.$$

Hence, if $P_1 + P_2 \equiv \mathcal{O} \pmod{p}$, then

$$x_1 \equiv x_2 \pmod{p}.$$

As p does not divide the denominator of the coordinates of $P_1 + P_2$, if follows that $y_1 \equiv y_2 \pmod p$ and hence $P_1 \equiv P_2 \pmod p$. In this case

$$y_1 \equiv y_2 \equiv 0 \pmod p.$$

Now write $x_2 = x_1 + p^k x$ with $k \in \mathbb{N}$ such that neither the denominator nor the numerator of x is divisible by p. It is easy to see that then $y_2 = y_1 + p^k y$ is of the same form. The equation

$$\begin{aligned} y_2^2 &= (x_1 + p^k x)^3 + a_4(x_1 + p^k x) + a_6 \\ &\equiv x_1^3 + a_4 x_1 + a_6 + p^k(3x x_1^2 + a_4 x) \pmod p \\ &\equiv y_1^2 + p^k x(3x_1^2 + a_4) \pmod p \end{aligned}$$

yields

$$y_2^2 - y_1^2 \equiv p^k x(3x_1^2 + a_4) \pmod p.$$

As $y_2^2 - y_1^2 = (y_2 - y_1)(y_2 + y_1)$ is divisible by p^{k+1}, it follows that $3x_1^2 + a_4$ is divisible by p. As in the case $P_1 = P_2$ one can conclude that p divides the discriminant Δ, which is a contradiction to the hypothesis.

Now let p be a prime divisor of n and suppose that $P_1 + P_2 \not\equiv \mathcal{O} \pmod p$. We show that in this case p does not divide the denominator of the coordinates of $P_1 + P_2$.

If $x_1 \not\equiv x_2 \pmod p$ this is trivial. Hence suppose $x_1 \equiv x_2 \pmod p$. Then $y_1 \equiv \pm y_2 \pmod p$. From $P_1 + P_2 \neq \mathcal{O} \pmod p$ it follows that

$$y_2 \equiv y_1 \not\equiv 0 \pmod p.$$

If $P_1 = P_2$, it then follows directly that the denominator of the coefficients of $P_1 + P_2 = 2P_1$ is not divisible by p. If $P_1 \neq P_2$ use the same equation as above to find

$$\frac{y_2^2 - y_1^2}{x_2 - x_1} \equiv 3x_1^2 + a \pmod p.$$

Since p does not divide $y_2 + y_1 \equiv 2y_1 \pmod p$, it follows that p does not divide the denominator of

$$\frac{y_2 - y_1}{x_2 - x_1} = \frac{y_2^2 - y_1^2}{(y_2 + y_1)(x_2 - x_1)},$$

and hence does not divide the denominator of the coordinates of $P_1 + P_2$. □

The factorization method of Lenstra uses this proposition to find a nontrivial divisor of the integer n. For this method we choose an elliptic curve $E|\mathbb{Q}$ as in the proposition. If we find one with $1 < \gcd(\Delta, n) < n$, we get a nontrivial divisor of n.

We take some bounds $B, C \in \mathbb{N}$ and

$$k = \prod_{l \le B} l^{\alpha_l},$$

where α_l is the largest exponent such that $l^{\alpha_l} \leq C$. After choosing a point $P \in E(\mathbb{Q})$ where the denominators of the coordinates are prime to n, we compute $k_1 P$ for $2 \leq k_1 \leq k$. Denote by $\sharp\widetilde{E}(\mathbb{F}_p)$ the number of points on the mod p reduced curve E (see Chapter 3, Section 3.2). If there exists a prime $p \mid n$ such that $p + 1 + 2\sqrt{p} < C$ (see Theorem 3.3) and $\sharp\widetilde{E}(\mathbb{F}_p)$ is not divisible by any prime $> B$, then k is a multiple of $\sharp\widetilde{E}(\mathbb{F}_p)$, so there exists a $2 \leq k_1 \leq k$ with $k_1 P \equiv \mathcal{O} \pmod p$. From Proposition 1.26 it follows that the denominators of the coordinates of $k_1 P$ are not prime to n. By computing the gcd of these denominators and n we find a divisor $d > 1$ of n.

It can happen that $d = n$. This happens, if $k_1 P \equiv \mathcal{O} \pmod p$ for all $p \mid n$, what is relatively unlikely if n has two or more large prime divisors. If the curve is bad, that means if for all prime numbers $p \mid n$ the integer $\sharp\widetilde{E}(\mathbb{F}_p)$ is divisible by a large prime ($> B$), we choose another elliptic curve.

Algorithm 1.27 (Lenstra Jr. [129]).

INPUT: A composite integer n.
OUTPUT: A nontrivial divisor d of n (if possible).
1. Choose an elliptic curve $E : y^2 = x^3 + a_4 x + a_6$ with $a_4, a_6 \in \mathbb{Z}$ and $\gcd(\Delta, n) = 1$.
2. Choose a point $P \in E(\mathbb{Q})$ such that the denominators of the coordinates are prime to n.
3. Choose $B, C \in \mathbb{N}$ and $k \leftarrow \prod_{l \leq B} l^{\alpha_l}$ with $\alpha_l = \left[\frac{\log C}{\log l}\right]$.
4. $Q \leftarrow P$.
5. For $k_1 = 2$ to k do
6. $\quad$ $Q \leftarrow Q + P$ with the representation $Q = \left(\frac{x_Q}{z_Q^2}, \frac{y_Q}{z_Q^3}\right)$.
7. $\quad$ If $1 < \gcd(z_Q, n) < n$ then return d.
8. Go to Step 1.

The primality test using elliptic curves is based on the following proposition.

Proposition 1.28. *Let $n \in \mathbb{N}$ be coprime to 6 and > 1. Let E be an elliptic curve over $\mathbb{Q}$ with integral coefficients and $\gcd(\Delta, n) = 1$. Let $m \in \mathbb{N}$ with a prime divisor $q > (n^{1/4} + 1)^2$. If there is a point $P \in E(\mathbb{Q})$ with*

$$mP \equiv \mathcal{O} \pmod n \quad \text{and} \quad \frac{m}{q} P \not\equiv \mathcal{O} \pmod n,$$

then n is prime.

Proof. See the article of Goldwasser and Kilian [83]. □

If we find during the computation that the multiplication is not possible, then using Proposition 1.26 we have found a divisor $d > 1$ of the integer n. If $d < n$ then n is not prime.

Algorithm 1.29 (Goldwasser, Kilian).

INPUT: An integer $n > 1$ coprime to 6.
OUTPUT: prime, not prime.

1. Choose an elliptic curve $E : y^2 = x^3 + a_4x + a_6$ with $a_4, a_6 \in \mathbb{Z}$ and $\gcd(\Delta, n) = 1$. If $1 < \gcd(\Delta, n) < n$ then return 0.
2. Compute $m = \sharp E(\mathbb{Z}/n\mathbb{Z})$ (assuming n is prime) with one of the algorithms from Chapter 3, Section 3.2 If m can not be computed, n is not prime, hence return 'not prime'.
3. Try to factorize $m = uq$ where u has small prime divisors and q satisfies some primality or pseudo-primality tests. If $q \leq (n^{1/4} + 1)^2$ go to Step 1.
4. Choose $x \in \mathbb{Z}/n\mathbb{Z}$ for which the Jacobi-symbol $\left(\frac{x^3+a_4x+a_6}{n}\right)$ is 0 or 1. Find $y \in \mathbb{Z}/n\mathbb{Z}$ with $y^2 \equiv x^3 + a_4x + a_6 \pmod{n}$. If this is not possible, n is not prime, hence return 'not prime'.
5. Consider the point $P = (x, y)$ and try to compute $mP \pmod{n}$ and $\frac{m}{q}P \pmod{n}$.
   If the multiplication is not possible then do:
6. Determine the gcd of n and of the denominators of the coordinates in question.
7. If this gcd is $< n$ then return 'not prime'.
8. If $mP \equiv \mathcal{O} \pmod{n}$ and $\frac{m}{q}P \not\equiv \mathcal{O} \pmod{n}$ then return 'prime'.
9. Go to Step 1.

We remark here that Agrawal, Kayal and Saxena [2] announced recently a deterministic polynomial-time algorithm to determine whether an integer is prime or composite (see `http://www.cse.iitk.ac.in/primality.pdf`).

1.5 Isogenies and endomorphisms of elliptic curves

In this section we give a short introduction to the theory of isogenies and endomorphisms of elliptic curves. More information can be found for example in Hasse [91] and in Silverman [204], Chapter III.

Definition 1.30. Let E_1, E_2 be elliptic curves over a field $\mathbb{K}$.

a) A *morphism* from E_1 to E_2 is a rational map which is regular at every point of E_1.

b) An *isogeny* from E_1 to E_2 is a morphism which maps $\mathcal{O}$ on E_1 to $\mathcal{O}$ on E_2. The *zero isogeny* is the only constant isogeny, it is the morphism which maps every point on E_1 to $\mathcal{O}$ on E_2.

c) A non-constant isogeny $\phi : E_1 \to E_2$ leads to an injection of function fields

$$\phi^* : \overline{\mathbb{K}}(E_2) \to \overline{\mathbb{K}}(E_1), \quad f \mapsto f \circ \phi.$$

where $\overline{\mathbb{K}}$ as always denotes the algebraic closure of $\mathbb{K}$. The isogeny ϕ is called *separable* (*purely inseparable*) if and only if the finite extension $\mathbb{K}(E_1)|\phi^*(\mathbb{K}(E_2))$ is separable (purely inseparable).

d) For a non-constant isogeny $\phi : E_1 \to E_2$ define the *degree* of ϕ as

$$\deg \phi := [\mathbb{K}(E_1) : \phi^*(\mathbb{K}(E_2))],$$

the *separable degree* of ϕ as

$$\deg_s \phi := [\mathbb{K}(E_1) : \phi^*(\mathbb{K}(E_2))]_s,$$

and the *inseparable degree* of ϕ as

$$\deg_i \phi := [\mathbb{K}(E_1) : \phi^*(\mathbb{K}(E_2))]_i.$$

For the zero isogeny 0 we define $\deg(0) := 0$.

Definition 1.31. Let $E|\mathbb{K}$ be an elliptic curve over a field $\mathbb{K}$. The set of all isogenies from E to E forms the *ring of endomorphisms* $\mathrm{End}(E)$ of E.

Proposition 1.32. *Let $\phi : E_1 \to E_2$ be an isogeny of degree d. There exists a unique isogeny $\hat{\phi} : E_2 \to E_1$ with*

$$\hat{\phi} \circ \phi = d \quad (= \textit{multiplication with } d \textit{ on } E_1),$$

$$\phi \circ \hat{\phi} = d \quad (= \textit{multiplication with } d \textit{ on } E_2).$$

The isogeny $\hat{\phi}$ is called the *dual isogeny* to ϕ. One has

$$\deg(\hat{\phi}) = \deg(\phi).$$

Proof. Silverman [204], Chapter III, Theorem 6.1.a) and Theorem 6.2.e). □

Proposition 1.33. *Let $\phi : E_1 \to E_2$ be a non-constant isogeny. Then for every $Q \in E_2$,*

$$\sharp\phi^{-1}(Q) = \deg_s \phi,$$

in particular

$$\sharp \ker \phi = \deg_s \phi.$$

If ϕ is separable then

$$\sharp \ker \phi = \deg \phi.$$

Proof. Silverman [204], Chapter III, Theorem 4.10. □

Examples for endomorphisms are the multiplication maps.

Proposition 1.34. *Let $E|\mathbb{K}$ be an elliptic curve and $m \in \mathbb{Z}$. The multiplication map corresponding to m is defined as*

$$m : E \to E, \quad P \mapsto mP.$$

a) *The map m is an endomorphism on E. Therefore the ring of integers $\mathbb{Z}$ is a subring of the endomorphism ring* $\mathrm{End}(E)$.

b) *If* $\gcd(m, \mathrm{char}(\mathbb{K})) = 1$ *or if* $\mathrm{char}(\mathbb{K}) = 0$ *then the endomorphism m is separable.*

c) *For the degree one has*

$$\deg(m) = m^2.$$

d) *The endomorphism m is its own dual isogeny:*

$$\hat{m} = m.$$

Proof. See Silverman [204], Chapter III, Section 4; Corollary 5.4; and Theorem 6.2. □

There are three classes of endomorphism rings of elliptic curves.

Theorem 1.35. *Let $E|\mathbb{K}$ be an elliptic curve over a field $\mathbb{K}$. Then the endomorphism ring of E is one of the following rings:*

- $\mathrm{End}(E) = \mathbb{Z}$,
- $\mathrm{End}(E)$ *is an order in an imaginary quadratic field,*
- $\mathrm{End}(E)$ *is an order in a quaternion algebra.*

Proof. See the article of Deuring [48], where a complete description of $\mathrm{End}(E)$ is given. □

Definition 1.36. An elliptic curve with endomorphism ring strictly larger than $\mathbb{Z}$ has *complex multiplication.*

Examples. 1) Let $\mathbb{K}$ be a field which contains the primitive third root of unity ζ. Then an elliptic curve of the form

$$E : Y^2 = X^3 + b, \quad 0 \neq b \in \mathbb{K}$$

has complex multiplication. One gets the endomorphism

$$\zeta : E \to E$$

defined by

$$\zeta P = \begin{cases} \mathcal{O} & \text{if } P = \mathcal{O}, \\ (\zeta \cdot x, y) & \text{if } P = (x, y). \end{cases}$$

It is an easy exercise to show that

$$\zeta^2 = -\zeta - 1 \quad \text{and} \quad \zeta^3 = 1$$

in terms of endomorphisms.

2) Let $\mathbb{K}$ be a field which contains the fourth root of unity $i = \sqrt{-1}$. Then an elliptic curve of the form

$$E : Y^2 = X^3 + aX, \quad 0 \neq a \in \mathbb{K}$$

has complex multiplication. One gets the endomorphism

$$i : E \to E$$

defined by

$$iP = \begin{cases} \mathcal{O} & \text{if } P = \mathcal{O}, \\ (-x, iy) & \text{if } P = (x, y). \end{cases}$$

It is an easy exercise to show that

$$i^2 = -1$$

in terms of endomorphisms.

1.6 Exercises

1) a) Homogenise the following polynomials:
   (i) $X^4 + 2X^2Y - 7Y^3$
   (ii) $-4XY^4 + Y^3 + 7X^2$

   b) Dehomogenise the following polynomials:
   (i) $XYZ^4 + 2X^2Y^3Z - 9X^5Z$
   (ii) $5XZ^2 + 3XYZ - 2Y^3$

2) Compute the Tate values, the discriminant and the j-invariant of the following curves in Weierstraß form:

a) $Y^2 + XY - 2Y = X^3 + 3$

b) $Y^2 + 4XY - 2Y = X^3 + 3X^2$

c) $Y^2 = X^3 - 5X^2$

3) Show the identities

$$4b_8 = b_2b_6 - b_4^2 \quad \text{and} \quad 12^3\Delta = c_4^3 - c_6^2.$$

4) Let $E|\mathbb{K}$ be given by an equation in long Weierstraß normal form.

a) Use the homogeneous equation to show that the point $\mathcal{O}$ is not singular.

b) Assume that E is singular at the point $P = (0, 0)$. Show that this implies that $a_3 = a_4 = a_6 = 0$. Compute c_4 and Δ.

c) Show that E has a node if and only if $\Delta = 0$ and $c_4 \neq 0$, and that E has a cusp if and only if $\Delta = c_4 = 0$.

d) Complete the proof of Proposition 1.5 for $\operatorname{char}(\mathbb{K}) \neq 2, 3$. (Hint: Use a short Weierstraß normal form.)

4) Test if the following equations define an elliptic curve.

a) $Y^2 = X^3 - 6X^2 + 9X$ over $\mathbb{Q}$

b) $Y^2 + XY = X^3 + 2X + 1$ over $\mathbb{Q}$

c) $Y^2 = X^3 - 15$ over $\mathbb{F}_2$, over $\mathbb{F}_3$, and over $\mathbb{F}_7$

5) Prove Theorem 1.7.

6) Why are the given birational transformations the only variable transformations which leave the Weierstraß normal form fixed?

7) Find a birational transformation to short Weierstraß normal form for

$$E : Y^2Z + 4XYZ = X^3 + 2X^2Z + XZ^2 + 3Z^3.$$

8) Prove Proposition 1.11

9) Check the different addition formulas for the different representations of points on elliptic curves.

10) a) Consider the elliptic curve $E : Y^2 = X^3 + 8$ over $\mathbb{Q}$. Let

$$P = (1, 3),\ T = (-2, 0) \in E(\mathbb{Q}).$$

Compute

$$2P,\ P + T,\ 2T,\ -P,\ -T,\ 2P + T.$$

b) Consider the elliptic curve $E : Y^2Z = X^3 + XZ^2 + 2Z^3$ over $\mathbb{F}_5$. Let $P = [1 : 3 : 1] \in E(\mathbb{F}_5)$. Compute

$$-P,\ 2P,\ 3P,\ 4P.$$

11) Let $E|\mathbb{Q}$ be an elliptic curve of the form

$$E : Y^2 = X^3 + a_4 X + a_6$$

with $a_4, a_6 \in \mathbb{Z}$ and $P = (x, y) \in E(\mathbb{Q})$. Show

$$x = \frac{u}{w^2}, \quad y = \frac{v}{w^3}$$

with $u, v \in \mathbb{Z}$, $w \in \mathbb{N}$, $\gcd(u, w) = \gcd(v, w) = 1$.

12) Proof Proposition 1.20.

13) Let $n, k \in \mathbb{N}$. Consider the multiplication formulas.

a) Show that

$$\frac{\phi_{nk}(X)}{\psi_{nk}^2(X)} = \frac{\phi_n\left(\frac{\phi_k(X)}{\psi_k^2(X)}\right)}{\psi_n^2\left(\frac{\phi_k(X)}{\psi_k^2(X)}\right)}$$

b) Compute the degrees and the first coefficients of the polynomials

- $\phi_{nk}(X)$,
- $\phi_n\left(\frac{\phi_k(X)}{\psi_k^2(X)}\right)\psi_n^{2n^2}(X)$,
- $\psi_{nk}^2(X)$,
- $\psi_n^2\left(\frac{\phi_k(X)}{\psi_k^2(X)}\right)\psi_n^{2n^2}(X)$.

c) Show the formulas (1.10) and (1.9) by considering the degrees and the first coefficients of the numerators and the denominators in the equation

$$\frac{\phi_{nk}(X)}{\psi_{nk}^2(X)} = \frac{\phi_n\left(\frac{\phi_k(X)}{\psi_k^2(X)}\right)}{\psi_n^2\left(\frac{\phi_k(X)}{\psi_k^2(X)}\right)} \cdot \frac{\psi_k^{2n^2}(X)}{\psi_k^{2n^2}(X)}.$$

Chapter 2

Elliptic curves over the complex numbers

Elliptic curves over the complex numbers are at the origin of the whole theory. They can be parametrized by the Weierstraß $\wp$-function and its derivative to define an analytic group homomorphic to the quotient of the additive group of complex numbers $\mathbb{C}$ by a lattice. In this connection we refer in particular to the paper of C. Meyer [146].

In the first section we define lattices. The Weierstraß $\wp$-function is presented in the second section. At the end of this chapter we develop the main theorem of complex multiplication.

In this chapter let $\mathbb{H} := \{\tau \in \mathbb{C} : \mathrm{Im}(\tau) > 0\}$ be the *upper half plane*.

2.1 Lattices

In this section we give a short introduction to lattices.

Definition 2.1. a) A *lattice* L (in $\mathbb{C}$) is the additive group

$$L := \omega_1\mathbb{Z} + \omega_2\mathbb{Z} = \langle \omega_1, \omega_2 \rangle_{\mathbb{Z}}$$

in $\mathbb{C}$, generated by $\omega_1, \omega_2 \in \mathbb{C}$ which are supposed to be linearly independent over $\mathbb{R}$. The generators ω_1, ω_2 are called *fundamental periods* of the lattice. A *period parallelogram* associated to L is given by

$$\{m_1\omega_1 + m_2\omega_2 : m_1, m_2 \in \mathbb{R}, 0 \leq m_1, m_2 < 1\}.$$

b) Let L, L' be two lattices in $\mathbb{C}$ and $\lambda \in \mathbb{C}$ with $\lambda L \subset L'$. Then λ induces a *homomorphism*

$$z \bmod L \mapsto \lambda z \bmod L',$$

also denoted by λ:

$$\lambda : \mathbb{C}/L \to \mathbb{C}/L'.$$

Either $\lambda = 0$ or $\lambda : \mathbb{C}/L \to \mathbb{C}/L'$ is surjective with kernel isomorphic to $L'/\lambda L$. Such a lattice homomorphism is called an *isogeny*.

c) Two lattices L and L' in $\mathbb{C}$ are *linearly equivalent* , if there exists a non-zero complex number $\lambda \in \mathbb{C}$ such that

$$\lambda L = L'.$$

This complex number induces an *isomorphism*

$$\lambda \colon \mathbb{C}/L \to \mathbb{C}/L'.$$

d) For two lattices L and L' in $\mathbb{C}$ with $L' \subset L$ the *index* of L' in L is defined as

$$[L : L'] := \frac{a(L')}{a(L)},$$

where $a(L)$ is the area of a period parallelogram associated to L.

Proposition 2.2. *Let $L = \omega_1\mathbb{Z} + \omega_2\mathbb{Z}$ be a lattice in $\mathbb{C}$ with $\mathrm{Im}(\frac{\omega_2}{\omega_1}) > 0$. Then L is linearly equivalent to a lattice $L' = \mathbb{Z} + \tau\mathbb{Z}$ in $\mathbb{C}$ with $\tau = \frac{\omega_2}{\omega_1} \in \mathbb{H}$.*

A lattice of the form $L' = \mathbb{Z} + \tau\mathbb{Z}$ with $\tau \in \mathbb{H}$ is called an *inhomogeneous lattice*, a lattice of the form $L = \omega_1\mathbb{Z} + \omega_2\mathbb{Z}$ as above is called a *homogeneous lattice*.

Proof. This is trivial, if we use the lattice isomorphism $\alpha = \frac{1}{\omega_1}$. □

In this section we write (as usual)

$$\sum_{\omega}{}' \quad \text{for} \quad \sum_{\omega \in L\setminus\{0\}} \qquad \text{and} \qquad \prod_{\omega}{}' \quad \text{for} \quad \prod_{\omega \in L\setminus\{0\}}.$$

We define several constants associated to a lattice in $\mathbb{C}$.

Definition 2.3. Let L be a lattice in $\mathbb{C}$.

a) We define the (convergent) quantities

$$g_2(L) := 60\sum_{\omega}{}' \frac{1}{\omega^4}, \quad g_3(L) := 140\sum_{\omega}{}' \frac{1}{\omega^6}.$$

b) The *discriminant* of the lattice is

$$\Delta(L) = g_2^3(L) - 27g_3^2(L).$$

c) The *j-invariant* of the lattice is

$$j(L) = 12^3 \frac{g_2^3(L)}{\Delta(L)}.$$

d) If the lattice L is given as $L = \mathbb{Z} + \tau\mathbb{Z}$ with $\tau \in \mathbb{H}$, we write

$$g_2(\tau) := g_2(L), \quad g_3(\tau) := g_3(L), \quad \Delta(\tau) := \Delta(L), \quad j(\tau) := j(L).$$

Proposition 2.4. *Let L and L' be two linearly equivalent lattices, $L = \lambda L'$, in $\mathbb{C}$ with $\lambda \in \mathbb{C}$. Then*

$$\begin{aligned} g_2(L) &= \lambda^{-4} g_2(L'), \\ g_3(L) &= \lambda^{-6} g_3(L'), \\ \Delta(L) &= \lambda^{-12} \Delta(L'), \\ j(L) &= j(L'). \end{aligned}$$

Proof. Exercise 1). □

As we want to compute the function $j(\tau)$ for $\tau \in \mathbb{H}$ (see section 3.3), we introduce the following functions.

Definition 2.5. For $\tau \in \mathbb{H}$ set $q = e^{2\pi i\tau}$.

a) The *Dedekind η-function* is given by

$$\eta(\tau) = q^{1/24}\prod_{n=1}^{\infty}(1-q^n) = q^{1/24}\sum_{n=-\infty}^{\infty}(-1)^n q^{(3n^2+n)/2}.$$

(Such a sum is called *q-expansion*.)

b) The *Weber functions* are

$$\begin{aligned}\mathfrak{f}(\tau) &= e^{-i\pi/24}\frac{\eta((\tau+1)/2)}{\eta(\tau)},\\ \mathfrak{f}_1(\tau) &= \frac{\eta(\tau/2)}{\eta(\tau)},\\ \mathfrak{f}_2(\tau) &= \sqrt{2}\frac{\eta(2\tau)}{\eta(\tau)}.\end{aligned}$$

The η-function satisfies the fundamental relations

$$\begin{aligned}\eta(\tau+1) &= q_1^{\frac{1}{24}}\eta(\tau), \quad \text{where } q_1 := e^{2\pi i},\\ \eta\left(-\frac{1}{\tau}\right) &= \sqrt{-i\tau}\,\eta(\tau).\end{aligned}$$

(See Weber [231] and Lang [113].) The Weber functions are not independent. They can be shown to satisfy the relations

$$\begin{aligned}\mathfrak{f}(\tau)\mathfrak{f}_1(\tau)\mathfrak{f}_2(\tau) &= \sqrt{2},\\ \mathfrak{f}(\tau)^8 &= \mathfrak{f}_1(\tau)^8 + \mathfrak{f}_2(\tau)^8.\end{aligned}$$

(See Weber [231], §34.)

For more information about the Weber functions see for example the (already quoted) book of Lang [118] or the article of Schertz [185]. (See also Lay [123].)

The following formulas can be used to compute the value of $j(\tau)$.

Proposition 2.6. *Let $\tau \in \mathbb{H}$. Then*

$$j(\tau) = \frac{(\mathfrak{f}^{24}(\tau)-16)^3}{\mathfrak{f}^{24}(\tau)} = \frac{(\mathfrak{f}_1^{24}(\tau)+16)^3}{\mathfrak{f}_1^{24}(\tau)} = \frac{(\mathfrak{f}_2^{24}(\tau)+16)^3}{\mathfrak{f}_2^{24}(\tau)}.$$

Proof. See Weber [231], §54 and the relations in §34. □

There is also a q-expansion of the function j, but because of the rapid growth of the coefficients c_n it is not recommendable to use this formula for the computation of j.

Proposition 2.7. *For $\tau \in \mathbb{H}$ there exist coefficients $c_n \in \mathbb{Z}$ with*

$$j(\tau) = \frac{1}{q} + 744 + \sum_{n=1}^{\infty} c_n q^n.$$

The values $c_n \in \mathbb{Z}$ grow rapidly with the asymptotic formula

$$c_n \sim \frac{e^{4\pi\sqrt{n}}}{\sqrt{2}n^{3/4}} \quad \text{as } n \to \infty.$$

Proof. See Silverman [207], Chapter I, Proposition 7.4.b), and Lang [118]. The asymptotic formula for c_n is given in [207] Chapter I, Remark 7.4.4. □

2.2 Weierstraß ℘-function

The connection between lattices and elliptic curves over the complex numbers is given by the Weierstraß ℘-function.

Definition 2.8. Let L be a lattice in $\mathbb{C}$. The (*classical*) *Weierstraß ℘-function* associated to L is the function

$$\wp : \mathbb{C} \to \mathbb{C} \cup \{\infty\}$$

with

$$\wp(L; z) := \wp(z) := \frac{1}{z^2} + \sum_{\omega}{}' \left(\frac{1}{(z-\omega)^2} - \frac{1}{\omega^2} \right).$$

The Weierstraß ℘-function has the following properties:

Proposition 2.9. *Let L be a lattice in $\mathbb{C}$ and $\wp$ the Weierstraß ℘-function associated to L.*

a) *The first derivative (with respect to z) is*

$$\wp'(z) = -2 \sum_{\omega \in L} \frac{1}{(z-\omega)^3}.$$

b) *Both functions $\wp$ and $\wp'$ are elliptic functions with period lattice L, i.e. they are meromorphic functions on $\mathbb{C}$ with*

$$\wp(z) = \wp(z+\omega), \quad \wp'(z) = \wp'(z+\omega)$$

for all $z \in \mathbb{C}$ and all $\omega \in L$. The latter relations mean that $\wp$, $\wp'$ are periodic with L.

c) *The Weierstraß $\wp$-function satisfies the differential equation*

$$\wp'(z)^2 = 4\wp(z)^3 - g_2(L)\wp(z) - g_3(L).$$

d) *The differential equation can be written as*

$$\wp'(z)^2 = 4\left(\wp(z) - \wp\left(\frac{\omega_1}{2}\right)\right)\left(\wp(z) - \wp\left(\frac{\omega_2}{2}\right)\right)\left(\wp(z) - \wp\left(\frac{\omega_1+\omega_2}{2}\right)\right).$$

e) *The Laurent series expansion of the Weierstraß $\wp$-function is*

$$\wp(z) = \frac{1}{z^2} + \sum_{n=1}^{\infty}(2n+1)G_{n+1}z^{2n}$$

with

$$G_k = \sum_{\omega}{}' \frac{1}{\omega^{2k}}$$

for $k \in \mathbb{N}$. Note that $g_2 = 60G_2$, $g_3 = 140G_3$.

Proof. See e.g. Fischer–Lieb [63], Chapter VII, §6, or Ahlfors [3], Chapter 7, Section 3. □

Now we generalize the definition of the Weierstraß $\wp$-function (see Folz [64]).

Definition 2.10. Let L be a lattice in $\mathbb{C}$. The *generalized Weierstraß functions* associated to L are functions $\mathfrak{p}, \tilde{\mathfrak{p}}$ over $\mathbb{C}$ with the property that

$$\mathfrak{p}(z) := \wp(z) + d_2$$

and

$$\begin{aligned}\tilde{\mathfrak{p}}(z) &:= \frac{1}{2}\wp'(z) + d_1\wp(z) + d_3 \\ &= \frac{1}{2}\mathfrak{p}'(z) + d_1\mathfrak{p}(z) + d_3 - d_1 d_2.\end{aligned}$$

where $\wp$ is the classical Weierstraß $\wp$-function associated to L and $d_1, d_2, d_3 \in \mathbb{C}$ are complex numbers.

Note. For sake of simplicity we do not use a notation like $\mathfrak{p}_{d_1,d_2,d_3}$ and $\tilde{\mathfrak{p}}_{d_1,d_2,d_3}$; hence we assume that the constants d_1, d_2, d_3 are known in the context.

From the differential equation of the Weierstraß $\wp$-function we get an equation for the generalized Weierstraß functions.

Proposition 2.11. *Let L be a lattice in $\mathbb{C}$ and $\mathfrak{p}$, $\tilde{\mathfrak{p}}$ the generalized Weierstraß functions associated to L with constants $d_1, d_2, d_3 \in \mathbb{C}$. Then*

$$\tilde{\mathfrak{p}}(z)^2 + a_1\mathfrak{p}(z)\tilde{\mathfrak{p}}(z) + a_3\tilde{\mathfrak{p}}(z) = \mathfrak{p}(z)^3 + a_2\mathfrak{p}(z)^2 + a_4\mathfrak{p}(z) + a_6$$

with

$$\begin{aligned} a_1 &= -2d_1, \\ a_2 &= -d_1^2 - 3d_2, \\ a_3 &= 2(d_1d_2 - d_3), \\ a_4 &= -\frac{1}{4}g_2(L) + 3d_2^2 + 2d_1(d_1d_2 - d_3), \\ a_6 &= -\frac{1}{4}g_3(L) + \frac{1}{4}d_2g_2(L) - d_2^3 - (d_1d_2 - d_3)^2. \end{aligned}$$

Proof. Exercise 2) or Folz [64]. □

Using the values a_1, a_2, a_3, a_4, a_6 we can compute the Tate values

$$\begin{aligned} b_2 &= a_1^2 + 4a_2 = -12d_2, \\ b_4 &= a_1a_3 + 2a_4 = -\frac{1}{2}g_2(L) + 6d_2^2, \\ b_6 &= a_3^2 + 4a_6 = -g_3(L) + d_2g_2(L) - 4d_2^3, \\ b_8 &= a_1^2a_6 + 4a_2a_6 - a_1a_3a_4 + a_2a_3^2 - a_4^2 = \frac{1}{4}(b_2b_6 - b_4^2) \\ &= -\frac{1}{16}g_2(L)(g_2(L) + 24d_2^2) + 3d_2(g_3(L) + d_2^3). \end{aligned}$$

The generalized Weierstraß function $\mathfrak{p}(z)$ is also connected with the classical Weierstraß $\wp$-function via the following differential equation.

Proposition 2.12. *Let L be a lattice in $\mathbb{C}$ with periods $\omega_1, \omega_2 \in \mathbb{C}$, associated to the functions $\wp$ and $\mathfrak{p}$ as above. Then*

$$\begin{aligned} \wp'(z)^2 &= 4\left(\mathfrak{p}(z) - \mathfrak{p}\left(\frac{\omega_1}{2}\right)\right)\left(\mathfrak{p}(z) - \mathfrak{p}\left(\frac{\omega_2}{2}\right)\right)\left(\mathfrak{p}(z) - \mathfrak{p}\left(\frac{\omega_1+\omega_2}{2}\right)\right) \\ &= 4\mathfrak{p}(z)^3 + b_2\mathfrak{p}(z)^2 + 2b_4\mathfrak{p}(z) + b_6. \end{aligned}$$

Proof. This follows from the definition of the generalized Weierstraß functions and the differential equations in 2.9. □

From complex analysis one obtains addition theorems for the classical Weierstraß ℘-function.

Theorem 2.13 (Classical addition theorem). *Let L be a lattice in $\mathbb{C}$ with corresponding Weierstraß ℘-function ℘.*

a) *For $z_1, z_2 \in \mathbb{C}$ with $z_1 \not\equiv z_2 \pmod{L}$ the addition formulas are*

$$\wp(z_1+z_2) = -(\wp(z_1)+\wp(z_2)) + \frac{1}{4}\left(\frac{\wp'(z_2)-\wp'(z_1)}{\wp(z_2)-\wp(z_1)}\right)^2$$

and

$$\wp'(z_1+z_2) = \left(\frac{\wp'(z_2)-\wp'(z_1)}{\wp(z_2)-\wp(z_1)}\right)(\wp(z_1)-\wp(z_1+z_2)) - \wp'(z_1).$$

b) *For $z \in \mathbb{C}$ with $2z \not\equiv 0 \pmod{L}$ the duplication formulas are*

$$\wp(2z) = -2\wp(z) + \frac{1}{4}\left(\frac{6\wp(z)^2 - \frac{1}{2}g_2(L)}{\wp'(z)}\right)^2$$

and

$$\wp'(2z) = \left(\frac{6\wp(z)^2 - \frac{1}{2}g_2(L)}{\wp'(z)}\right)(\wp(z)-\wp(2z)) - \wp'(z).$$

Proof. See for example Hurwitz–Courant [102], II.1, §8 or Folz [64]. □

From these classical addition formulas we can deduce the generalized addition formulas (see also Chapter 1, Section 1.2, Theorem 1.16).

Theorem 2.14 (Generalized addition theorem). *Let L be a lattice in $\mathbb{C}$ with corresponding generalized Weierstraß functions $\mathfrak{p}$ and $\tilde{\mathfrak{p}}$ for the complex constants d_1, d_2, d_3.*

a) *For $z_1, z_2 \in \mathbb{C}$ with $z_1 \not\equiv z_2 \pmod{L}$ the addition formulas are*

$$\begin{aligned}\mathfrak{p}(z_1+z_2) &= -(\mathfrak{p}(z_1)+\mathfrak{p}(z_2)) + \left(\frac{\tilde{\mathfrak{p}}(z_2)-\tilde{\mathfrak{p}}(z_1)}{\mathfrak{p}(z_2)-\mathfrak{p}(z_1)}\right)^2 \\ &\quad + a_1\left(\frac{\tilde{\mathfrak{p}}(z_2)-\tilde{\mathfrak{p}}(z_1)}{\mathfrak{p}(z_2)-\mathfrak{p}(z_1)}\right) - a_2\end{aligned}$$

and

$$\tilde{\mathfrak{p}}(z_1+z_2) = \left(\frac{\tilde{\mathfrak{p}}(z_2)-\tilde{\mathfrak{p}}(z_1)}{\mathfrak{p}(z_2)-\mathfrak{p}(z_1)}\right)(\mathfrak{p}(z_1)-\mathfrak{p}(z_1+z_2)) \\ - a_1\mathfrak{p}(z_1+z_2) - a_3 - \tilde{\mathfrak{p}}(z_1).$$

b) *For* $z \in \mathbb{C}$ *with* $2z \not\equiv 0 \pmod{L}$ *the duplication formulas are*

$$\mathfrak{p}(2z) = -2\mathfrak{p}(z) + \frac{1}{4}\left(\frac{6\mathfrak{p}(z)^2+b_2\mathfrak{p}(z)+b_4}{2\tilde{\mathfrak{p}}(z)+a_1\mathfrak{p}(z)+a_3}\right)^2 - \frac{1}{4}b_2$$

and

$$\tilde{\mathfrak{p}}(2z) = \frac{1}{2}\left(\frac{6\mathfrak{p}(z)^2+b_2\mathfrak{p}(z)+b_4}{2\tilde{\mathfrak{p}}(z)+a_1\mathfrak{p}(z)+a_3}\right)(\mathfrak{p}(z)-\mathfrak{p}(2z)) \\ - \frac{1}{2}a_1(\mathfrak{p}(z)+\mathfrak{p}(2z)) - a_3 - \tilde{\mathfrak{p}}(z).$$

Proof. Inserting $\mathfrak{p}(z) = \wp(z) + d_2$ in the equation for $\wp(z_1+z_2)$, we have (for $z_1 \not\equiv z_2 \pmod{L}$)

$$\mathfrak{p}(z_1+z_2) = -(\mathfrak{p}(z_1)+\mathfrak{p}(z_2)) + \frac{1}{4}\left(\frac{\wp'(z_2)-\wp'(z_1)}{\mathfrak{p}(z_2)-\mathfrak{p}(z_1)}\right)^2 + 3d_2.$$

We also have

$$\wp'(z) = 2\tilde{\mathfrak{p}}(z) - 2d_1\mathfrak{p}(z) + 2(d_1d_2-d_3) = 2\tilde{\mathfrak{p}}(z) + a_1\mathfrak{p}(z) + a_3. \tag{2.1}$$

In sum we obtain for $z_1 \not\equiv z_2 \pmod{L}$

$$\begin{aligned}
&\mathfrak{p}(z_1+z_2) \\
&= -(\mathfrak{p}(z_1)+\mathfrak{p}(z_2)) + \frac{1}{4}\left(\frac{2(\tilde{\mathfrak{p}}(z_2)-\tilde{\mathfrak{p}}(z_1))+a_1(\mathfrak{p}(z_2)-\mathfrak{p}(z_1))}{\mathfrak{p}(z_2)-\mathfrak{p}(z_1)}\right)^2 + 3d_2 \\
&= -(\mathfrak{p}(z_1)+\mathfrak{p}(z_2)) + \frac{1}{4}\left(2\left(\frac{\tilde{\mathfrak{p}}(z_2)-\tilde{\mathfrak{p}}(z_1)}{\mathfrak{p}(z_2)-\mathfrak{p}(z_1)}\right)+a_1\right)^2 + 3d_2 \\
&= -(\mathfrak{p}(z_1)+\mathfrak{p}(z_2)) + \left(\frac{\tilde{\mathfrak{p}}(z_2)-\tilde{\mathfrak{p}}(z_1)}{\mathfrak{p}(z_2)-\mathfrak{p}(z_1)}\right)^2 + a_1\left(\frac{\tilde{\mathfrak{p}}(z_2)-\tilde{\mathfrak{p}}(z_1)}{\mathfrak{p}(z_2)-\mathfrak{p}(z_1)}\right) - a_2
\end{aligned}$$

because

$$\frac{1}{4}a_1^2 + 3d_2 = d_1^2 + 3d_2 = -a_2.$$

The other equations are left to the reader as an exercise (see Exercise 3)). □

With the Weierstraß functions we get isomorphisms between the complex plane $\mathbb{C}$ modulo a lattice and an elliptic curve over the complex numbers $\mathbb{C}$.

Theorem 2.15. *Let L be a lattice in $\mathbb{C}$.*

a) *The Weierstraß $\wp$-function associated to L induces an isomorphism from $\mathbb{C}/L$ to the elliptic curve*

$$E : Y^2 = X^3 - \frac{g_2(L)}{4}X - \frac{g_3(L)}{4}$$

over $\mathbb{C}$ by the function

$$z \pmod{L} \mapsto \left(\wp(z), \frac{1}{2}\wp'(z)\right).$$

b) *The generalized Weierstraß functions $\mathfrak{p}$, $\tilde{\mathfrak{p}}$ associated to L and the constants d_1, $d_2, d_3 \in \mathbb{C}$ induce an isomorphism from $\mathbb{C}/L$ to the elliptic curve*

$$E : Y^2 + a_1XY + a_3Y = X^3 + a_2X^2 + a_4X + a_6$$

over $\mathbb{C}$ by the function

$$z \pmod{L} \mapsto (\mathfrak{p}(z), \tilde{\mathfrak{p}}(z)).$$

The constants a_1, a_2, a_3, a_4, a_6 are given in Proposition 2.11

This is the parametrization of the curve $E|\mathbb{C}$ by the Weierstraß $\wp$-function (and its derivative $\wp'$).

Proof. a) The function is a homomorphism because of the addition formulas given above. As $\wp$ has only $0 \pmod{L}$ as a pole in $\mathbb{C}/L$, the kernel of the function is trivial, so the function is injective. Further $\wp(z)$ attains every value in $\mathbb{C}$ twice (counted with multiplicity). So if $(x, y) \in E(\mathbb{C})$, there is a complex number $z \pmod{L}$ with $\wp(z) = x$. From the differential equation it follows that $\frac{1}{2}\wp'(z) = \pm y$. As $\wp$ is an even function and $\wp'$ is an odd function, it follows that

$$\wp(-z) = x \quad \text{and} \quad \frac{1}{2}\wp'(-z) = -\frac{1}{2}\wp'(z) = \mp y.$$

Hence there is exactly one value $z \pmod{L}$ with $(\wp(z), \frac{1}{2}\wp'(z)) = (x, y)$.

b) This is proved in the same way as above. To show that the function is surjective, we need the fact that $\mathfrak{p}$ is an even function which attains every value in $\mathbb{C}$ twice. This follows from the corresponding property of the Weierstraß $\wp$-function. We also need the fact that

$$\begin{aligned}
\tilde{\mathfrak{p}}(-z) &= \frac{1}{2}\wp'(-z) + d_1\wp(-z) + d_3 \\
&= -\frac{1}{2}\wp'(z) + d_1\wp(z) + d_3 \\
&= -\tilde{\mathfrak{p}}(z) + 2d_1\wp(z) + 2d_3 \\
&= -\tilde{\mathfrak{p}}(z) - a_1\mathfrak{p}(z) - 2(d_1d_2 - d_3) \\
&= -\tilde{\mathfrak{p}}(z) - a_1\mathfrak{p}(z) - a_3.
\end{aligned}$$

□

The isomorphism between an elliptic curve $E|\mathbb{C}$ and $\mathbb{C}/L$, where L is a lattice of the form $L = \mathbb{Z} + \tau\mathbb{Z}$, leads to the fact that

$$j(E) = j(\tau).$$

In the last part of this section we prove the generalized multiplication formulas. These are formulas for the values of $\mathfrak{p}(nz)$, $\tilde{\mathfrak{p}}(nz)$ for $n \in \mathbb{N}$. As a result we get the multiplication formulas from Chapter 1, Section 1.3. The proofs are adopted from Folz [64] (see also Cassels [25] or Zimmer [255]).

Remark. We have changed the notation here (see Definition 1.18): If $P = (x, y)$ is a point in the group $E(\mathbb{K})$ for an elliptic curve E in short Weierstraß form over a number field $\mathbb{K}$, we have $x = \wp(z)$ and $y = \frac{1}{2}\wp'(z)$ (see Theorem 2.15). The number field $\mathbb{K}$ can be regarded as a subfield of the field $\mathbb{C}$ of complex numbers in which the variable z varies.

We fix until the end of this section a lattice L in $\mathbb{C}$ with fundamental periods ω_1, ω_2 and the constants $d_1, d_2, d_3 \in \mathbb{C}$.

Lemma 2.16. *For $n \in \mathbb{N}$ there exists an elliptic function ψ_n such that*

$$\psi_n(z)^2 = n^2 \prod_{\substack{r,s=0 \\ (r,s)\neq(0,0)}}^{n-1} \left(\mathfrak{p}(z) - \mathfrak{p}\left(\frac{r\omega_1 + s\omega_2}{n}\right)\right).$$

The Laurent series expansion of the function ψ_n begins according to

$$\psi_n(z) = \frac{(-1)^{n+1}n}{z^{n^2-1}} + \cdots .$$

Proof. (See also the book of Fricke [71].) We have to show, among other things, that the product on the right side is a square. If n is odd, then because of

$$\frac{r\omega_1 + s\omega_2}{n} \equiv -\frac{(n-r)\omega_1 + (n-s)\omega_2}{n} \pmod{L}$$

and therefore

$$\begin{aligned} \mathfrak{p}\left(\frac{r\omega_1 + s\omega_2}{n}\right) &= \mathfrak{p}\left(-\frac{(n-r)\omega_1 + (n-s)\omega_2}{n}\right) \\ &= \mathfrak{p}\left(\frac{(n-r)\omega_1 + (n-s)\omega_2}{n}\right) \end{aligned}$$

(the function $\mathfrak{p}$ is even), every factor of the product appears twice, so that the product is a square.

If n is even, every factor of the product shows up twice, except for a factor for which

$$\begin{aligned}
\frac{r\omega_1 + s\omega_2}{n} &\equiv \frac{(n-r)\omega_1 + (n-s)\omega_2}{n} \pmod{L} \\
\Leftrightarrow 2\frac{r\omega_1 + s\omega_2}{n} &\equiv 0 \pmod{L} \\
\Leftrightarrow \quad \frac{r\omega_1 + s\omega_2}{n} &\equiv \frac{\omega_1}{2}, \frac{\omega_2}{2}, \frac{\omega_1 + \omega_2}{2} \pmod{L}.
\end{aligned}$$

Therefore we have to show that

$$\left(\mathfrak{p}(z) - \mathfrak{p}\left(\frac{\omega_1}{2}\right)\right)\left(\mathfrak{p}(z) - \mathfrak{p}\left(\frac{\omega_2}{2}\right)\right)\left(\mathfrak{p}(z) - \mathfrak{p}\left(\frac{\omega_1+\omega_2}{2}\right)\right)$$

is a square. This follows from Proposition 2.12.

From the Laurent series expansion of the Weierstraß $\wp$-function it follows that the Laurent series expansion of the generalized Weierstraß function begins according to

$$\mathfrak{p}(z) = \frac{1}{z^2} + \cdots,$$

and the Laurent series expansion of $\wp'$ begins according to

$$\wp'(z) = \frac{(-2)}{z^3} + \cdots.$$

If n is odd, then ψ_n is a polynomial in $\mathfrak{p}(z)$ of degree $\frac{n^2-1}{2}$ with first coefficient n, so its Laurent series expansion begins according to

$$\psi_n(z) = \frac{n}{z^{n^2-1}} + \cdots.$$

If n is even, it follows from the above discussion that

$$\psi_n(z) = \frac{1}{2}\wp'(z) P_n(\mathfrak{p}(z)),$$

where P_n is a polynomial in $\mathfrak{p}(z)$ of degree $\frac{n^2-4}{2}$ with first coefficient n. Therefore, in both cases, the Laurent series expansion of ψ_n begins according to

$$\psi_n(z) = \frac{(-1)^{n+1} n}{z^{n^2-1}} + \cdots.$$

Of course, we consider here only Laurent series expansions around 0. □

Definition 2.17. Let L be a lattice in $\mathbb{C}$.

a) The product

$$\sigma(z) := z \prod_{\omega}{}' \left(\left(1 - \frac{z}{\omega}\right) \cdot \exp\left(\frac{z}{\omega} + \frac{1}{2}\left(\frac{z}{\omega}\right)^2\right)\right)$$

defines a holomorphic function on $\mathbb{C}$, which has the points in L as simple zeros and no other zeros. This function is called the *Weierstraß σ-function* (cf. Hurwitz–Courant [102], II.1.§13).

b) The *Weierstraß ζ-function* associated to L is

$$\zeta(z) = \frac{1}{z} + \sum_{\omega}{}' \left(\frac{1}{z-\omega} + \frac{1}{\omega} + \frac{z}{\omega^2} \right)$$

(cf. Hurwitz–Courant [102], II.1.§11).

c) The *quasi-period map* η associated to L is the map

$$\eta : L \to \mathbb{C}$$

with

$$\zeta(z+\omega) = \zeta(z) + \eta(\omega) \quad \text{for all } z \in \mathbb{C},\ \omega \in L,$$

where ζ is the Weierstraß ζ-function associated to L.

The connection between the Weierstraß ζ-function and the Weierstraß $\wp$-function is given by the formula

$$\frac{d}{dz}\zeta(z) = -\wp(z).$$

The quasi-period map η should not be confused with the Dedekind η-function (see Definition 2.5). Let the lattice L be given by the periods ω_1, ω_2. We define $\eta_j := 2\zeta\left(\frac{\omega_j}{2}\right)$ for $j = 1, 2$. Then for $\omega = m_1\omega_1 + m_2\omega_2 \in L$ the quasi-period map is

$$\eta(\omega) = m_1\eta_1 + m_2\eta_2.$$

The numbers η_j satisfy the *Legendre-relation*

$$\eta_1\omega_2 - \eta_2\omega_1 = 2\pi i.$$

For more information about these functions see for example the book of Hurwitz and Courant [102] II.1.

The Weierstraß σ-function has the following properties.

Proposition 2.18. a) *Let $\omega = m_1\omega_1 + m_2\omega_2 \in L$ with $m_1, m_2 \in \mathbb{Z}$. Then*

$$\sigma(z+\omega) = \varepsilon(\omega) e^{\eta(\omega)(z+\frac{\omega}{2})}\sigma(z)$$

with

$$\varepsilon(\omega) = \begin{cases} 1 & \text{if } \frac{\omega}{2} \in L, \\ -1 & \text{if } \frac{\omega}{2} \notin L, \end{cases}$$

where $\eta(\omega)$ is the quasi-period map associated to L.

b) *We have, for* $z_1, z_2 \in \mathbb{C}$, *the relation*

$$\wp(z_1) - \wp(z_2) = -\frac{\sigma(z_1+z_2)\sigma(z_1-z_2)}{\sigma(z_1)^2\sigma(z_2)^2}.$$

c) *We have furthermore*

$$\wp'(z) = -\frac{\sigma(2z)}{\sigma(z)^4}.$$

d) *For* $n \in \mathbb{N}$, *the formula*

$$\psi_n(z) = (-1)^{n+1}\frac{\sigma(nz)}{\sigma(z)^{n^2}}.$$

is valid. In particular,

$$\psi_1(z) = \frac{\sigma(z)}{\sigma(z)} = 1.$$

Proof. We only give the proof of Part d). The proof of the other properties can be found in Hurwitz, Courant [102], II.1, §13 and §14. (It was not the intention in this book to give all the proofs in detail.)

For the proof of Part d) (see Folz [64]), we first show that the function on the right hand side is an elliptic function, i.e. that it is meromorphic with period lattice L. Let $\omega \in L$. From a) it follows that

$$\begin{aligned}\frac{\sigma(n(z+\omega))}{\sigma(z+\omega)^{n^2}} &= \frac{\varepsilon(n\omega)e^{\eta(n\omega)(nz+\frac{n\omega}{2})}\sigma(nz)}{\varepsilon(\omega)^{n^2}e^{\eta(\omega)(z+\frac{\omega}{2})n^2}\sigma(z)^{n^2}}\\ &= \frac{\varepsilon(\omega)^{n}e^{\eta(\omega)(z+\frac{\omega}{2})n^2}\sigma(nz)}{\varepsilon(\omega)^{n^2}e^{\eta(\omega)(z+\frac{\omega}{2})n^2}\sigma(z)^{n^2}}\\ &= \frac{\sigma(nz)}{\sigma(z)^{n^2}}.\end{aligned}$$

Because $\sigma(z)$ has a simple zero at $z \equiv 0 \pmod{L}$, the function $\frac{\sigma(nz)}{\sigma(z)^{n^2}}$ has a pole of order n^2-1 at $z \equiv 0 \pmod{L}$ and a simple zero at

$$z = \frac{r\omega_1 + s\omega_2}{n}$$

for $r, s = 0, \ldots, n-1$, $(r, s) \neq (0, 0)$. These are exactly the same poles and zeros (in the period parallelogram) of the function $\psi_n(z)$. (This can be seen in Lemma 2.16.) Therefore these functions differ only by a constant c,

$$\psi_n(z) = c\frac{\sigma(nz)}{\sigma(z)^{n^2}}.$$

The Laurent series expansion around $z = 0$ of these two elliptic functions begin as follows:

$$\psi_n(z) = \frac{(-1)^{n+1}n}{z^{n^2-1}} + \cdots$$

(see Lemma 2.16) and

$$\frac{\sigma(nz)}{\sigma(z)^{n^2}} = \frac{n}{z^{n^2-1}} + \cdots$$

(which becomes clear from the definition of the Weierstraß σ-function), so that the constant is $c = (-1)^{n+1}$. □

We now obtain the following representation for $\mathfrak{p}(nz) - \mathfrak{p}(mz)$:

Lemma 2.19. *Let $z \in \mathbb{C}$, $m, n \in \mathbb{N}$ with $m \neq n$. Then*

$$\mathfrak{p}(nz) - \mathfrak{p}(mz) = -\frac{\psi_{n+m}(z)\psi_{n-m}(z)}{\psi_n(z)^2\psi_m(z)^2}.$$

Proof. From Part b) of Proposition 2.18 we obtain

$$\mathfrak{p}(nz) - \mathfrak{p}(mz) = -\frac{\sigma((n+m)z)\sigma((n-m)z)}{\sigma(nz)^2\sigma(mz)^2}.$$

On the other hand we obtain with Part d) of Proposition 2.18

$$\begin{aligned}
&\frac{\psi_{n+m}(z)\psi_{n-m}(z)}{\psi_n(z)^2\psi_m(z)^2} \\
&\quad = \frac{(-1)^{n+m+1}\sigma((n+m)z)(-1)^{n-m+1}\sigma((n-m)z)(\sigma(z)^{n^2})^2(\sigma(z)^{m^2})^2}{\sigma(z)^{(n+m)^2}\sigma(z)^{(n-m)^2}\sigma(nz)^2\sigma(mz)^2} \\
&\quad = \frac{\sigma((n+m)z)\sigma((n-m)z)}{\sigma(nz)^2\sigma(mz)^2}.
\end{aligned}$$

□

Corollary 2.20. *Let $z \in \mathbb{C}$ and $n \in \mathbb{N}$, $n > 1$. Then*

$$\mathfrak{p}(nz) = \mathfrak{p}(z) - \frac{\psi_{n+1}(z)\psi_{n-1}(z)}{\psi_n(z)^2}.$$

Proof. We put $m = 1$ in the preceding lemma. □

We also need the following lemma.

Lemma 2.21. *Let $z \in \mathbb{C}$ and $n \in \mathbb{N}$. Then*

$$\wp'(nz) = 2\tilde{\mathfrak{p}}(nz) + a_1\mathfrak{p}(nz) + a_3 = \frac{\psi_{2n}(z)}{\psi_n(z)^4}.$$

Proof. The first equation is equation (2.1). From Part c) and d) of Proposition 2.18 we get

$$\wp'(nz) = -\frac{\sigma(2nz)}{\sigma(nz)^4} = \frac{(-1)^{2n+1}\frac{\sigma(2nz)}{\sigma(z)^{(2n)^2}}}{\left(\frac{\sigma(nz)}{\sigma(z)^{n^2}}\right)^4} = \frac{\psi_{2n}(z)}{\psi_n(z)^4}.$$

□

We now define the functions ψ_n for all integers n.

Definition 2.22. For $z \in \mathbb{C}$ and for $n \in \mathbb{N}$ put

$$\psi_0(z) := 0,$$
$$\psi_{-n}(z) := -\psi_n(z).$$

As $\sigma(z)$ is an odd function, this definition can be regarded as a consequence of the representation

$$\psi_n(z) = (-1)^{n+1}\frac{\sigma(nz)}{\sigma(z)^{n^2}}$$

for $n \in \mathbb{N}$, when we take $-n$ instead of n on the right side of the equation.

Lemma 2.23. *For $z \in \mathbb{Z}$ and $k, l, m \in \mathbb{Z}$, the relation*

$$\psi_{k+l}(z)\psi_{k-l}(z)\psi_m(z)^2 + \psi_{l+m}(z)\psi_{l-m}(z)\psi_k(z)^2 + \psi_{m+k}(z)\psi_{m-k}(z)\psi_l(z)^2 = 0$$

holds.

Proof. From Lemma 2.19 we have

$$\begin{aligned} 0 &= (\mathfrak{p}(kz) - \mathfrak{p}(lz)) + (\mathfrak{p}(lz) - \mathfrak{p}(mz)) + (\mathfrak{p}(mz) - \mathfrak{p}(kz)) \\ &= -\frac{\psi_{k+l}(z)\psi_{k-l}(z)}{\psi_k(z)^2\psi_l(z)^2} - \frac{\psi_{l+m}(z)\psi_{l-m}(z)}{\psi_l(z)^2\psi_m(z)^2} - \frac{\psi_{m+k}(z)\psi_{m-k}(z)}{\psi_m(z)^2\psi_k(z)^2}. \end{aligned}$$

Multiplying by $-\psi_k(z)^2\psi_l(z)^2\psi_m(z)^2$ we obtain the asserted equation. □

Corollary 2.24. *For $z \in \mathbb{C}$ and $k, l \in \mathbb{Z}$, we have*

$$\psi_{k+l}(z)\psi_{k-l}(z) = \psi_{k+1}(z)\psi_{k-1}(z)\psi_l(z)^2 - \psi_{l+1}(z)\psi_{l-1}(z)\psi_k(z)^2.$$

Proof. Put $m = 1$ in Lemma 2.23 and use the identity $\psi_{1-k}(z) = -\psi_{k-1}(z)$ (see Definition 1.18). □

For the proof of the multiplication formulas we first develop recursion formulas for the functions ψ_n.

Theorem 2.25. *For $n \in \mathbb{N}$, the functions ψ_n are polynomials in $\mathfrak{p}$ and $\tilde{\mathfrak{p}}$ with coefficients in $\mathbb{Z}[a_1, a_2, a_3, a_4, a_6]$. These polynomials are given by the recursion formulas*

$$\begin{aligned}
\psi_1(z) &= 1, \\
\psi_2(z) &= 2\tilde{\mathfrak{p}}(z) + a_1\mathfrak{p}(z) + a_3, \\
\psi_3(z) &= 3\mathfrak{p}(z)^4 + b_2\mathfrak{p}(z)^3 + 3b_4\mathfrak{p}(z)^2 + 3b_6\mathfrak{p}(z) + b_8, \\
\psi_4(z) &= \psi_2(z)(2\mathfrak{p}(z)^6 + b_2\mathfrak{p}(z)^5 + 5b_4\mathfrak{p}(z)^4 + 10b_6\mathfrak{p}(z)^3 + 10b_8\mathfrak{p}(z)^2 \\
&\qquad + (b_2b_8 - b_4b_6)\mathfrak{p}(z) + b_4b_8 - b_6^2),
\end{aligned}$$

and for $n \geq 2$:

$$\begin{aligned}
\psi_{2n+1}(z) &= \psi_{n+2}(z)\psi_n(z)^3 - \psi_{n+1}(z)^3\psi_{n-1}(z), \\
\psi_{2n}(z)\psi_2(z) &= \psi_n(z)(\psi_{n-1}(z)^2\psi_{n+2}(z) - \psi_{n-2}(z)\psi_{n+1}(z)^2).
\end{aligned}$$

Proof. The formula for ψ_1 follows from Proposition 2.18 d). For ψ_2, we see with the same proposition and equation (2.1) that

$$\psi_2(z) = \frac{-\sigma(2z)}{\sigma(z)^4} = \wp'(z) = 2\tilde{\mathfrak{p}}(z) + a_1\mathfrak{p}(z) + a_3.$$

From Lemma 2.19, we then get

$$\begin{aligned}
\mathfrak{p}(2z) - \mathfrak{p}(z) &= \frac{-\psi_3(z)}{\psi_2(z)^2} \\
\Leftrightarrow \psi_3(z) &= -\wp'(z)^2(\mathfrak{p}(2z) - \mathfrak{p}(z)) \\
&= -\wp'(z)^2\left(-2\mathfrak{p}(z) + \frac{1}{4}\left(\frac{6\mathfrak{p}(z)^2 + b_2\mathfrak{p}(z) + b_4}{2\tilde{\mathfrak{p}}(z) + a_1\mathfrak{p}(z) + a_3}\right)^2 - \frac{1}{4}b_2 - \mathfrak{p}(z)\right)
\end{aligned}$$

where we applied the duplication formula for the function $\mathfrak{p}$ (Theorem 2.14). With the definition of $\tilde{\mathfrak{p}}$ (see Equation (2.1)) it follows that

$$\begin{aligned}
\psi_3(z) &= -\wp'(z)^2\left(- 3\mathfrak{p}(z) + \frac{1}{4}\left(\frac{6\mathfrak{p}(z)^2 + b_2\mathfrak{p}(z) + b_4}{\wp'(z)}\right)^2 - \frac{1}{4}b_2\right) \\
&= 3\wp'(z)^2\mathfrak{p}(z) - \frac{1}{4}(6\mathfrak{p}(z)^2 + b_2\mathfrak{p}(z) + b_4)^2 + \frac{1}{4}b_2\wp'(z)^2.
\end{aligned}$$

Using the differential equation from Proposition 2.12 one gets after further computation

$$\psi_3(z) = 3\mathfrak{p}(z)^4 + b_2\mathfrak{p}(z)^3 + 3b_4\mathfrak{p}(z)^2 + 3b_6\mathfrak{p}(z) + b_8.$$

For the computation of ψ_4 we have from Lemma 2.21 and Theorem 2.13

$$
\begin{aligned}
\wp'(2z) &= \frac{\psi_4(z)}{\psi_2(z)^4} \\
\Leftrightarrow \ \psi_4(z) &= \wp'(2z)\psi_2(z)^4 \\
&= \psi_2(z)\wp'(2z)\wp'(z)^3 \\
&= \psi_2(z)\left(\left(\frac{6\wp(z)^2 - \frac{1}{2}g_2}{\wp'(z)}\right)(\wp(z) - \wp(2z)) - \wp'(z)\right)\wp'(z)^3 \\
&= \psi_2(z)\left(\left(6\wp(z)^2 - \frac{1}{2}g_2\right)(\wp(z) - \wp(2z)) - \wp'(z)^2\right)\wp'(z)^2
\end{aligned}
$$

with $g_2 = g_2(L)$. (Here we used the duplication formula for $\wp'$.) Employing the duplication formula for $\wp$, we obtain for $\psi_4(z)$ the equation:

$$
\psi_2(z)\left(\left(6\wp(z)^2 - \frac{1}{2}g_2\right)\left(3\wp(z)\wp'(z)^2 - \frac{1}{4}\left(6\wp(z)^2 - \frac{1}{2}g_2\right)^2\right) - \wp'(z)^4\right).
$$

Further computation using the formula $\wp(z) = \mathfrak{p}(z) + \frac{1}{12}b_2$ and the formulas on page 38 leads to the desired equation

$$
\begin{aligned}
\psi_4(z) = \ &\psi_2(z)\big(2\mathfrak{p}(z)^6 + b_2\mathfrak{p}(z)^5 + 5b_4\mathfrak{p}(z)^4 + 10b_6\mathfrak{p}(z)^3 + 10b_8\mathfrak{p}(z)^2 \\
&+ (b_2b_8 - b_4b_6)\mathfrak{p}(z) + b_4b_8 - b_6^2\big).
\end{aligned}
$$

The formula for ψ_{2n+1} can be obtained by using Corollary 2.24 for $k = n + 1$ and $l = n$. Using the same corollary for $k = n + 1$ and $l = n - 1$ leads to the formula for ψ_{2n}. (See Exercise 5.) □

Now we can prove the multiplication formulas (see Definition 1.18).

Theorem 2.26. *Let $n \in \mathbb{N}$. There are elliptic functions ψ_n, ϕ_n, Ω_n, and $\tilde{\Omega}_n$, such that*

$$
\begin{aligned}
\mathfrak{p}(nz) &= \frac{\phi_n(z)}{\psi_n(z)^2}, \\
\tilde{\mathfrak{p}}(nz) &= \frac{\Omega_n(z)}{\psi_n(z)^3}, \\
\frac{1}{2}\wp'(nz) &= \frac{\tilde{\Omega}_n(z)}{\psi_n(z)^3}.
\end{aligned}
$$

Here ψ_n, ϕ_n, Ω_n are polynomials in $\mathfrak{p}$ and $\tilde{\mathfrak{p}}$, and $\tilde{\Omega}_n$ is a polynomial in $\mathfrak{p}$ and $\wp'$. All polynomials have coefficients in $\mathbb{Z}[a_1, a_2, a_3, a_4, a_6]$. The polynomials ψ_n are given *by the recursion formulas in Theorem 2.25. The other polynomials are given by the recursion formulas*

$$\begin{aligned}
\phi_1(z) &= \mathfrak{p}(z), \\
\Omega_1(z) &= \tilde{\mathfrak{p}}(z), \\
\tilde{\Omega}_1(z) &= \frac{1}{2}\wp'(z),
\end{aligned}$$

and for $n \geq 2$:

$$\begin{aligned}
\phi_n(z) &= \mathfrak{p}(z)\psi_n(z)^2 - \psi_{n-1}(z)\psi_{n+1}(z), \\
2\psi_2(z)\Omega_n(z) &= \psi_{n-1}(z)\psi_{n+2}(z)^2 - \psi_{n-2}(z)\psi_{n+1}(z)^2 \\
&\quad - \psi_2(z)\psi_n(z)(a_1\phi_n(z) + a_3\psi_n(z)^2) \\
&= 2\psi_2(z)\tilde{\Omega}_n(z) - \psi_2(z)\psi_n(z)(a_1\phi_n(z) + a_3\psi_n(z)^2), \\
2\psi_2(z)\tilde{\Omega}_n(z) &= \psi_{n-1}(z)^2\psi_{n+2}(z) - \psi_{n-2}(z)\psi_{n+1}(z)^2.
\end{aligned}$$

Proof. The recursion formulas for ψ_n are already stated in Theorem 2.25. The formulas for ϕ_1, Ω_1, and $\tilde{\Omega}_1$ are trivial.

For ϕ_n, consider the formula from Corollary 2.20:

$$\begin{aligned}
\mathfrak{p}(nz) &= \mathfrak{p}(z) - \frac{\psi_{n+1}(z)\psi_{n-1}(z)}{\psi_n(z)^2} \\
&= \frac{\mathfrak{p}(z)\psi_n(z)^2 - \psi_{n-1}(z)\psi_{n+1}(z)}{\psi_n(z)^2}.
\end{aligned}$$

For Ω_n, we need Lemma 2.21:

$$\begin{aligned}
\tilde{\mathfrak{p}}(nz) &= \frac{1}{2}\left(\frac{\psi_{2n}(z)}{\psi_n(z)^4} - a_1\mathfrak{p}(nz) - a_3\right) \\
&= \frac{1}{2}\left(\frac{\psi_{2n}(z)}{\psi_n(z)^4} - a_1\frac{\phi_n(z)}{\psi_n(z)^2} - a_3\right) \\
&= \frac{1}{2}\left(\frac{\psi_{2n}(z)}{\psi_n(z)} - a_1\phi_n(z)\psi_n(z) - a_3\psi_n(z)^3\right)\frac{1}{\psi_n(z)^3}
\end{aligned}$$

Using the recursion formula for $\psi_{2n}(z)$ and multiplying by $\psi_n(z)^3$, we obtain

$$\begin{aligned}
\psi_n(z)^3\tilde{\mathfrak{p}}(nz) &= \frac{1}{2}\Bigg(\frac{\psi_{n-1}(z)^2\psi_{n+2}(z) - \psi_{n-2}(z)\psi_{n+1}(z)^2}{\psi_2(z)} \\
&\qquad - a_1\phi_n(z)\psi_n(z) - a_3\psi_n(z)^3\Bigg)
\end{aligned}$$

and so we obtain the asserted formula for Ω_n:

$$\Omega_n(z) = \psi_n(z)^3\tilde{\mathfrak{p}}(nz) = \frac{1}{2\psi_2(z)}(\psi_{n-1}(z)^2\psi_{n+2}(z) - \psi_{n-2}(z)\psi_{n+1}(z)^2 - \psi_2(z)\psi_n(z)(a_1\phi_n(z) + a_3\psi_n(z)^2)).$$

For $\tilde{\Omega}_n$ we also use Lemma 2.21 and the recursion formula for $\psi_{2n}(z)$ and get

$$\begin{aligned}\wp'(nz) &= \frac{\psi_{2n}(z)}{\psi_n(z)^4}\\ &= \frac{1}{\psi_2(z)}\frac{\psi_{n-1}(z)^2\psi_{n+2}(z) - \psi_{n-2}(z)\psi_{n+1}(z)^2}{\psi_n(z)^3}\end{aligned}$$

and we obtain the recursion formula for $\tilde{\Omega}_n$. The rest of the theorem is trivial. (See Exercise 6.) □

From the above equations it follows that, for $n \geq 2$:

$$\psi_{2n}(z) = 2\psi_n(z)\tilde{\Omega}_n(z).$$

Using the fact that an elliptic curve over $\mathbb{C}$ is analytically isomorphic to $\mathbb{C}/L$, where L is a lattice in $\mathbb{C}$, we can describe the points of finite order. This can also be used for elliptic curves over number fields.

Proposition 2.27. a) *Let $E|\mathbb{C}$ be an elliptic curve over $\mathbb{C}$ isomorphic to $\mathbb{C}/L$ with the lattice $L = \omega_1\mathbb{Z} + \omega_2\mathbb{Z}$ in $\mathbb{C}$, and $n \in \mathbb{N}$. Then the set of n-division points is*

$$E[n] = E(\mathbb{C})[n] \cong \left\{\frac{r\omega_1 + s\omega_2}{n} : 0 \leq r, s < n\right\} \cong \mathbb{Z}/n\mathbb{Z} \times \mathbb{Z}/n\mathbb{Z}.$$

In particular

$$\sharp E[n] = n^2.$$

b) *Let $E|\mathbb{K}$ an elliptic curve over the number field $\mathbb{K}$ and $n \in \mathbb{N}$. Then the set of n-division points is*

$$E[n] = E(\overline{\mathbb{K}})[n] \cong \mathbb{Z}/n\mathbb{Z} \times \mathbb{Z}/n\mathbb{Z}.$$

In particular

$$\sharp E[n] = n^2.$$

Proof. a) The elements of finite order in $\mathbb{C}/L$ are the elements

$$\frac{r\omega_1 + s\omega_2}{n} \quad \text{with } 0 \leq r, s < n.$$

The proposition follows with the analytic isomorphism $E(\mathbb{C}) \cong \mathbb{C}/L$.

b) The number field $\mathbb{K}$ can be embedded in $\mathbb{C}$, so $E|\mathbb{K}$ can be regarded as an elliptic curve E over $\mathbb{C}$. Then we apply Part a). □

2.3 Periods of elliptic curves

Definition 2.28. Let $E|\mathbb{C}$ be an elliptic curve over the complex numbers. If $E(\mathbb{C})$ is isomorphic to $\mathbb{C}/L$, where $L = \omega_1\mathbb{Z} + \omega_2\mathbb{Z}$ with $\omega_1, \omega_2 \in \mathbb{C}$, then ω_1, ω_2 are the *periods of the elliptic curve* E.

Theorem 2.29. *Let $E|\mathbb{C}$ be an elliptic curve given in Legendre normal form (see Definition 1.10)*

$$E = E_\lambda : y^2 = x(x-1)(x-\lambda)$$

with $\lambda \in \mathbb{C}\backslash\{0, 1\}$. We assume that $|\lambda| < 1$ and $|\lambda - 1| < 1$. Then the periods of the curve E are

$$\omega_1 = i\pi F(1-\lambda), \quad \omega_2 = \pi F(\lambda),$$

where

$$F(z) := \sum_{n=0}^{\infty} \binom{-1/2}{n}^2 z^n.$$

Proof. Let $L = \omega_1\mathbb{Z} + \omega_2\mathbb{Z}$ be the lattice such that $\mathbb{C}/L \cong E(\mathbb{C})$. Then

$$\omega_1 = \int_{-\infty}^{0} \frac{dx}{\sqrt{x(x-1)(x-\lambda)}}, \qquad \omega_2 = \int_{1}^{\infty} \frac{dx}{\sqrt{x(x-1)(x-\lambda)}}$$

(see for example Husemöller [103], 9.6).

Using the change of variable $x = 1 - y$ for ω_1, we get

$$\omega_1 = \int_1^{\infty} \frac{dy}{\sqrt{(1-y)(-y)(1-y-\lambda)}} = -i \int_1^{\infty} \frac{dy}{\sqrt{y(y-1)(y-(1-\lambda))}}.$$

Now substituting $y = \frac{1}{s^2}$ leads to

$$\omega_1 = 2i \int_0^1 \frac{ds}{\sqrt{(1-s^2)(1-(1-\lambda)s^2)}}.$$

With $s = \sin\theta$ we get [1]

$$\omega_1 = 2i \int_0^{\frac{\pi}{2}} \frac{d\theta}{\sqrt{(1-(1-\lambda)\sin^2\theta)}}.$$

Using the binomial series (here we need that $|1 - \lambda| < 1$)

$$(1-(1-\lambda)\sin^2\theta)^{-1/2} = \sum_{n=0}^{\infty} \binom{-1/2}{n} (-(1-\lambda)\sin^2\theta)^n$$

[1] Such integrals occur during the computation of the arc length of an ellipse. Studying those elliptic integrals over $\mathbb{C}$ leads to a certain Riemann surface which turns out to be an elliptic curve. This is the origin of the name *elliptic*.

and integrating term by term, we get

$$\omega_1 = 2i \sum_{n=0}^{\infty} \binom{-1/2}{n} (-(1-\lambda))^n \int_0^{\frac{\pi}{2}} \sin^{2n} \theta d\theta.$$

Now we substitute $t = \sin^2 \theta$. Then

$$d\theta = \frac{1}{2\sqrt{t}\sqrt{1-t}} dt.$$

We get thus

$$\omega_1 = i \sum_{n=0}^{\infty} \binom{-1/2}{n} (-(1-\lambda))^n \int_0^1 t^{n-1/2} (1-t)^{-1/2} dt.$$

From Bronstein et al. [21], we see that

$$\int_0^1 t^{n-1/2}(1-t)^{-1/2} dt = \frac{\Gamma\left(n+\frac{1}{2}\right)\Gamma\left(\frac{1}{2}\right)}{\Gamma(n+1)}$$

with the ordinary Γ-function. We get

$$\begin{aligned} \frac{\Gamma\left(n+\frac{1}{2}\right)\Gamma\left(\frac{1}{2}\right)}{\Gamma(n+1)} &= \frac{\left(n-\frac{1}{2}\right)\left(n-\frac{3}{2}\right)\cdots\frac{1}{2}\Gamma\left(\frac{1}{2}\right)\Gamma\left(\frac{1}{2}\right)}{n!} \\ &= \pi(-1)^n \frac{\left(-\frac{1}{2}\right)\left(-\frac{3}{2}\right)\cdots\left(-\frac{1}{2}-(n-1)\right)}{n!} \\ &= \pi(-1)^n \binom{-1/2}{n}. \end{aligned}$$

Together this gives

$$\omega_1 = i\pi F(1-\lambda).$$

The proof for ω_2 is analogous. □

There is a simpler formula for computing the periods, which uses the arithmetic-geometric mean of Gauss.

Definition 2.30. Let $a, b \in \mathbb{C}$. The *arithmetic-geometric mean of* a, b is

$$\operatorname{AGM}(a, b) = \lim_{n\to\infty} a_n = \lim_{n\to\infty} b_n,$$

where the a_n and b_n are defined as follows:

$$\begin{aligned} (a_0, b_0) &= (a, b) \\ (a_{n+1}, b_{n+1}) &= \left(\frac{a_n + b_n}{2}, \sqrt{a_n b_n}\right). \end{aligned}$$

During the computation of the AGM, one has to choose always the same complex square root. We use the choice specified in the book of Smart [213] XIII.2, that is, b_{n+1} should satisfy

$$|a_{n+1} - b_{n+1}| \leq |a_{n+1} + b_{n+1}|$$

and if these two numbers are equal, then $\mathrm{Im}\left(\frac{b_{n+1}}{a_{n+1}}\right) > 0$.

Theorem 2.31. *Let $E|\mathbb{C}$ an elliptic curve given by an equation of the form*

$$E : \left(Y + \frac{a_1 X + a_3}{2}\right)^2 = X^3 + \frac{b_2}{4}X^2 + \frac{b_4}{2}X + \frac{b_6}{4} = (X - e_1)(X - e_2)(X - e_3)$$

with $e_i \in \mathbb{C}$. Then the periods of E are

$$\begin{aligned} \omega_1 &= \frac{\pi}{\mathrm{AGM}(\sqrt{e_3 - e_1}, \sqrt{e_3 - e_2})} \\ \omega_2 &= \frac{\pi i}{\mathrm{AGM}(\sqrt{e_3 - e_1}, \sqrt{e_2 - e_1})}. \end{aligned}$$

Proof. See the book of Cremona [42]. □

Examples. 1) Let

$$E : y^2 = x^3 - x = x(x-1)(x+1).$$

We take $e_1 < e_2 < e_3$ and have

$$\begin{aligned} \omega_1 &= \frac{\pi}{\mathrm{AGM}(\sqrt{2}, 1)}, \\ \omega_2 &= \frac{\pi i}{\mathrm{AGM}(\sqrt{2}, 1)} = i\omega_1. \end{aligned}$$

2) Let

$$E : y^2 = x^3 - 1 = (x-1)(x-\zeta_3)(x-\zeta_3^2),$$

where ζ_3 is a third root of unity. Computing the periods with the AGM gives

$$\begin{aligned} \omega_1 &\approx 2.4286506479 \\ \omega_2 &= \frac{(1+\sqrt{-3})}{2}\omega_1 \\ &\approx 1.2143253239 + 2.1032731580 \cdot i. \end{aligned}$$

2.4 Complex multiplication

The notion of isogenies can be redefined in terms of lattices. Let L, L' be two lattices in $\mathbb{C}$. We recall that a complex number $\lambda \neq 0$ satisfying

$$\lambda L \subseteq L'$$

induces an isogeny

$$\lambda : \mathbb{C}/L \to \mathbb{C}/L'$$

(see Definition 2.1). Either $\lambda = 0$ or the above isogeny is surjective with kernel

$$\ker(\lambda) = \lambda^{-1}L'/L.$$

Definition 2.32. a) Let L be a lattice in $\mathbb{C}$ and λ a complex number satisfying $\lambda L \subseteq L$. The isogeny

$$\lambda : \mathbb{C}/L \to \mathbb{C}/L$$

induced by λ is called an *endomorphism* of L.

b) Let L be a lattice in $\mathbb{C}$. Define

$$\mathrm{End}(\mathbb{C}/L) := \{\lambda \in \mathbb{C} : \lambda L \subseteq L\}.$$

c) Let L, L' be two lattices in $\mathbb{C}$ and $\lambda \neq 0$ be an isogeny between L and L'. The *degree* $\deg(\lambda)$ is

$$\deg(\lambda) := \sharp \ker(\lambda) = \sharp \lambda^{-1}L'/L = [L' : \lambda L] \quad (< \infty).$$

The *dual isogeny* to λ is the isogeny $\hat{\lambda}$ given by

$$\hat{\lambda} := \frac{\deg(\lambda)}{\lambda} : \mathbb{C}/L' \to \mathbb{C}/L.$$

We define (as before)

$$\deg(0) := 0, \quad \hat{0} := 0.$$

We now list some properties of isogenies.

Proposition 2.33. *Let L, L' be two lattices in $\mathbb{C}$ and λ be an isogeny of degree n between L and L'. Then*

a) $\hat{\hat{\lambda}} = \lambda$.

b) $\deg(\hat{\lambda}) = \deg(\lambda) = n$.

c) *As complex numbers* $\lambda\hat{\lambda} = \deg(\lambda) = n$.

Proof. a) For the area $a(\lambda L)$ of a period parallelogram associated to the lattice λL, we have

$$a(\lambda L) = |\lambda|^2 a(L) = \lambda\bar{\lambda} a(L),$$

where $a(L)$ is the area of a period parallelogram associated to L and $\bar{\lambda}$ is the complex conjugate of λ. Then it follows that

$$\deg(\lambda) = [L' : \lambda L] = \frac{a(\lambda L)}{a(L')} = \lambda\bar{\lambda}\frac{a(L)}{a(L')}.$$

With the definition of $\hat{\lambda}$ we obtain

$$\hat{\lambda} = \frac{\deg(\lambda)}{\lambda} = \bar{\lambda}\frac{a(L)}{a(L')},$$

such that

$$\hat{\hat{\lambda}} = \bar{\hat{\lambda}}\frac{a(L')}{a(L)} = \bar{\lambda}\overline{\frac{a(L)}{a(L')}} \cdot \frac{a(L')}{a(L)} = \lambda.$$

b) We have

$$\lambda = \hat{\hat{\lambda}} = \frac{\deg(\hat{\lambda})}{\hat{\lambda}} = \frac{\deg(\hat{\lambda})}{\deg(\lambda)} \cdot \lambda,$$

such that $\deg(\hat{\lambda}) = \deg(\lambda)$.

c) This follows directly from the definition of $\hat{\lambda}$. □

Proposition 2.34. *Let L be a lattice in $\mathbb{C}$ and $n \in \mathbb{N}$. Then,*

a) $\deg(n) = n^2$.

b) $\hat{n} = n$.

Proof. a) Let ω_1, ω_2 be fundamental periods of L. Then

$$\ker(n) = \left\{\frac{r\omega_1 + s\omega_2}{n} : 0 \le r, s < n\right\} \cong \mathbb{Z}/n\mathbb{Z} \times \mathbb{Z}/n\mathbb{Z},$$

hence $\deg(n) = \sharp \ker(n) = n^2$.

b) This can be easily derived from Part a). □

Definition 2.35. Let L be a lattice in $\mathbb{C}$ and λ be an endomorphism of L. The *trace* $\mathrm{Tr}(\lambda)$ is defined by

$$\mathrm{Tr}(\lambda) := \lambda + \hat{\lambda}.$$

and the *norm* by

$$\mathrm{N}(\lambda) := \deg(\lambda) = \lambda\hat{\lambda}.$$

The *characteristic polynomial* of λ is

$$\mathcal{F}_\lambda(u) := u^2 - \mathrm{Tr}(\lambda)u + \mathrm{N}(\lambda).$$

Proposition 2.36. *Let L be a lattice in $\mathbb{C}$ and λ an endomorphism of L. Then*

a) $\mathrm{Tr}(\lambda) \in \mathbb{Z}$.

b) $\mathcal{F}_\lambda(u) \in \mathbb{Z}[u]$.

c) $\mathcal{F}_\lambda(\lambda) = 0$.

Proof. a) In the proof of Part a) of Proposition 2.33 we have

$$\hat{\lambda} = \bar{\lambda}\frac{a(L)}{a(L')}.$$

Therefore $\hat{}$ is a group endomorphism on the multiplicative group of endomorphisms $\neq 0$ of L. It follows that

$$\begin{aligned} \mathrm{N}(1+\lambda) &= (1+\lambda)\widehat{(1+\lambda)} \\ &= (1+\lambda)(1+\hat{\lambda}) \\ &= 1+\mathrm{Tr}(\lambda)+\mathrm{N}(\lambda). \end{aligned}$$

Then $\mathrm{Tr}(\lambda)$ must be an integer, because

$$\mathrm{Tr}(\lambda) = \mathrm{N}(1+\lambda) - \mathrm{N}(\lambda) - 1 \in \mathbb{Z}.$$

b) This follows from Part a) and the definition of the degree.
c) We have

$$\mathcal{F}_\lambda(\lambda) = \lambda^2 - (\lambda+\hat{\lambda})\lambda + \lambda\hat{\lambda} = 0.$$ □

Proposition 2.37 (Hasse [91]). *Let L be a lattice in $\mathbb{C}$ and λ be an endomorphism of L. For the characteristic polynomial of λ we have*

$$\mathcal{F}_\lambda(u) \geq 0$$

for all $u \in \mathbb{R}$. Furthermore

$$|\mathrm{Tr}(\lambda)| \leq 2\sqrt{\mathrm{N}(\lambda)}.$$

Proof. For $u = \frac{m}{n} \in \mathbb{Q}, m, n \in \mathbb{Z}, n > 0$, we have

$$\begin{aligned} 0 &\leq \mathrm{N}(m-n\lambda) \\ &= (m-n\lambda)(m-n\hat{\lambda}) \\ &= m^2 - mn(\lambda+\hat{\lambda}) + n^2\lambda\hat{\lambda} \\ &= m^2 - \mathrm{Tr}(\lambda)mn + \mathrm{N}(\lambda)n^2 \\ &= n^2\mathcal{F}_\lambda(u). \end{aligned}$$

As $\mathcal{F}_\lambda(u) \geq 0$ for all $u \in \mathbb{Q}$ the quadratic equation

$$\mathcal{F}_\lambda(u) = u^2 - \mathrm{Tr}(\lambda)u + \mathrm{N}(\lambda)$$

is nonnegative in the real numbers, since $\mathbb{Q}$ is dense in $\mathbb{R}$. Therefore, we have for the discriminant of this equation:

$$(\lambda - \hat{\lambda})^2 = \mathrm{Tr}(\lambda)^2 - 4\mathrm{N}(\lambda) \leq 0.$$

It follows that

$$\mathrm{Tr}(\lambda)^2 \leq 4\mathrm{N}(\lambda) \Leftrightarrow |\mathrm{Tr}(\lambda)| \leq 2\sqrt{\mathrm{N}(\lambda)}.$$ □

In the next theorem we show that isogenies of lattices and isogenies of elliptic curves are essentially the same.

Theorem 2.38. *Let E_1, E_2 be elliptic curves over $\mathbb{C}$ corresponding to the lattices L_1, L_2. Then there is a bijection*

$$\{\phi\colon E_1 \to E_2 : \phi \textit{ is an isogeny}\} \to \{\lambda\colon \mathbb{C}/L_1 \to \mathbb{C}/L_2 : \lambda L_1 \subseteq L_2\}.$$

Proof. An isogeny $\phi : E_1 \to E_2$ induces a holomorphic map $\phi : \mathbb{C}/L_1 \to \mathbb{C}/L_2$ simply by using the isomorphism in Theorem 2.15. This map can be lifted to a holomorphic map $\phi' : \mathbb{C} \to \mathbb{C}$ such that for any $\omega \in L_1$

$$\phi'(z + \omega) = \phi'(z) \bmod L_2$$

for all $z \in \mathbb{C}$. As L_2 is discrete, it follows that $\phi'(z + \omega) - \phi'(z)$ is independent of z and hence

$$\phi'(z + \omega) = \phi'(z)$$

for all $z \in \mathbb{C}$ and $\omega \in L_1$. That means that ϕ' is a holomorphic elliptic function, hence constant. Then we can write $\phi(z)$ as $\phi(z) = \lambda z + \nu$ with $\lambda, \nu \in \mathbb{C}$. With $\phi(0) = 0$ we see that $\phi(z) = \lambda z$.

From this construction it follows easily that the map $\phi \mapsto \lambda$ is injective. To show the surjectivity, suppose that $\lambda \in \mathbb{C}$ is a complex number with $\lambda L_1 \subseteq L_2$. Considering the elliptic curves, we see that the map corresponding to λ is given as

$$\begin{aligned} E_1 &\to E_2 \\ (\mathfrak{p}_1(z), \tilde{\mathfrak{p}}_1(z)) &\mapsto (\mathfrak{p}_2(\lambda z), \tilde{\mathfrak{p}}_2(\lambda z)), \end{aligned}$$

where $\mathfrak{p}_i$, $\tilde{\mathfrak{p}}_i$ are the generalized Weierstraß functions associated to E_i (see Theorem 2.15). It suffices now to show that $\mathfrak{p}_2(\lambda z)$ and $\tilde{\mathfrak{p}}_2(\lambda z))$ can be expressed as rational functions of $\mathfrak{p}_1(z)$ and $\tilde{\mathfrak{p}}_1(z)$.

For any $\omega \in L_1$ we get

$$\mathfrak{p}_2(\lambda(z + \omega)) = \mathfrak{p}_2(\lambda z + \lambda\omega) = \mathfrak{p}_2(\lambda z),$$

as $\lambda L_1 \subseteq L_2$. A similar relation holds for $\tilde{\mathfrak{p}}_2(\lambda z)$. Hence $\mathfrak{p}_2(\lambda z)$ and $\tilde{\mathfrak{p}}_2(\lambda z)$ are elliptic functions with period lattice L_1. Therefore they can be represented as rational functions of $\wp_1(z)$ and $\wp_1'(z)$, where $\wp_1(z)$ is the Weierstraß $\wp$-function associated to L_1. The proof is finished using the definition of the generalized Weierstraß functions. □

Proposition 2.39. a) *Let $E|\mathbb{C}$ be an elliptic curve over the complex numbers. Then* End(E) *is either $\mathbb{Z}$ or an order in an imaginary quadratic field.*

b) *Let $E|\mathbb{K}$ be an elliptic curve over a number field $\mathbb{K}$. Then* End(E) *is either $\mathbb{Z}$ or an order in an imaginary quadratic field.*

If End(E) is an order in an imaginary quadratic field, the curve has *complex multiplication.*

Proof. As the number field $\mathbb{K}$ in Part b) can be embedded in $\mathbb{C}$, it is enough to prove Part a).

From Proposition 2.36 it follows that every element of End(E) satisfies a monic quadratic equation over $\mathbb{Z}$, so it is of degree 1 or 2 over $\mathbb{Q}$. If every element of End(E) is of degree 1 over $\mathbb{Q}$, then with $\mathbb{Z} \subseteq \mathrm{End}(E)$ it follows that

$$\mathrm{End}(E) = \mathbb{Z}.$$

Otherwise, as every endomorphism is of degree ≤ 2 over $\mathbb{Q}$, considering the possible endomorphism rings (Theorem 1.35), one sees that End(E) must be a subset of a quadratic number field. From Proposition 2.37 it follows that the characteristic equations have no roots in $\mathbb{R}$, hence the number field which contains End(E) must be an imaginary quadratic number field. Since the characteristic polynomials have integer coefficients, End(E) is a subring of the ring of integers in this imaginary quadratic field. Because $\mathbb{Z} \subseteq \mathrm{End}(E)$ and $\mathbb{Z} \neq \mathrm{End}(E)$, the ring of endomorphisms is an order in this field. □

Theorem 2.40. *Let $E|\mathbb{C}$ be an elliptic curve over the complex numbers which is isomorphic to $\mathbb{C}/L$ with the lattice $L = \mathbb{Z} + \tau\mathbb{Z}$ with $\tau \in \mathbb{H}$. The curve has complex multiplication if and only if τ is an imaginary quadratic number. In the case of complex multiplication* End(E) *is an order in the imaginary quadratic field $\mathbb{Q}(\tau)$.*

Proof. If E has complex multiplication, then there exists $\lambda \in \mathrm{End}(E)$ with $\lambda \notin \mathbb{Z}$. Then, because of $\lambda L \subset L$, it follows that $\lambda, \lambda\tau \in L$, so that

$$\lambda = a + \tau b, \qquad \text{with } a, b \in \mathbb{Z}, b \neq 0,$$

and

$$\lambda\tau = c + \tau d, \qquad \text{with } c, d \in \mathbb{Z}.$$

Together, we obtain

$$c + d\tau = a\tau + b\tau^2,$$

hence τ satisfies a quadratic equation. On the other hand, if τ satisfies a quadratic equation

$$a\tau^2 + b\tau + c = 0$$

with $a, b, c \in \mathbb{Z}$, $a \neq 0$, then

$$a\tau^2 = -c - b\tau \in L,$$

so that, for all $m + n\tau \in L$, we get

$$(a\tau)(m + n\tau) = ma\tau + n(a\tau^2) = -cn + (am - bn)\tau \in L.$$

Therefore, there is an endomorphism $a\tau \notin \mathbb{Z}$, hence the curve has complex multiplication.

The last assertion follows from the quadratic relation for τ over $\mathbb{Q}$. □

Examples. 1) Let

$$E : Y^2 = X^3 - X.$$

We have seen that this curve has the periods $\omega_2 = i\omega_1$, hence

$$\tau = \frac{\omega_2}{\omega_1} = i.$$

As this is an imaginary quadratic number, it follows that E has complex multiplication. Indeed, we have an endomorphism

$$\begin{aligned} i \colon E &\to E \\ \mathcal{O} &\mapsto \mathcal{O}, \\ (x, y) &\mapsto (-x, iy) = (i^2 x, iy). \end{aligned}$$

It is easy to show that $i \circ i = -1$, i.e. applying two times the endomorphism i is the same as multiplication by -1.

This is also true for all elliptic curves of the form

$$E : Y^2 = X^3 + aX$$

with non-zero $a \in \mathbb{C}$, because they are all isomorphic over $\mathbb{C}$.

2) Let

$$E : Y^2 = X^3 - 1.$$

We have seen that this curve has the periods

$$\omega_2 = \frac{(1 + \sqrt{-3})}{2}\omega_1 = -\zeta_3\omega_1,$$

where ζ_3 is the third root of unity given above. This curve has complex multiplication. The third root of unity induces the endomorphism

$$\begin{aligned} \zeta_3 : E &\to E \\ \mathcal{O} &\mapsto \mathcal{O}, \\ (x, y) &\mapsto (\zeta_3 x, y). \end{aligned}$$

This curve has also an endomorphism induced by the sixth root of unity $\zeta_6 = -\zeta_3$:

$$\begin{aligned} \zeta_6 \colon E &\to E \\ \mathcal{O} &\mapsto \mathcal{O}, \\ (x, y) &\mapsto (\zeta_3 x, -y) = (\zeta_6^4 x, \zeta_6^3 y). \end{aligned}$$

This is also true for all elliptic curves of the form

$$E : Y^2 = X^3 + b$$

with non-zero $b \in \mathbb{C}$, because they are all isomorphic over $\mathbb{C}$.

Theorem 2.41. *Let $E|\mathbb{C}$ (or $E|\mathbb{K}$) be an elliptic curve over the complex numbers $\mathbb{C}$ (or over a number field $\mathbb{K}$) with complex multiplication. Then*

$$E \cong \mathbb{C}/\mathfrak{a},$$

where $\mathfrak{a}$ is an ideal of the order End(E).

Proof. If E is given over a number field $\mathbb{K}$, we embed this field in $\mathbb{C}$ and consider the curve over $\mathbb{C}$.

The elliptic curve $E|\mathbb{C}$ is isomorphic to $\mathbb{C}/L$ with a lattice L of the form

$$L = \mathbb{Z} + \tau\mathbb{Z}.$$

Here τ is an imaginary quadratic number. The endomorphism ring End(E) is an order $\mathcal{O}$ of the field $\mathbb{Q}(\tau)$. From the definition of endomorphisms, it follows that for all $x \in \mathcal{O} = \mathrm{End}(E)$ we have

$$xL \subseteq L,$$

hence L is an $\mathcal{O}$-ideal. □

An ideal $\mathfrak{a}$ in an imaginary quadratic field has a representation as a $\mathbb{Z}$-module of $\mathbb{Z}$-rank 2 not contained in $\mathbb{R}$ (see Cohen [34]), hence it defines a lattice in $\mathbb{C}$. Therefore we can define the j-invariant $j(\mathfrak{a})$ as the j-invariant of the associated lattice.

Theorem 2.42 (First main theorem of complex multiplication). *Let $\mathcal{O}$ be an order in an imaginary quadratic field $\mathbb{K}$ with ideal class group $\mathcal{C}l(\mathcal{O})$ and $E|\mathbb{C}$ an elliptic curve with* $\mathrm{End}(E) \cong \mathcal{O}$. *Then $j(E)$ is an algebraic integer with minimal polynomial*

$$W_{\mathcal{O}}(x) := \prod_{\mathfrak{a} \in \mathcal{C}l(\mathcal{O})} (x - j(\mathfrak{a}))$$

and the field

$$\mathbb{H}_{\mathcal{O}} := \mathbb{K}(j(\mathfrak{a}))$$

for an $\mathfrak{a} \in \mathcal{C}l(\mathcal{O})$ is the Hilbert ring class field for the order $\mathcal{O}$ in $\mathbb{K}$. As $j(\mathfrak{a})$ only depends on the class of $\mathfrak{a}$, this field is independent of the choice of $\mathfrak{a}$ within its class. Furthermore the other $j(\mathfrak{a}')(\mathfrak{a}' \in \mathcal{C}l(\mathcal{O}))$ are the conjugates of $j(\mathfrak{a})$.

Proof. From Theorem 2.41 we see that $E(\mathbb{C}) \cong \mathbb{C}/\mathfrak{a}$ for an $\mathcal{O}$-ideal $\mathfrak{a}$. Two ideals $\mathfrak{a}_1, \mathfrak{a}_2$ are in the same class, if and only if there exists an $x \in \mathcal{O}$ with $x\mathfrak{a}_1 = \mathfrak{a}_2$, which is the case if and only if $j(\mathfrak{a}_1) = j(\mathfrak{a}_2)$ by Proposition 2.4. Hence the elliptic curves over $\mathbb{C}$ with endomorphism ring equal to $\mathcal{O}$ considered modulo isomorphisms, correspond to the ideal class group $\mathcal{C}l(\mathcal{O})$.

Since $j(E)$ depends only on the isomorphism class of E, it follows that it has minimal polynomial $W_{\mathcal{O}}(x)$.

For a proof of the rest of the theorem, we refer for example to Deuring [54] or Lang [116] (see also Lay [123]). □

2.5 Exercises

1) Prove Proposition 2.4.
2) Use the definition of the generalized Weierstraß functions and the differential equation of the Weierstraß $\wp$-function to prove Proposition 2.11
3) Finish the proof of Theorem 2.14.
4) Take linear equivalent lattices L, L' over $\mathbb{C}$ and $d_1, d_2, d_3, d_1', d_2', d_3' \in \mathbb{C}$. Define the generalized Weierstraß functions $\mathfrak{p}_1, \widetilde{\mathfrak{p}_1}$ associated to L and d_1, d_2, d_3, and $\mathfrak{p}_2, \widetilde{\mathfrak{p}_2}$ associated to L' and d_1', d_2', d_3'. These functions satisfy the equations
$$\widetilde{\mathfrak{p}_1}^2 + a_1\mathfrak{p}_1\widetilde{\mathfrak{p}_1} + a_3\widetilde{\mathfrak{p}_1} = \mathfrak{p}_1^3 + a_2\mathfrak{p}_1^2 + a_4\mathfrak{p}_1 + a_6$$
and
$$\widetilde{\mathfrak{p}_2}^2 + a_1'\mathfrak{p}_2\widetilde{\mathfrak{p}_2} + a_3'\widetilde{\mathfrak{p}_2} = \mathfrak{p}_2^3 + a_2'\mathfrak{p}_2^2 + a_4'\mathfrak{p}_2 + a_6'.$$
As the lattices are linear equivalent, these two elliptic curves are isomorphic. Use the formulas for the generalized Weierstraß functions to express the coefficients a_i' in terms of the a_i. The result should be the formulas on page 6.
5) Complete the proof of Theorem 2.25.
6) Complete the proof of Theorem 2.26.

Chapter 3
Elliptic curves over finite fields

In this chapter we investigate elliptic curves over finite fields. First we explain the Frobenius endomorphism. After that we present special curves and some methods for computing the number of rational points. Then we introduce an algorithm which can lead to the construction of elliptic curves with given number of points. In the last sections we consider an important application of the theory of elliptic curves over finite fields: elliptic curve cryptosystems.

Let $\mathbb{P}$ be the set of prime numbers, $p \in \mathbb{P}$, $k \in \mathbb{N}$, $q = p^k$, and $\mathbb{F}_q$ be the finite field with q elements.

3.1 Frobenius endomorphism and supersingular curves

Definition 3.1. Let $E|\mathbb{F}_q$ be an elliptic curve over the finite field $\mathbb{F}_q$. The *q-Frobenius endomorphism* $\varphi_q \colon E \to E$ is given by

$$\varphi_q(x, y) = (x^q, y^q), \quad \varphi_q(\mathcal{O}) = \mathcal{O}.$$

As $\varphi_q \in \operatorname{End}(E)$ for elliptic curves $E|\mathbb{F}_q$, it follows that elliptic curves over finite fields always have complex multiplication (see also Deuring [48]).

Theorem 3.2. *Let $E|\mathbb{F}_q$ be an elliptic curve and φ_q the q-Frobenius endomorphism.*

a) *Let $P \in E$. Then*
$$P \in E(\mathbb{F}_q) \Leftrightarrow \varphi_q(P) = P.$$

b) *The endomorphism φ_q is purely inseparable.*

c) *The degree is* $\deg(\varphi_q) = q$.

d) *There exists an integer $t = t_q$ such that*
$$\varphi_q^2 - t\varphi_q + q = 0,$$
that is to say that, for all $P \in E$, we have the equation
$$\varphi_q^2(P) - t\varphi_q(P) + qP = \mathcal{O}.$$
(The integer t is called the *trace* of the q-Frobenius endomorphism.)

e) *The trace t of the q-Frobenius endomorphism is linked with the number of rational points by the relation:*

$$\sharp E(\mathbb{F}_q) = q + 1 - t.$$

Proof. a) The q-Frobenius automorphism ϕ_q of the field $\overline{\mathbb{F}}_q$ is defined as

$$\phi_q : \overline{\mathbb{F}}_q \to \overline{\mathbb{F}}_q, \quad X \mapsto X^q.$$

It has the property that $x \in \mathbb{F}_q \Leftrightarrow \phi_q(x) \in \mathbb{F}_q$. From this the assertion follows easily.

b) We here consider the function field $\mathbb{F}_q(E)$. The elements of this function field are quotients of homogeneous polynomials of the same degree. The q-Frobenius endomorphism induces an endomorphism φ_q^* on the function field (see Definition 1.30). Consider $\frac{F}{G} \in \mathbb{F}_q(E)$. Then

$$\varphi_q^* \left(\frac{F}{G}\right) = \frac{F(X^q, Y^q)}{G(X^q, Y^q)} = \frac{F(X, Y)^q}{G(X, Y)^q},$$

where the last equality follows from the fact that for every $x \in \mathbb{F}_q$ we have $x^q = x$. Therefore $\varphi_q^*(\mathbb{F}_q(E)) = \mathbb{F}_q(E)^q$ and the field extension $\mathbb{F}_q(E)|\mathbb{F}_q(E)^q$ is purely inseparable.

c) Here one has to show that the degree of the field extension $\mathbb{F}_q(E)|\mathbb{F}_q(E)^q$ is equal to q. This can not be proved with the background given by this book, but see for example the article of Hasse [91].

d) We first compute the trace of the q-Frobenius endomorphism. Consider the endomorphism $\varphi_q - 1$.

$$\begin{aligned} \deg(\varphi_q - 1) &= (\varphi_q - 1)(\widehat{\varphi_q} - 1) \\ &= \varphi_q \widehat{\varphi_q} - (\varphi_q + \widehat{\varphi_q}) + 1 \\ &= \deg(\varphi_q) - (\varphi_q + \widehat{\varphi_q}) + 1 \end{aligned}$$

and hence with $\deg(\varphi_q) = q$:

$$t := \mathrm{Tr}(\varphi_q) = \varphi_q + \widehat{\varphi_q} = q + 1 - \deg(\varphi_q - 1) \in \mathbb{Z}.$$

Then we compute (with $\mathrm{N}(\varphi_q) = \varphi_q \widehat{\varphi_q}$)

$$0 = (\varphi_q - \varphi_q)(\varphi_q - \widehat{\varphi_q}) = \varphi_q^2 - t\varphi_q + q.$$

e) One can show that the endomorphism $\varphi_q - 1$ is separable (see for example Silverman [204] Chapter III, Corollary 5.5). Hence

$$\deg(\varphi_q - 1) = \sharp(\ker(\varphi_q - 1)) = \sharp(E(\mathbb{F}_q)).$$

This proves the theorem. □

There is an estimate for the trace of the Frobenius endomorphism which was proved by Hasse.

Theorem 3.3 (Hasse, Analogue of the Riemann Hypothesis). *Let $E|\mathbb{F}_q$ be an elliptic curve. Then*

$$|\sharp E(\mathbb{F}_q) - (q+1)| = |t| \leq 2\sqrt{q}.$$

Proof. This is proved in the article of Hasse [91]. See also Chapter 2, Proposition 2.37. An elementary proof, which contains some errors, is given in an article of Manin [134]. See also Elistratov [58] or Zimmer [247]. Proofs can also be found in the books of Chahal [31] or Knapp [109]. □

There are elliptic curves over finite fields with special properties.

Definition 3.4. Let $E|\mathbb{F}_q$ be an elliptic curve over a finite field of characteristic p with $\sharp E(\mathbb{F}_q) = q + 1 - t$. The curve is called *supersingular*, if $p \mid t$. A curve which is not supersingular is called *ordinary*.

For more information about supersingular curves see Elkies [60].

The following characterizations for supersingular curves are important for the application in cryptography.

Theorem 3.5. *Let $E|\mathbb{F}_q$ be an elliptic curve.*

a) *The curve is supersingular if and only if* $\mathrm{End}(E)$ *is the maximal order in a quaternion algebra.*

b) *If* $\mathrm{char}(\mathbb{F}_q) = 2$ *or* 3, *the curve is supersingular if and only if* $j(E) = 0$.

Proof. This result goes back to Deuring [48]. See the book of Blake, Seroussi, and Smart [18], III.6. □

3.2 Computing the number of points

The estimate of Hasse (Theorem 3.3) can be rewritten to obtain the following estimate for the number of points of an elliptic curve E over $\mathbb{F}_q$:

$$-2\sqrt{q} \leq \sharp E(\mathbb{F}_q) - (q+1) \leq 2\sqrt{q} \;\Leftrightarrow\; (\sqrt{q}-1)^2 \leq \sharp E(\mathbb{F}_q) \leq (\sqrt{q}+1)^2.$$

In this section we present three different methods for the computation of the number of points:

- naive counting,
- the method of Shanks and Mestre,
- the method of Schoof.

For the naive counting we run through a representation system for the finite field $\mathbb{F}_q$ and test if an element is the first coordinate of a point of the elliptic curve. If this point is a point of order > 2, the element leads to two different points. We have thereby to take into account the point at infinity.

Algorithm 3.6 (Naive counting).

INPUT: q and $E|\mathbb{F}_q$ with coefficients a_1, a_2, a_3, a_4, a_6.
OUTPUT: The number $\sharp E(\mathbb{F}_q)$.
1. $n \leftarrow 1$. /* $\mathcal{O}$ */
2. Find a representation system R_q for $\mathbb{F}_q$.
3. For every $x \in R_q$ do:
4.   If there is an $y \in R_q$ with
    $y^2 + a_1xy + a_3y = x^3 + a_2x^2 + a_4x + a_6$ then do:
5.   If $y = -y - a_1x - a_3$ then $n \leftarrow n + 1$,
6.   else $n \leftarrow n + 2$.
7. Return n.

Example. We consider the elliptic curve

$$E : Y^2 + XY = X^3 + X$$

over $\mathbb{F}_2$. For the computation of the number of points we first set $x = 0$ and determine all $y \in \mathbb{F}_2$ with

$$y^2 \equiv 0 \pmod 2.$$

The only solution is $y \equiv 0 \pmod 2$. Then we take $x = 1$ and determine all $y \in \mathbb{F}_2$ with

$$y^2 + y \equiv 0 \pmod 2.$$

This is true for $y \equiv 0 \pmod 2$ and for $y \equiv 1 \pmod 2$. Therefore

$$\sharp E(\mathbb{F}_2) = 1 + 1 + 2 = 4.$$

(The theorem of Hasse is of course satisfied.)

If $\operatorname{char}(\mathbb{F}_q) \neq 2$ the elliptic curve can be given in the form

$$E : Y^2 = X^3 + a_2X^2 + a_4X + a_6.$$

In this case we precompute the squares in $\mathbb{F}_q$ and get the following algorithm.

Algorithm 3.7 (Naive counting, $\operatorname{char}(\mathbb{F}_q) \neq 2$).

INPUT: q, $E|\mathbb{F}_q$ with coefficients a_2, a_4, a_6.
OUTPUT: The number $\sharp E(\mathbb{F}_q)$.
1. $n \leftarrow 1$. /* $\mathcal{O}$ */
2. Find a representation system R_q for $\mathbb{F}_q$.

```
    Let S_q be the set of squares in F_q.
    For every x ∈ R_q do:
        If x^3 + a_2x^2 + a_4x + a_6 ∈ S_q then do:
            If x^3 + a_2x^2 + a_4x + a_6 = 0 then n ← n + 1,
            else n ← n + 2.
    Return n.
```

As an example we take the curve

$$E : Y^2 = X^3 + \xi X$$

over $\mathbb{K} = \mathbb{F}_5(\xi) \cong \mathbb{F}_{25}$, where ξ is a root of $X^2 - 3$. A system of representatives for $\mathbb{K}$ is

$$\mathbb{K} = \{a + b\xi : 0 \le a, b \le 4\}.$$

First we compute all squares in $\mathbb{K}$ and represent them in the above system. Note that we have to compute the squares only for the half of $\mathbb{K}$.

$x = a + b\xi$	0	1	2	ξ	$1+\xi$	$2+\xi$	$3+\xi$	$4+\xi$
x^2	0	1	4	3	$4+2\xi$	$2+4\xi$	$2+\xi$	$4+3\xi$

$x = a + b\xi$	2ξ	$1+2\xi$	$2+2\xi$	$3+2\xi$	$4+2\xi$
x^2	2	$3+4\xi$	$1+3\xi$	$1+2\xi$	$3+\xi$

Now we compute for all $x \in \mathbb{K}$ the number $x^3 + \xi x$ in the above representation and test if this is a square or not. We only have to test half of the elements because

$$(-x)^3 + \xi(-x) = -(x^3 + \xi x)$$

and $-1 \equiv 4 \pmod 5$ is a square in $\mathbb{K}$. We then double the results for $x \neq 0$.

$x = a + b\xi$	0	1	2	ξ	$1+\xi$	$2+\xi$	$3+\xi$
$x^3 + \xi x$	0	$1+\xi$	$3+2\xi$	$3+3\xi$	$3+2\xi$	$4+2\xi$	$2+3\xi$
points	1	0	0	0	0	4	0

$x = a + b\xi$	$4+\xi$	2ξ	$1+2\xi$	$2+2\xi$	$3+2\xi$	$4+2\xi$	
$x^3 + \xi x$	3	$1+4\xi$	$3+\xi$	1	$1+\xi$	$4+4\xi$	
points	4	0	4	4	0	0	1

It follows that $\sharp E(\mathbb{K}) = 18$.

If $q = p$ with $p > 2$ we can use the Legendre-symbol to test if an element of $\mathbb{F}_p$ is a square.

Proposition 3.8. *Let $E|\mathbb{F}_p$ be an elliptic curve in the form*

$$E : Y^2 = X^3 + a_2X^2 + a_4X + a_6$$

over $\mathbb{F}_p$, $p > 2$. Then

$$\sharp E(\mathbb{F}_p) = p + 1 + \sum_{x=0}^{p-1} \left(\frac{x^3 + a_2x^2 + a_4x + a_6}{p} \right).$$

Here $\left(\frac{0}{p}\right) = 0$.

Proof. Exercise 1). □

Algorithm 3.9 (Legendre-symbol method, $p \neq 2$).

INPUT: The prime p, $E|\mathbb{F}_p$ with coefficients a_2, a_4, a_6.
OUTPUT: The number $\sharp E(\mathbb{F}_p)$.
1. $n \leftarrow p + 1$.
2. For $x = 0$ to $p - 1$ do $n \leftarrow n + \left(\frac{x^3+a_2x^2+a_4x+a_6}{p}\right)$.
3. Return n.

Using Proposition 3.8 we are able to compute the number of points on certain elliptic curves.

Lemma 3.10. *Let $k \in \mathbb{Z}$, $k \neq 0$ without a 6-th power, $E_k : Y^2 = X^3 + k$. Further let $p \in \mathbb{P}$ with $p \nmid 6k$ and $p \equiv 2 \pmod 3$. Consider E_k as an elliptic curve over $\mathbb{F}_p$ (a so-called Mordell curve). Then $E_k|\mathbb{F}_p$ is supersingular and*

$$\sharp E_k(\mathbb{F}_p) = p + 1.$$

Proof. For the first part, see Definition 3.4 and Theorem 3.5. For the second, one has

$$\sharp E_k(\mathbb{F}_p) = p + 1 + \sum_{x=0}^{p-1} \left(\frac{x^3 + k}{p} \right).$$

We first show the following lemma.

Lemma 3.11. *Let $p \in \mathbb{P}$ be a prime number with $p \equiv 2 \pmod 3$. Then the map*

$$X \mapsto X^3$$

is an automorphism of the multiplicative group of $\mathbb{F}_p^*$.

Proof. The map is a homomorphism, because

$$(ab)^3 \equiv a^3 b^3 \pmod{p}$$

for all $a, b \in \mathbb{F}_p^*$. We show that the map is injective. Then it is automatically surjective.

Let $x \in \mathbb{F}_p^*$ be in the kernel of the above function:

$$x^3 \equiv 1 \pmod{p}.$$

We assume that $x \not\equiv 1 \pmod{p}$. Then x is of order 3. This implies that $3 \mid (p-1)$, which is a contradiction to $p \equiv 2 \pmod{3}$. □

With this result, it follows that the map $X \mapsto X^3$ is an automorphism of $\mathbb{F}_p^*$. In particular it is bijective on $\mathbb{F}_p$. Then the map $X \mapsto X^3 + k$ is also bijective on $\mathbb{F}_p$. Hence we have

$$\sharp E_k(\mathbb{F}_p) = p + 1 + \sum_{x=0}^{p-1} \left(\frac{x^3+k}{p}\right) = p + 1 + \sum_{x=0}^{p-1} \left(\frac{x}{p}\right).$$

Here again $\left(\frac{0}{p}\right) = 0$. Further there are exactly $\frac{p-1}{2}$ quadratic residues in $\mathbb{F}_p$ and $\frac{p-1}{2}$ quadratic non-residues in $\mathbb{F}_p$. Thus

$$\sharp E_k(\mathbb{F}_p) = p + 1 + \frac{p-1}{2} + \frac{p-1}{2}(-1) = p + 1.$$

This completes the proof of Lemma 3.10. □

Lemma 3.12. *Let $p \in \mathbb{P}$ with $p > 2$ and consider the curve $E : Y^2 = X^3 + X$ over $\mathbb{F}_p$. Then*

$$4 \mid \sharp E(\mathbb{F}_p).$$

Proof. First let $p \equiv 1 \pmod{4}$. Then -1 is a square mod p, meaning that there exists an integer $n \pmod{p}$ with $n^2 \equiv -1 \pmod{p}$. Then the curve is given modulo p by

$$E : Y^2 \equiv X(X^2 - n^2) \equiv X(X-n)(X+n) \pmod{p}.$$

There are 3 points of exact order 2 in $E(\mathbb{F}_p)$:

$$(0,0), (n,0), (-n,0).$$

Further there are an even number, say $2k$, of points $(x, \pm y)$ with $1 \le x \le \frac{p-1}{2}$ and $y \neq 0$. For such a point $(x, \pm y)$, also $(-x, \pm yn)$ is a point on the curve, because

$$(-x)((-x)^2 + 1) \equiv n^2(x(x^2+1)) \equiv n^2 y^2 \pmod{p}.$$

Hence $1 \leq x \leq \frac{p-1}{2}$ corresponds either to no point or to the 4 points

$$(x, \pm y), (-x, \pm yn).$$

With this construction we got all points with $x \neq 0$, $y \neq 0$. Thus there are

$$\sharp E(\mathbb{F}_p) = 1 + 3 + 4k = 4(k+1)$$

points in the modulo p reduced curve E over $\mathbb{F}_p$.

Now let $p \equiv 3 \pmod 4$. Then -1 is not a square mod p, which implies that the only point of order 2 on E modulo p is the point $(0, 0)$.

Let $1 \leq x \leq \frac{p-1}{2}$. If there is a point of the form $(x, \pm y)$, then there is no point with the first coordinate $-x$, because

$$(-x)^3 + (-x) \equiv -(x^3 + x) \equiv -y^2 \pmod p$$

is no square modulo p. On the other hand, if there is no point with the first coordinate x, then

$$\left(\frac{x(x^2+1)}{p}\right) = -1$$

and

$$\left(\frac{(-x)^3 + (-x)}{p}\right) = \left(\frac{-1}{p}\right)\left(\frac{x^3 + x}{p}\right) = (-1)(-1) = 1.$$

This implies that one gets the 2 points

$$(-x, \pm y).$$

So every $1 \leq x \leq \frac{p-1}{2}$ leads to exactly 2 points, which means that one has

$$\sharp E(\mathbb{F}_p) = 1 + 1 + 2 \cdot \frac{p-1}{2} = p + 1 \equiv 0 \pmod 4. \qquad \square$$

This method of naive counting is only convenient for small values of q. Cohen [34] suggests for example to use the Legendre-symbol method only for $p \leq 10000$. For large values of q, the method of Shanks worked out by Mestre is more suitable. This is a baby step-giant step method combined with the estimate of Hasse.

From the Hasse estimate it follows that $\sharp E(\mathbb{F}_q) = q + 1 - t$ with $|t| \leq 2\sqrt{q}$. Let $P \in E(\mathbb{F}_q)$ be a random point. It is assumed that the order of P is greater than $4\sqrt{q}$. If this is not the case, we choose another random point P. Such a point exists because trivially $(q-1)^2 > 0$. We set $Q = (q + 1 + \lfloor 2\sqrt{q} \rfloor)P$. Then,

$$Q = (q + 1 - t + t + \lfloor 2\sqrt{q} \rfloor)P = (t + \lfloor 2\sqrt{q} \rfloor)P.$$

During the algorithm we compute an integer $0 \leq n \leq 4\sqrt{q}$ with $Q = nP$. This integer exists because of $0 \leq t + \lfloor 2\sqrt{q} \rfloor \leq 4\sqrt{q}$ by Hasse's Theorem 3.3. As the order of the point P is greater than $4\sqrt{q}$, we obtain the identity $t = n - \lfloor 2\sqrt{q} \rfloor$.

The integer n is computed using the baby step-giant step method. We know that $0 \leq n \leq 4\sqrt{q}$. Taking $m = \lceil 2q^{1/4} \rceil$ we write

$$n = im + j$$

with $0 \leq i \leq m, 0 \leq j \leq m-1$. The baby steps consist now in computing the points jP for $0 \leq j \leq m-1$. The giant steps are then to compute the points $Q - i(mP)$ for $0 \leq i \leq m$ and to test if there exists a j with $Q - i(mP) = jP$. If this is the case, the $n = im + j$ has been found.

Algorithm 3.13 (Shanks–Mestre method).

INPUT: The prime power q and $E|\mathbb{F}_q$.
OUTPUT: The number $t = q + 1 - \sharp E(\mathbb{F}_q)$.
1. $m \leftarrow \lceil 2q^{1/4} \rceil$.
2. Let $P \in E(\mathbb{F}_q)$ be a random point with order $> 4\sqrt{q}$.
3. $Q \leftarrow (q + 1 + \lfloor 2\sqrt{q} \rfloor)P$.
4. *Baby steps:* For $j = 0$ to $m - 1$ do:
5. Compute (and store) jP.
6. *Giant steps:* For $i = 0$ to m do:
7. If $Q - i(mP) = jP$ for a $0 \leq j \leq m-1$ then
$t \leftarrow im + j - \lfloor 2\sqrt{q} \rfloor$. /* Here n=im+j */
8. Return t.

We give an example for the Shanks–Mestre method. Consider $q = p = 163$ and the elliptic curve

$$E : Y^2 = X^3 + 5X + 3$$

over $\mathbb{F}_{163}$. Here $m = 8$. The random point is

$$P = (1, 3).$$

Then we have

$$Q = (163 + 1 + 25)P = (53, 0).$$

The baby steps lead to the following table:

j	0	1	2	3	4
jP	$\mathcal{O}$	(1, 3)	(36, 59)	(109, 85)	(23, 95)

j	5	6	7
jP	(11, 133)	(157, 88)	(46, 101)

We only have to carry out the giant steps for $i = 0$ and $i = 1$. Then we get

$$Q - 0(8P) = (53, 0), \quad Q - (8P) = (1, 3) = P.$$

So we find

$$\begin{aligned} n &= 8+1 = 9, \\ t &= 8+1-25 = -16, \end{aligned}$$

and

$$\sharp E(\mathbb{F}_{163}) = 163+1-(-16) = 180.$$

The algorithm of Shanks and Mestre has complexity $O(q^{1/4+\varepsilon})$, where ε is a positive constant that can be made arbitrarily small (see Blake, Seroussi, Smart [18], VI.3). In cryptographic applications, a method with complexity $O(\log^8 q)$ is used. This improvement is due to work of Schoof ([193], [194]). The idea is to compute the order of the group modulo small primes and then to use the Chinese Remainder Theorem to obtain the exact order. For the computation of the order modulo small primes $l > 2$ we apply the q-Frobenius endomorphism.

Algorithm 3.14 (Schoof).

INPUT: The prime power q and $E|\mathbb{F}_q$.
OUTPUT: The integer $t = q+1-\sharp E(\mathbb{F}_q)$.
1. $l_{\max} \leftarrow \min\left\{p \in \mathbb{P} : \prod_{l\in\mathbb{P}, l\le p} l > 4\sqrt{q}\right\}$.
2. If $2\,|\,q$ then do:
3.   If $j(E) = 0$ then $t_2 \leftarrow 1$, else $t_2 \leftarrow 0$,
4. else do:
5.   If $\sharp E(\mathbb{F}_q)[2] = 1$ then $t_2 \leftarrow 1$, else $t_2 \leftarrow 0$.
6. For all $l \in \mathbb{P}, 3 \le l \le l_{\max}$, do:
7.   Take a random point $P \in E[l]\backslash\{\mathcal{O}\}$.
8.   Compute $\varphi_q^2(P) + q_l P$ with $0 \le q_l < l, q_l \equiv q \pmod{l}$.
9.   For $\tau = 0$ to l do:
10.   Compute $\tau\varphi_q(P)$.
11.   If $\tau\varphi_q(P) = \varphi_q^2(P) + q_l P$ then do:
12. $t_l \leftarrow \tau$.
13.   Go to the next prime in Step 6.
14. Use the Chinese Remainder Theorem to determine t with $|t| \le 2\sqrt{q}$ and $t \equiv t_l \pmod{l}$ for all $l \in \mathbb{P}$, $2 \le l \le l_{\max}$.
15. Return t.

To understand the algorithm, observe that, for $q = p^k$,

$$\sharp E(\mathbb{F}_q) = q+1-t$$

with $|t| \le 2\sqrt{q}$. If we can determine $t_l := t \pmod{l}$ for all $2 \le l \le l_{\max}, l \in \mathbb{P}$, where

$$l_{\max} = \min\left\{p' \in \mathbb{P} : \prod_{l\in\mathbb{P}, l\le p'} l > 4\sqrt{q}\right\},$$

then we can determine t by the Chinese Remainder Theorem. The number $l_{\max}$ is thus chosen in such a way that the product of all primes $\leq l_{\max}$ is equal to a number $> 4\sqrt{q}$.

If $l = 2$ and q is divisible by 2, it follows from Theorem 3.5 that $t_2 = 0$ if and only if the curve is supersingular, that is $j(E) = 0$. When 2 does not divide q then $t_2 = 0$ if and only if there exists a nontrivial point of order 2.

Let $l > 2$ and φ_q be the q-th Frobenius endomorphism. Then for all $P \in E(\overline{\mathbb{F}}_q)$, we have

$$\varphi_q^2(P) - t\varphi_q(P) + qP = \mathcal{O}.$$

If there exists a $\tau \in \{0, 1, \ldots, l-1\}$ such that for $P \in E[l]\backslash\{\mathcal{O}\}$ we have

$$\varphi_q^2(P) + q_l P = \tau\varphi_q(P),$$

where $q_l \equiv q \pmod{l}$, then $t_l \equiv \tau \pmod{l}$. For the computation of points in $E[l]$, see Proposition 1.25, which admits many improvements (see Lay [123]).

There are many modifications of this algorithm. We mention the methods of Atkin and Elkies.

The method of Schoof depends on the determination of t_l. This constant does also appear in the characteristic polynomial of the Frobenius map

$$\mathcal{F}_l(U) := U^2 - t_l U + q_l = 0 \pmod{l}.$$

Definition 3.15. A prime number l is an *Elkies prime*, if $\mathcal{F}_l(u)$ has two roots in $\mathbb{F}_l$, i.e. if $t^2 - 4q \equiv t_l^2 - 4q_l \pmod{l}$ is a square in $\mathbb{F}_l$. Otherwise it is an *Atkin prime*.

The splitting type of a certain lth modular polynomial $\Phi_l(x, y)$ over the ground field $\mathbb{F}_q$, with j-invariant of the curve substituted for one of the variables, determines whether l is an Elkies or an Atkin prime (see Blake, Seroussi, Smart [18] VII.3.).

If the prime l is an Elkies prime, one uses a polynomial constructed in [18] (see [18] VII.2.1, VII.7) to efficiently searching a point $P = (x, y)$ and a value $\lambda \in \{1, 2, \ldots, l-1\}$ such that

$$(x^q, y^q) = \lambda(x, y).$$

Then λ is an eigenvalue of the Frobenius map, hence a root of the polynomial $\mathcal{F}_l(u)$. Therefore

$$t \equiv t_l \equiv \lambda + \frac{q}{\lambda} \pmod{l}.$$

If the prime is an Atkin prime, the method developed by Atkin does not produce the exact value of t_l, but a small set of possible values for t_l. For more details see the book of Blake, Seroussi, and Smart [18], VII.

For more improvements of Schoof's algorithm see for example the articles of Couveignes and Morain [39] or Lehmann, Maurer, Müller, and Shoup [125].

3.3 Construction of elliptic curves with given group order

The construction of elliptic curves over finite fields with given group order is a fundamental task in computational number theory. It is of great importance for cryptography, where elliptic curves which have an order divisible by a large prime number, are needed.

From the estimate of Hasse it follows that for a given finite field there is only a finite set of possible group orders for elliptic curves over that field. Moreover, there is a result of Waterhouse [230]:

Lemma 3.16. *Let $q = p^k$. There exists an elliptic curve $E|\mathbb{F}_q$ such that $\sharp E(\mathbb{F}_q) = q + 1 - t$ if and only if one of the following conditions hold:*

(i) $t \not\equiv 0 \pmod{p}$ *and* $t^2 \leq 4q$.

(ii) *k is odd and one of the following is true:*

- a) $t = 0$.
- b) $p = 2$ *and* $t^2 = 2q$.
- c) $p = 3$ *and* $t^2 = 3q$.

(iii) *k is even and one of the following is true:*

- a) $t^2 = 4q$.
- b) $p \not\equiv 1 \pmod{3}$ *and* $t^2 = q$.
- c) $p \not\equiv 1 \pmod{4}$ *and* $t = 0$.

Proof. See the article of Waterhouse [230] □

The following theorem gives the possible abelian groups for elliptic curves over finite fields.

Theorem 3.17. *Let $q = p^k$. Let*

$$\sharp E(\mathbb{F}_q) = n = \prod_{l \in \mathbb{P}} l^{n_l}$$

be a possible order of an elliptic curve E over $\mathbb{F}_q$. Then all possible groups $E(\mathbb{F}_q)$ with $\sharp E(\mathbb{F}_q) = n$ are the following (up to isomorphism):

$$\mathbb{Z}/p^{n_p}\mathbb{Z} \times \prod_{l \neq p} (\mathbb{Z}/l^{a_l}\mathbb{Z} \times \mathbb{Z}/l^{n_l - a_l}\mathbb{Z}) .$$

Here, for $l \neq p$,

a) *in case* (iii) a) *of Lemma* 3.16: *Each* a_l *is equal to* $\frac{n_l}{2}$.

b) *in all other cases of Lemma* 3.16: a_l *is an arbitrary integer satisfying* $0 \le a_l \le \min\left\{v_l(q-1), \left\lfloor \frac{n_l}{2} \right\rfloor\right\}$.

Proof. See the article of Rück [182]. □

There is a simple method for the construction of elliptic curves with given group order. First check that the desired order is possible over the given field, according to Lemma 3.16. Then determine the group order for randomly generated curves over the field until a curve with desired group order is found.

There is another method towards determining the group order which employs the theory of complex multiplication (see Lay and Zimmer [123], [124]). This method enables us to construct non-supersingular elliptic curves almost with given group order and is based on the following theorem.

Theorem 3.18 (Deuring). *Let* $\mathbb{K}$ *be an imaginary quadratic field and* $\mathbb{H}_{\mathcal{O}}$ *be the ring class field associated to an order* $\mathcal{O}$ *in* $\mathbb{K}$. *Denote by* p *a rational prime which splits completely in* $\mathbb{K}$ *and by* $\mathfrak{p}$ *a prime in* $\mathbb{H}_{\mathcal{O}}$ *above* p *with residue degree* $f = f_{\mathfrak{p}|p}$ *and such that* $[\mathcal{O}_{\mathbb{K}} : \mathcal{O}] \notin \mathfrak{p}$. *Let* $\mathcal{E}$ *be an elliptic curve over* $\mathbb{H}_{\mathcal{O}}$ *which has complex multiplication by* $\mathcal{O}$ *and let* $E|\mathbb{F}_q$ *be the curve* $\mathcal{E}$ *reduced modulo* $\mathfrak{p}$, *which should be an ordinary elliptic curve. Then there is an element* $\pi \in \mathcal{O}\backslash p\mathcal{O}$ *satisfying the system of norm equations*

$$\begin{aligned} q &= N_{\mathbb{K}}(\pi) \\ m = \sharp E(\mathbb{F}_q) &= N_{\mathbb{K}}(1-\pi), \end{aligned}$$

where $q = p^f$ *and* $N_{\mathbb{K}}$ *is the norm of* $\mathbb{K}$ *over* $\mathbb{Q}$. *The endomorphism ring of* $\mathcal{E}$ *is stable under the reduction map* $\mathcal{E} \mapsto E$, *i.e.* $\mathrm{End}(\mathcal{E}) = \mathrm{End}(E) = \mathcal{O}$. *Moreover, every elliptic curve over* $\mathbb{F}_q$ *with endomorphism ring* $\mathcal{O}$ *arises in this way.*

Proof. See Deuring [54], Lang [116], or also Lay [123]. □

Let $\mathbb{K} = \mathbb{Q}(\sqrt{\delta})$ with $\delta \in \mathbb{Q}$, $\delta < 0$. We write the norm equations using $\pi = a + b\sqrt{\delta}$ with $a, b \in \mathbb{Q}$. As π is integral, we can assume that the denominators of a and b divide 2. For given q and m, we want to construct an elliptic curve E defined over F_q with $\sharp E(F_q) = m$. The corresponding norm equations are

$$\begin{aligned} q &= a^2 - \delta b^2 \\ m &= (1-a)^2 - \delta b^2. \end{aligned}$$

Eliminating $-\delta b^2$ leads to $a = \frac{1}{2}((q+1) - m)$. Inserting this in the first equation gives

$$4\delta b^2 = (q+1-m)^2 - 4q.$$

Therefore, the imaginary quadratic field in the theorem is

$$\mathbb{K} = \mathbb{Q}\big(\sqrt{(q+1-m)^2 - 4q}\,\big).$$

We consider positive definite binary quadratic forms

$$Q = (a, b, c) : \ (X, Y) \mapsto aX^2 + bXY + cY^2$$

with $a, b, c \in \mathbb{Z}$. The discriminant of such a form is $\delta = b^2 - 4ac$. To every quadratic form Q we associate the number $\tau_Q = (-b + \sqrt{\delta})/2a$. Let $\mathcal{QF}(\delta)$ be the set of all such quadratic forms with discriminant δ. We consider the $SL_2(\mathbb{Z})$-equivalence class group of this set, which is called $\mathcal{C}(\delta)$.

We need the following proposition.

Proposition 3.19. *Let $\mathcal{O}$ be the order of discriminant δ in the imaginary quadratic field $\mathbb{K} = \mathbb{Q}(\sqrt{\delta})$ and let $\mathcal{C}l(\mathcal{O})$ be the ideal class group of $\mathcal{O}$. The map*

$$\Psi : \mathcal{QF}(\delta) \to \mathcal{C}l(\mathcal{O}), \quad Q = (a, b, c) \mapsto [1, \tau_Q]$$

induces an isomorphism between $\mathcal{C}(\delta)$ and $\mathcal{C}l(\mathcal{O})$.

Proof. See Cox [40], §7.B. (See also Zagier [242].) □

From the theory of complex multiplication (see Theorem 2.42) it follows that in the context of the above theorem

$$\mathbb{H}_{\mathcal{O}} = \mathbb{K}(j(\mathcal{E})).$$

On the other hand

$$\mathbb{H}_{\mathcal{O}} = \mathbb{K}[X]/W_\delta(X)\mathbb{K}[X]$$

where $W_\delta = W_{\mathcal{O}}$ is the polynomial

$$W_\delta(x) = \prod_{[Q]\in\mathcal{C}(\delta)} (X - j(\tau_Q))$$

with the notation of the proposition (see Theorem 2.42). Here $[Q]$ is the class of the quadratic form $Q \in \mathcal{QF}(\delta)$.

This polynomial can be computed by approximating the values $j(\tau_Q)$ by the formulas of Proposition 2.6 to a sufficiently high precision. Since the resulting polynomial has coefficients in $\mathbb{Z}$, it then suffices to round off the coefficients (see Lay, Zimmer [124]).

In the situation of Theorem 3.18 we have that $\mathcal{O}_{\mathbb{H}_{\mathcal{O}}}/\mathfrak{p} \cong \mathbb{F}_q$. We want to compute the j-invariant $j(\mathcal{E})$, which is a root of the polynomial W_δ. Because we are interested only in the reduction of $\mathcal{E}$ (mod $\mathfrak{p}$), it suffices to compute the roots of W_δ (mod $\mathfrak{p}$) in $\mathbb{F}_q$. Each root of W_δ yields the image of the j-invariant of an elliptic curve E over

$\mathbb{F}_q$ with group order m (for the theory described above see also the article of Yui and Zagier [241]).

Now we have to find the desired elliptic curve. In Proposition 1.9 we carried through constructions for elliptic curves with given j-invariant. We have to take into account that the isomorphism class of elliptic curves over $\mathbb{F}_q$ is not uniquely determined by the j-invariant. We only consider the case of characteristic $\neq 3$. The case of characteristic 3 needs special consideration similar to the case of characteristic 2, but this is not important for cryptographic applications.

Definition 3.20. a) Let $E|\mathbb{F}_q$ be an elliptic curve given by a short Weierstraß equation

$$E : Y^2 = X^3 + a_4X + a_6,$$

with $q = p^k$ and $p > 3$. For any fixed non-square $c \in \mathbb{F}_q^*$, the *c-twist* of E is the elliptic curve

$$E_c : Y^2 = X^3 + a_4c^2X + a_6c^3.$$

b) Let $E|\mathbb{F}_{2^k}$ be an elliptic curve given by the Weierstraß normal form

$$E : Y^2 + XY = X^3 + a_2X^2 + a_6$$

with $a_2 \in \{0, \gamma\}$, where $\gamma \in \mathbb{F}_{2^k}$ is an element of trace $\mathrm{Tr}(\gamma) = 1$. The *γ-twist* of E is the elliptic curve

$$E_\gamma : Y^2 + XY = X^3 + (a_2 + \gamma)X^2 + a_6.$$

Proposition 3.21. *Let $E|\mathbb{F}_q$ be an ordinary elliptic curve and E' be a twist. Then*

1) $j(E) = j(E')$,

2) $\sharp E(\mathbb{F}_q) + \sharp E'(\mathbb{F}_q) = 2q + 2.$

Proof. 1) This is an easy computation.

2) Consider first $\mathrm{char}(\mathbb{F}_q) > 3$. An element in $\mathbb{F}_q$ is a square if and only if it is a square in $\mathbb{F}_p$ with $p \in \mathbb{P}, q = p^k$. There are q elements $x \in \mathbb{F}_q$. If $x^3 + a_4x + a_6$ is a square, there are two points $(x, \pm y) \in E(\mathbb{F}_q)$. Further in this case, since c is not a square in $\mathbb{F}_q^*$,

$$\left(\frac{(cx)^3 + a_4c^2(cx) + a_6c^3}{p}\right) = \left(\frac{c}{p}\right)\left(\frac{x^3 + a_4x + a_6}{p}\right) = -1,$$

so that there is no point on $E'(\mathbb{F}_q)$ with first coordinate cx. If $x^3 + a_4x + a_6$ is not a square in $\mathbb{F}_q$, then there is no point on $E(\mathbb{F}_q)$ with first coordinate x, but

$$\left(\frac{(cx)^3 + a_4c^2(cx) + a_6c^3}{p}\right) = \left(\frac{c}{p}\right)\left(\frac{x^3 + a_4x + a_6}{p}\right) = 1.$$

Hence there are two points $(cx, \pm y)$ on $E'(\mathbb{F}_q)$ with first coordinate cx.

Altogether, every $x \in \mathbb{F}_q$ gives two points, either on E or on E'. Adding the two points at infinity gives the result.

The case of $\mathrm{char}(\mathbb{F}_q) = 2$ is an exercise, as is the case of $\mathrm{char}(\mathbb{F}_q) = 3$. □

Hence, for the construction of the curve $E|\mathbb{F}_q$ with given number of points we construct a curve and the twists for this j-invariant. The right choice between the curves is made by trial and error. (However, the number of points obtained might differ a little from what was intended.)

We summarize the algorithm:

Algorithm 3.22 (Construction of an elliptic curve).

```
INPUT: An integer m and a prime power q.
OUTPUT: An ordinary elliptic curve E|F_q with ♯E(F_q) = m.
1.  If m is no possible order for an ordinary elliptic curve
    over F_q (see Lemma 3.16 then return a message.
2.  δ ← (q + 1 − m)^2 − 4q, K ← Q(√δ).
3.  Compute the polynomial W_δ.
4.  Compute the roots of this polynomial in F_q.  These roots
    correspond to j-invariants of elliptic curves.
5.  For each root construct an elliptic curve E|F_q together
    with its twist.
6.  Decide which elliptic curve has the desired group order.
7.  Return E|F_q.
```

We give a simple example to understand the algorithm. Consider the field $\mathbb{F}_5$, hence $p = q = 5$. We want to construct an elliptic curve $E|\mathbb{F}_5$ with $\sharp E(\mathbb{F}_5) = 8$, that means $m = 8$. The trace t of the q-Frobenius endomorphism is then $t = -2$. Lemma 3.16 (i) shows that an elliptic curve over $\mathbb{F}_5$ with order $m = 8$ exists.

Then we compute δ:

$$\delta = (q + 1 - m)^2 - 4q = -16.$$

That means that we consider the number field $\mathbb{K} = \mathbb{Q}(\sqrt{\delta}) = \mathbb{Q}(i)$, where $i = \sqrt{-1}$. The class number of this field is 1, hence there is only one equivalence class of quadratic forms with discriminant -16.

We take the form

$$Q = (2, 0, 2) : \ (X, Y) \mapsto 2X^2 + 2Y^2.$$

The discriminant of this form is equal to $\delta = -16$. It maps to the complex number

$$\tau_Q = \frac{\sqrt{\delta}}{4} = i.$$

Using the formula for $j(\tau)$ (see Proposition 2.6) we get

$$j(\tau_Q) = j(i) = 1728 = 12^3.$$

The polynomial is then

$$W_\delta(X) = X - 1728 \equiv X - 3 \pmod 5.$$

The root of this polynomial modulo 5 is 3 (mod 5). We use Proposition 1.9 to construct an elliptic curve $E|\mathbb{F}_5$ with j-invariant equal to $1728 \equiv 3 \pmod 5$:

$$E : Y^2 = X^3 + X.$$

To construct a twist of this curve, we take the number $c \equiv 2 \pmod 5$, which is not a square modulo 5. The twist of E is then

$$E' : Y^2 = X^3 + 4X.$$

Then we compute

$$\sharp E(\mathbb{F}_5) = 4, \quad \sharp E'(\mathbb{F}_5) = 8.$$

Therefore we return E' as the desired elliptic curve.

3.4 Elliptic curves in cryptography

Cryptography means the enciphering and deciphering of messages in secret code. This field has become so important that nowadays there are special conferences devoted exclusively to this topic. We begin with a short introduction to cryptosystems and then explain the elliptic curve public key cryptosystems. A more detailed description of elliptic curves in cryptography is given in the book of Blake, Seroussi, and Smart [18].

The message which should be enciphered is called *plaintext*, the enciphered message *ciphertext*. Both texts are written in an *alphabet* which consists of N *letters*. The plaintext and the ciphertext are given in certain *text units* (arrangements of several letters). Let $\mathcal{P}$ be the set of all possible plaintext units and $\mathcal{C}$ the set of all possible ciphertext units. An *enciphering function* is a bijective function $f : \mathcal{P} \to \mathcal{C}$. The *deciphering function* is the function f^{-1}. Such a system

$$f : \mathcal{P} \to \mathcal{C}, \quad f^{-1} : \mathcal{C} \to \mathcal{P}$$

is called *cryptosystem*. Someone who sends a message is called the *sender*, someone who receives a message is called the *receiver*.

The enciphering function is a mathematical function. Therefore the text units have to be represented as mathematical objects. An easy method is for example to represent the blank by the number 0 and then to number the letters $A, \dots, Z$:

$$A \to 1, \ B \to 2, \ \dots, \ Z \to 26.$$

As we focus on elliptic curve cryptosystems, we now explain how to represent text units as points on an elliptic curve over a finite field.

We assume that a text unit consists of integers m such that $0 \leq m < M$, $M \in \mathbb{N}$. Let $\kappa \in \mathbb{N}$, $p \in \mathbb{P}$, $p > 2$, $q = p^k$, and $q > M\kappa$. We take an elliptic curve $E : Y^2 = X^3 + a_4X + a_6$ over $\mathbb{F}_q$ so that $j = 0$ if $p = 3$. The representation of the text units as points on $E(\mathbb{F}_q)$ is carried out in two steps:

a) Representation of the numbers $\{1, \ldots, M\kappa\}$ as elements of $\mathbb{F}_q$:

Let N be an element of $\{1, \ldots, M\kappa\}$. We use the p-adic representation of N:

$$N = \sum_{i=0}^{k-1} n_i p^i, \quad 0 \leq n_i < p.$$

Fixing an integer m satisfying $0 \leq m < M$, we write

$$N = m\kappa + j$$

with $1 \leq j < \kappa$.

The field $\mathbb{F}_q$ is given as $\mathbb{F}_q \cong \mathbb{F}_p[X]/(f(X))$ with a polynomial $f(X)$ which is irreducible over $\mathbb{F}_p$ and of degree k. We assign to $N = m\kappa + j$ the element

$$x_{m,j} := \sum_{i=0}^{k-1} n_i X^i \pmod{f(X)}.$$

Conversely, we get from an element $x \in \mathbb{F}_q$, which represents a number $1 \leq N = m\kappa + j \leq M\kappa$, the parameters m, j in the following way. Let

$$0 \neq x = \sum_{i=0}^{k-1} n_i X^i \pmod{f(X)}.$$

Then N is the number

$$N = \sum_{i=0}^{k-1} n_i p^i \quad (\text{so that } N < p^k = q).$$

By construction we have

$$1 \leq N \leq M\kappa \Leftrightarrow 0 \leq N - 1 < M\kappa \Leftrightarrow 0 \leq \left\lfloor \frac{N-1}{\kappa} \right\rfloor < M.$$

Hence we get the parameters

$$m := \left\lfloor \frac{N-1}{\kappa} \right\rfloor, \qquad j := N - m\kappa.$$

It remains to show that $1 \le j < \kappa$:

$$\begin{aligned} 1 &= N - \frac{N-1}{\kappa}\kappa \\ &\le N - \left\lfloor \frac{N-1}{\kappa} \right\rfloor \kappa \\ &= j \\ &= \frac{N-1}{\kappa}\kappa - \left\lfloor \frac{N-1}{\kappa} \right\rfloor \kappa + 1 \\ &< \left(\left\lfloor \frac{N-1}{\kappa} \right\rfloor + 1 \right) \kappa - \left\lfloor \frac{N-1}{\kappa} \right\rfloor \kappa + 1 \\ &= \kappa + 1. \end{aligned}$$

b) Representation of the numbers $0 \le m < M$ as points in $E(\mathbb{F}_q)$:

To represent a number m as a point in $E(\mathbb{F}_q)$, we first consider the number

$$N = m\kappa + 1.$$

In Part a) we determined the element $x_{m,1} \in \mathbb{F}_q$ corresponding to N. Now we test if $x_{m,1}^3 + a_4 x_{m,1} + a_6$ is a square in $\mathbb{F}_q$. If this is the case, there exists an $y_{m,1} \in \mathbb{F}_q$ such that

$$y_{m,1}^2 = x_{m,1}^3 + a_4 x_{m,1} + a_6.$$

Then m is assigned the point $(x_{m,1}, y_{m,1})$. Otherwise consider the number $N = m\kappa + 2$ and repeat the method.

In this way we consider successively all numbers $m\kappa + j$ for $j = 1, 2, \ldots$, until we have found a point $(x_{m,j}, y_{m,j})$. If $j > \kappa$ and no point is found, we have to pick another curve.

For any $x \in \mathbb{F}_q$, we have, in principle, q possibilities for the value

$$x^3 + a_4 x + a_6.$$

Only for ca. $\frac{1}{2}\sharp E(\mathbb{F}_q)$ of these values we get a point on $E(\mathbb{F}_q)$. So the probability that x leads to a point is

$$\frac{\sharp E(\mathbb{F}_q)}{2q} \simeq \frac{1}{2}.$$

because, by Hasse's estimate, we know that $(\sqrt{q} - 1)^2 \le \sharp E(\mathbb{F}_q) \le (\sqrt{q} + 1)^2$. Therefore, for a given m, the probability that $m\kappa + 1, m\kappa + 2, \ldots, m\kappa + \kappa$ lead to no point is about $2^{-\kappa}$.

From a point $P = (x, y) \in E(\mathbb{F}_q)$, which represents the number m, we obtain m by using the method of Part a) for x.

Public key cryptosystems are cryptosystems where the enciphering function is public. This has many advantages.

- Sender and receiver of a message need not exchange a key in secret.
- In a network of many users, there need not be different keys for all users.
- A *digital signature* is possible (see for example the book of Blake, Seroussi, and Smart [18]).

The enciphering functions for public key cryptosystems should be *one-way functions.* This are functions $f : \mathcal{P} \to \mathcal{C}$ such that for each plaintext unit $m \in \mathcal{P}$ the ciphertext unit $f(m)$ is easy to compute (for example computable in polynomial time). For most ciphertext units $c \in \mathcal{C}$ the plaintext unit $f^{-1}(c)$ should be hard to compute (infeasible using the best known algorithms and the best available computer technologies). If there is an extra information (*trapdoor*) with which the function f can be efficiently inverted, f is called a *trapdoor one-way function.*

The principle of public key cryptosystems using trapdoor one-way functions is the following. The receiver makes the enciphering function f public. Every sender can encipher the message $m \in \mathcal{P}$ and sends $c := f(m)$ to the receiver. As f is a one-way function, nobody can read the plaintext m from the ciphertext c, except the receiver who knows the trapdoor.

This system can also be used for a digital signature. If the sets $\mathcal{P}$ and $\mathcal{C}$ are the same, the receiver sends a message m enciphered with f^{-1}, that means he sends $f^{-1}(m)$. If someone wants to read this message, he can apply the publicly known function f to get $m = f(f^{-1}(m))$. This message has to come from the right person, because nobody else knows the function f^{-1}.

As an example for public key cryptosystems using elliptic curves, we look at the ElGamal method for elliptic curves [57].

Here let $E|\mathbb{F}_q$ be an elliptic curve over a finite field $\mathbb{F}_q$ and let $P \in E(\mathbb{F}_q)$ be a point of E over $\mathbb{F}_q$. The *discrete logarithm problem on E* is the question if, to a given point $Q \in E(\mathbb{F}_q)$, there exists an integer n with $Q = nP$ and if one can compute this n.

The principle of this ElGamal method is that the discrete logarithm problem on elliptic curves is difficult to solve.

One represents the text units $0 \le m < M$ as points P_m of $E(\mathbb{F}_q)$. The receiver makes a basis point $P \in E(\mathbb{F}_q)$ public. Further he chooses a secret integer n and makes the point nP public. The sender chooses a secret integer k and maps the text unit m to the pair

$$(kP, P_m + k(nP)).$$

The receiver can read the text unit m from this pair, because he knows the number n, that is, he can compute

$$P_m = (P_m + k(nP)) - n(kP).$$

A spy knows neither n nor k, so normally he cannot compute the point P_m.

3.5 The discrete logarithm problem on elliptic curves

We recall the *discrete logarithm problem* (DLP) on elliptic curves: Let $E|\mathbb{F}_q$ be an elliptic curve and $P, Q \in E(\mathbb{F}_q)$. The DLP on E is the question if there exists an integer n with $Q = nP$ and how to compute this n.

For cryptographic applications we need elliptic curves where the DLP is difficult to solve. In this section we present classes of elliptic curves where the DLP can be reduced and therefore solved easier than on a general elliptic curve. These classes of elliptic curves should *not* be used for cryptosystems.

Pohlig and Hellmann observed that the DLP in finite abelian groups can be reduced to the DLP in finite abelian groups of prime power order (see [168]). First the DLP for the subgroups of prime order is solved. This is applied to solve the DLP for the subgroups of prime power order. The DLP for the group can then be solved using the Chinese Remainder Theorem. To *avoid* this simplification, the group should have at least one subgroup of large prime order.

We will now explain the simplification method of Pohlig and Hellmann. Let G be a finite additive abelian group of order $\sharp G = m$, $P, Q \in G$. We want to solve the DLP $Q = nP$. Let p be a prime number which divides m. Set $m' = m/p$ and

$$Q' = m'Q, \quad P' = m'P.$$

These are elements of order dividing p. Instead of solving the DLP $Q = nP$, we first solve the DLP $Q' = nP' = n_0 P'$; here $n_0 \equiv n \pmod{p}$, $n_0 \leq n$ (say).

If $p^{i+1} \mid m$ for $i \geq 1$ and if we know the solution of the DLP modulo p^i, $n \equiv n_i \pmod{p^i}$, $n_i \leq n$, we can compute the solution modulo p^{i+1} with the same method as for the solution modulo p. Therefore we write $n = n_i + \lambda p^i$. We want to compute $\lambda \pmod{p}$. Suppose that we have

$$Q = (n_i + \lambda p^i)P.$$

This can be written as

$$R := Q - n_i P = \lambda(p^i P) =: \lambda S.$$

Thus we solve the DLP $R = \lambda S$ to get first $\lambda \pmod{p}$ and then $n \pmod{p^{i+1}}$.

With this method we can compute $n \pmod{p^k}$ for all prime powers p^k which divide m. Using the Chinese Remainder Theorem we can compute $n \pmod{m}$ and therefore solve the DLP.

Another method to solve the DLP when the group order is not too large is, for example, a baby step-giant step method. Therefore let G be a finite abelian group with order m as before, $P, Q \in G$. The initial DLP is given by $Q = nP$. Write

$$n = a\lceil\sqrt{m}\,\rceil + b \quad \text{with } 0 \leq a, b < \lceil\sqrt{m}\,\rceil.$$

The DLP can then be written as

$$R_b := Q - bP = a(\lceil\sqrt{m}\,\rceil P).$$

The baby steps are to compute and store the elements R_b for $0 \le b < \lceil \sqrt{m} \rceil$. The giant steps are then to compute $a(\lceil \sqrt{m} \rceil P)$ for $0 \le a < \lceil \sqrt{m} \rceil$. If one a is found with $R_b = a(\lceil \sqrt{m} \rceil P)$ for a $b \in \{0, \dots, \lceil \sqrt{m} \rceil - 1\}$, then the DLP is solved.

There is also a method suggested by Pollard [170] (λ-*method*) which works for finite groups, where the order is not too large. For this method, let G again be a finite additive abelian group of order m and $P, Q \in G$ with the DLP $Q = nP$. Further define both, a function

$$f : G \to \{1, \dots, s\}$$

for some $s \in \mathbb{N}$ and a set of multipliers

$$M_i := a_i P + b_i Q \quad \text{for } i = 1, \dots, s$$

with random $a_i, b_i \in \mathbb{Z}$. Consider the function

$$F : G \to G, \quad R \mapsto R + M_{f(R)}.$$

The λ-*method* of Pollard is the following: Take two elements

$$R_0 := x_0 P + x_0' Q, \quad S_0 := y_0 P + y_0' Q$$

at random and compute inductively

$$\begin{aligned} R_k &:= F(R_{k-1}) = x_k P + x_k' Q, \\ S_k &:= F(S_{k-1}) = y_k P + y_k' Q \end{aligned}$$

for $k > 0$. After some iterations there are $k, l \in \mathbb{N}_0$ with

$$R_k = S_l \Leftrightarrow x_k P + x_k' Q = y_l P + y_l' Q.$$

We then find

$$(x_k - y_l)P = (y_l' - x_k')Q = (y_l' - x_k')nP.$$

Furthermore, we can compute

$$n \equiv \frac{x_k - y_l}{y_l' - x_k'} \pmod{m}$$

provided that $\gcd(y_l' - x_k', m) = 1$.

If the iteration is done only for R_k until there are $k, l \in \mathbb{N}_0$ with $R_k = R_l$, the DLP can be solved in an analogous way. This is called Pollard's ρ-*method*.

The methods described to solve the DLP are general methods, which can be used for arbitrary finite abelian groups. There are some special methods for the DLP on elliptic curves over finite fields. One such method comes from results of Menezes, Okamoto, and Vanstone (MOV-attack, [142], see also the article of Frey and Rück [70]). This attack reduces the DLP on elliptic curves to that on finite fields, if the smallest value of l such that $q^l \equiv 1 \pmod{m}$ is not too large.

If the curve is supersingular, it follows from Lemma 3.16 that

$$m = \sharp E(\mathbb{F}_q) = q + 1 - t \quad \text{with } t^2 = 0, q, 2q, 3q, \text{or } 4q.$$

In the cases $t^2 = 0, q, 2q, 3q$ there exists a j, $1 \leq j \leq 6$, with $q^j \equiv 1 \pmod{m}$. Hence, for cryptographic applications, supersingular curves should not be used.

Another special method was proposed by Smart [214], and by Satoh and Araki [184]. Here the p-adic elliptic logarithm is used to directly solve the DLP for anomalous elliptic curves. An elliptic curve $E|\mathbb{F}_q$ is called *anomalous*, if the trace of the Frobenius endomorphism is 1:

$$t = q + 1 - \sharp E(\mathbb{F}_q) = 1.$$

This section leads to the following conclusion. Let $E|\mathbb{F}_q$ be an elliptic curve with $m = q+1-t = \sharp E(\mathbb{F}_q)$. For E to be used in cryptosystems, the following properties are required:

- The group $E(\mathbb{F}_q)$ should have a subgroup of large prime order.
- The curve E should not be anomalous (i.e. $q \neq m$).
- The smallest value of l such that $q^l \equiv 1 \pmod{m}$ should be large. (This condition removes curves with $t = 0, t = 2$ (for $q \in \mathbb{P}$), and supersingular curves.)

3.6 Exercises

1) Prove Proposition 3.8 and generalize the formula for $\sharp E(\mathbb{F}_q)$ for elliptic curves over a finite ring $R := \mathbb{Z}/n\mathbb{Z}$ for a suitable $n \in \mathbb{N}$.
2) Compute the number of points, using naive counting, of
   a) $E : Y^2 = X^3 + 8X + 9$ over $\mathbb{F}_3, \mathbb{F}_{13}, \mathbb{F}_{17}$
   b) $E : Y^2 + XY = X^3 + \xi X + 1$ over $\mathbb{F}_{2^3} = \mathbb{F}_2(\xi)$ with $\xi^3 + \xi + 1 = 0$.
   c) $E : Y^2 = X^3 - 2\xi$ over $\mathbb{F}_{7^2} = \mathbb{F}_7(\xi)$ with $\xi^2 + 3 = 0$.
3) Compute the number of points using the method of Shanks and Mestre of the curve $E : Y^2 = X^3 + 24X - 34$ over $\mathbb{F}_{157}$.
4) Show that either

$$E(\mathbb{F}_q) \cong \mathbb{Z}/d\mathbb{Z} \quad \text{with } d \mid q^2$$

or

$$E(\mathbb{F}_q) \cong \mathbb{Z}/d\mathbb{Z} \times \mathbb{Z}/d'\mathbb{Z} \quad \text{with } d \mid d' \text{ and } d \mid q - 1.$$

Here, E is an elliptic curve over the finite field $\mathbb{F}_q$, $q = p^k$, $p \in \mathbb{P}, k \in \mathbb{N}$.

5) Represent the blank by 0, and number the letters $A, \ldots, Z$ as $A \to 1, \ldots, Z \to 26$, so that $M = 27$. Take $p = 109$ and $\kappa = 4$. Then $M\kappa = 108 < p$. First represent the numbers $\{1, \ldots, 108\}$ as elements of $\mathbb{F}_{109}$. Then take the elliptic curve

$$E : Y^2 = X^3 + 8X$$

over $\mathbb{F}_{109}$. Represent the text units blank, $A, \ldots, Z$ as points on $E(\mathbb{F}_{109})$.

6) Using the curve of Exercise 4), encode the text

FERMATS LAST THEOREM .

Chapter 4
Elliptic curves over local fields

In this chapter we consider elliptic curves over local fields of characteristic 0. First we define the several reduction types. Then we consider the filtration of elliptic curves over local fields. In the last section we prove the local theorem of Nagell, Lutz, and Cassels, which, among other things, we employ later in Chapter 6 to compute the points of finite order of elliptic curves over number fields.

Let $\mathbb{K}$ be a local field with respect to the normalized discrete additive valuation ord. Further let

- $\mathcal{O}_{\mathbb{K}}$ be the ring of integers of $\mathbb{K}$,
- $\mathfrak{p}$ the prime ideal corresponding to ord,
- $\mathbb{F}_{\mathfrak{p}} = \mathcal{O}_{\mathbb{K}}/\mathfrak{p}$ the residue field of $\mathfrak{p}$,
- π a prime element in $\mathcal{O}_{\mathbb{K}}$ corresponding to $\mathfrak{p}$.

4.1 Reduction

Let $E|\mathbb{K}$ be an elliptic curve in long Weierstraß normal form with integral coefficients, that is

$$\operatorname{ord}(a_1) \geq 0, \quad \operatorname{ord}(a_2) \geq 0, \quad \operatorname{ord}(a_3) \geq 0, \quad \operatorname{ord}(a_4) \geq 0, \quad \operatorname{ord}(a_6) \geq 0.$$

Then, obviously the Tate values are also integral, and the discriminant is integral:

$$\operatorname{ord}(\Delta) \geq 0.$$

Definition 4.1. Let $E|\mathbb{K}$ be an elliptic curve in long Weierstraß normal form with integral coefficients over the local field $\mathbb{K}$. The equation is called *minimal*, if $\operatorname{ord}(\Delta)$ is minimal under this condition among all curves in the same isomorphism class.

(Reduction of elliptic curves is a topic of a large number of investigations.[1] When is an equation of an elliptic curve minimal?)

[1] See for example Stroeker, R. J., Reduction of elliptic curves over imaginary quadratic number fields. Pacific J. Math. **108,** 451–463, 1983.

Theorem 4.2. a) *For every elliptic curve $E|\mathbb{K}$ over a local field $\mathbb{K}$, there exists a minimal Weierstraß equation.*

b) *Let $E|\mathbb{K}$ be an elliptic curve in long Weierstraß normal form with integral coefficients. If* $\operatorname{ord}(\Delta) < 12$ *(or* $\operatorname{ord}(c_4) < 4$ *or* $\operatorname{ord}(c_6) < 6$*), then the equation is minimal. If* $\operatorname{char}(\mathbb{F}_\mathfrak{p}) \neq 2, 3$, *the converse is also true.*

Proof. a) Transform the Weierstraß equation, so that the coefficients $a_1, \dots, a_6$ get integral. Then, $\operatorname{ord}(\Delta) \geq 0$. This is true for all such equations with integral coefficients. Because $\operatorname{ord}(\Delta)$ takes values in $\mathbb{Z}$ and is ≥ 0, there is such an equation with minimal $\operatorname{ord}(\Delta)$, that is, a minimal equation.

b) Let E be given in long Weierstraß normal form with integral coefficients. If the equation is not minimal, there exists a transformation $u(\neq 0), r, s, t \in \mathbb{K}$, such that, for the new discriminant Δ':

$$\begin{aligned} 0 &\leq \operatorname{ord}(\Delta') \\ &= \operatorname{ord}(u^{-12}\Delta) \\ &= -12\operatorname{ord}(u) + \operatorname{ord}(\Delta) \end{aligned}$$

and

$$\operatorname{ord}(\Delta') < \operatorname{ord}(\Delta),$$

hence $\operatorname{ord}(u) > 0$. The valuation of Δ can only be changed by subtracting multiples of 12 from $\operatorname{ord}(\Delta)$. If $\operatorname{ord}(\Delta) < 12$ this is not possible, hence the equation is minimal for $\operatorname{ord}(\Delta) < 12$.

The investigation of c_4 and c_6 is analogous.

For the proof of the converse, assume that the curve is given in short Weierstraß normal form

$$E : Y^2 = X^3 + AX + B$$

with integral coefficients. Let the equation be minimal at $\mathfrak{p}$. One has

$$\begin{aligned} c_4 &= -2^4 3A, \\ c_6 &= -2^5 3^3 B. \end{aligned}$$

Furthermore,

$$\begin{aligned} c_4^3 &= \Delta j, \\ c_6^2 &= c_4^3 - 1728\Delta = \Delta(j - 12^3). \end{aligned}$$

If

$$\operatorname{ord}(c_4) \geq 4 \quad \text{and} \quad \operatorname{ord}(c_6) \geq 6,$$

the equation is not minimal, as in this case

$$A = \pi^4 A', \quad B = \pi^6 B'$$

with integral A', B' and the prime element π (remember $\mathrm{ord}(6) = 0$). The transformation with $r = s = t = 0$, $u = \pi$ yields a short Weierstraß normal form with integral coefficients, which has the Tate values c_4', c_6':

$$\mathrm{ord}(c_4') = \mathrm{ord}(c_4) - 4 \quad \text{and} \quad \mathrm{ord}(c_6') = \mathrm{ord}(c_6) - 6.$$

If these values are still too large, one repeats the method until $\mathrm{ord}(A) < 4$ or $\mathrm{ord}(B) < 6$ (new notation). Then, if $\mathrm{ord}(A) = \mathrm{ord}(c_4) < 4$,

$$\begin{aligned} \mathrm{ord}(\Delta) + \mathrm{ord}(j) &= 3\mathrm{ord}(c_4) \\ &< 12 + 3\mathrm{ord}(2^4 3) \\ &= 12. \end{aligned}$$

If $\mathrm{ord}(B) = \mathrm{ord}(c_6) < 6$,

$$\begin{aligned} \mathrm{ord}(\Delta) + \mathrm{ord}(j - 12^3) &= 2\mathrm{ord}(c_6) \\ &< 12 + 2\mathrm{ord}(2^5 3^3) \\ &= 12. \end{aligned}$$

In both cases

$$\begin{aligned} \mathrm{ord}(\Delta) &\leq \mathrm{ord}(\Delta) + \max\{\mathrm{ord}(j), 0\} \\ &= \mathrm{ord}(\Delta) + \max\{\mathrm{ord}(j), \mathrm{ord}(12^3)\} \\ &\leq \mathrm{ord}(\Delta) + \max\{\mathrm{ord}(j), \mathrm{ord}(j - 12^3)\} \\ &< 12. \end{aligned}$$

□

Definition 4.3. Let $E|\mathbb{K}$ be an elliptic curve over a local field $\mathbb{K}$ with minimal equation

$$E: \; Y^2 + a_1XY + a_3Y = X^3 + a_2X^2 + a_4X + a_6.$$

The *reduced curve* $\widetilde{E}$ is the curve

$$\widetilde{E}: \; Y^2 + \widetilde{a_1}XY + \widetilde{a_3}Y = X^3 + \widetilde{a_2}X^2 + \widetilde{a_4}X + \widetilde{a_6},$$

where $\widetilde{a_i}$ is the coefficient a_i considered modulo $\mathfrak{p}$, that is, the coefficient $\widetilde{a_i}$ is in the finite field $\mathbb{F}_\mathfrak{p}$.

The elliptic curve has *good reduction* if the reduced curve is an elliptic curve, that is, if it is nonsingular. If it is singular it has *bad reduction*. This latter behaviour is subdivided into *multiplicative reduction*, if the reduced curve has a node, and *additive reduction*, if the reduced curve has a cusp.

The multiplicative reduction is further divided into *split multiplicative reduction*, if the slopes of the tangents at the node lie in the field $\mathbb{F}_\mathfrak{p}$. If this is not the case then the curve is said to have *non-split multiplicative reduction*.

Proposition 4.4. *Let $E|\mathbb{K}$ be an elliptic curve over a local field $\mathbb{K}$ with minimal Weierstraß equation.*

a) *The curve has good reduction if and only if*

$$\operatorname{ord}(\Delta) = 0.$$

b) *The curve has multiplicative reduction if and only if*

$$\operatorname{ord}(\Delta) > 0 \quad \text{and} \quad \operatorname{ord}(c_4) = 0.$$

The reduction is split multiplicative, if

at $\operatorname{char}(\mathbb{F}_\mathfrak{p}) \neq 2, 3$*: $-c_4c_6$ is a square in $\mathbb{F}_\mathfrak{p}$,*
at $\operatorname{char}(\mathbb{F}_\mathfrak{p}) = 3$*: b_2 is a square in $\mathbb{F}_\mathfrak{p}$,*
at $\operatorname{char}(\mathbb{F}_\mathfrak{p}) = 2$*: The polynomial $X^2 + a_1X + (a_3a_1^{-1} + a_2)$ has a root in $\mathbb{F}_\mathfrak{p}$.*

Otherwise the reduction is non-split multiplicative.

c) *The curve has additive reduction if and only if*

$$\operatorname{ord}(\Delta) > 0 \quad \text{and} \quad \operatorname{ord}(c_4) > 0.$$

Proof. Proposition 1.5 for the modulo $\mathfrak{p}$ reduced curve in $\mathbb{F}_\mathfrak{p}$ yields the proof of everything but the subdivision between split and non-split multiplicative reduction. For the open proof we consider the polynomial

$$f(X, Y) = Y^2 + a_1XY + a_3Y - X^3 - a_2X^2 - a_4X - a_6.$$

Let the point $P = (x_0, y_0) \in E(\mathbb{K})$ reduce to a singular point modulo $\mathfrak{p}$, that is, let

$$\begin{aligned} f(x_0, y_0) &\equiv 0 \pmod{\mathfrak{p}}, \\ f_X(x_0, y_0) &\equiv 0 \pmod{\mathfrak{p}}, \\ f_Y(x_0, y_0) &\equiv 0 \pmod{\mathfrak{p}}. \end{aligned}$$

The Taylor series for f at P modulo $\mathfrak{p}$ is

$$f(X, Y) \equiv [(Y - y_0) - \alpha(X - x_0)][(Y - y_0) - \beta(X - x_0)] - (X - x_0)^3 \bmod \mathfrak{p}.$$

Here $\alpha, \beta \in \mathbb{F}_\mathfrak{p}$ are the slopes of the tangent at P modulo $\mathfrak{p}$, with $\alpha \neq \beta$ for multiplicative reduction.

If $\operatorname{char}(\mathbb{F}_\mathfrak{p}) \neq 2, 3$, then we can use an equation in short Weierstraß normal form, that is (cf. the proof of Theorem 1.7),

$$f(X, Y) \equiv Y^2 - X^3 + 27c_4X + 54c_6 \pmod{\mathfrak{p}}.$$

Taking into account the partial derivatives of f and the fact that P is singular modulo $\mathfrak{p}$ we see that

$$3x_0^2 - 27c_4 \equiv 0 \equiv 2y_0 \pmod{\mathfrak{p}}.$$

Since E has multiplicative reduction, $12^3\Delta = c_4^3 - c_6^2$, and $\operatorname{char}(\mathbb{F}_\mathfrak{p}) \neq 2, 3$, we see that

$$c_4^3 \equiv c_6^2 \pmod{\mathfrak{p}}.$$

Hence c_4 is a square modulo $\mathfrak{p}$ with square roots $\pm c_6 c_4^{-1} \pmod{\mathfrak{p}}$. Inserting the two possibilities for x_0 in the equation $f(x_0, y_0) \equiv 0 \bmod \mathfrak{p}$ and again observing that $\operatorname{char}(\mathbb{F}_\mathfrak{p}) \neq 2, 3$, we see that

$$P \equiv (-3c_6c_4^{-1}, 0) \pmod{\mathfrak{p}}.$$

Plugging this expression in the Taylor expansion and comparing the coefficients for XY and X^2 we see that

$$\alpha \equiv -\beta \pmod{\mathfrak{p}}$$

and

$$\alpha^2 \equiv -9c_6c_4^{-1} \pmod{\mathfrak{p}},$$

which proves the first part of Proposition 4.4 b).

If $\operatorname{char}(\mathbb{F}_\mathfrak{p}) = 3$, we use the equation (cf. the proof of Theorem 1.7)

$$Y^2 = 4X^3 + b_2X^2 + 2b_4X + b_6,$$

which leads to the polynomial

$$f(X, Y) \equiv Y^2 - X^3 - b_2X^2 + b_4X - b_6 \pmod{\mathfrak{p}}.$$

Taking into account the partial derivatives of f we see that

$$b_2x_0 + b_4 \equiv 0 \equiv 2y_0 \pmod{\mathfrak{p}}.$$

Because the reduction is multiplicative, we have $c_4 \equiv b_2^2 \not\equiv 0 \bmod \mathfrak{p}$ and get

$$P = (-b_4b_2^{-1}, 0) \pmod{\mathfrak{p}}.$$

Comparing the coefficients at XY and X^2 of the Taylor expansion yields

$$\alpha \equiv -\beta \pmod{\mathfrak{p}} \quad \text{and} \quad \alpha^2 \equiv b_2 \pmod{\mathfrak{p}},$$

which proves the second part of Proposition 4.4 b).

If $\operatorname{char}(\mathbb{F}_\mathfrak{p}) = 2$, we refer back to the long Weierstraß normal form. Because E has multiplicative reduction we have $a_1 \not\equiv 0 \bmod \mathfrak{p}$. Taking into account the partial derivatives of f we get

$$P \equiv (a_3a_1^{-1}, (a_3^2a_1^{-2} + a_4)a_1^{-1}) \pmod{\mathfrak{p}}.$$

Comparing the coefficients at XY and X^2 of the Taylor expansion yields

$$\alpha + \beta \equiv a_1 \pmod{\mathfrak{p}}$$

and

$$\alpha^2 + a_1\alpha + (a_3a_1^{-1} + a_2) \equiv 0 \pmod{\mathfrak{p}},$$

which proves the third part of Proposition 4.4 b). □

With the help of Theorem 4.2 and Proposition 4.4 one can in general determine a minimal equation of an elliptic curve and its reduction type. We only describe the algorithm for the case where $\mathrm{char}(\mathbb{F}_\mathfrak{p}) \neq 2, 3$.

Algorithm 4.5 (Minimal equation for $\mathrm{char}(\mathbb{F}_\mathfrak{p}) \neq 2, 3$).

INPUT: An elliptic curve in long Weierstraß normal form over $\mathbb{K}$.
OUTPUT: A $\mathfrak{p}$-minimal equation for the curve, $\mathfrak{p} \nmid 2, 3$.
1. Transform the curve so that the coefficients are $\mathfrak{p}$-integral.
2. While $\mathrm{ord}(\Delta) \geq 12$ and $\mathrm{ord}(c_4) \geq 4$ and $\mathrm{ord}(c_6) \geq 6$ do:
3. Transform the equation with $u = \pi$, $r = s = t = 0$, where π is a generator of $\mathfrak{p}$.
4. Return the equation.

Algorithm 4.6 (Reduction type).

INPUT: An elliptic curve with minimal equation over $\mathbb{K}$.
OUTPUT: The reduction type.
1. If $\mathrm{ord}(\Delta) = 0$ then return ‘‘good reduction’’.
2. If $\mathrm{ord}(\Delta) > 0$ and $\mathrm{ord}(c_4) = 0$ then
3. If $\mathrm{char}(\mathbb{F}_\mathfrak{p}) \neq 2, 3$ and $-c_4c_6$ is a square in $\mathbb{F}_\mathfrak{p}$ then return ‘‘split multiplicative reduction’’.
4. If $\mathrm{char}(\mathbb{F}_\mathfrak{p}) = 3$ and b_2 is a square in $\mathbb{F}_\mathfrak{p}$ then return ‘‘split multiplicative reduction’’.
5. If $\mathrm{char}(\mathbb{F}_\mathfrak{p}) = 2$ and $X^2 + a_1X + (a_3a_1^{-1} + a_2)$ has roots in $\mathbb{F}_\mathfrak{p}$ then return ‘‘split multiplicative reduction’’.
6. return ‘‘non split multiplicative reduction’’.
7. return ‘‘additive reduction’’.

Tate [220] subdivides the different reduction types into more cases. He gives an algorithm to determine the type of the singular fiber in an elliptic pencil. This algorithm can also be applied to compute a (local) minimal model of the elliptic curve, which is especially useful for prime ideals dividing 2 or 3 (see also the paper of Papadopoulos [159]). Descriptions of the algorithms can be found in the paper of Tate [220] as well as in the book of Silverman [207] Chapter IV, Section 9.

4.2 The filtration

Now we consider the points on the elliptic curve in dependence of to their behaviour under reduction.

Definition 4.7. Let $E|\mathbb{K}$ be an elliptic curve given in minimal equation over the local field $\mathbb{K}$. Define for $n \in \mathbb{Z}$ the set

$$E_n(\mathbb{K}) := \{P = (x, y) \in E(\mathbb{K}) : \operatorname{ord}(x) \leq -2n\} \cup \{\mathcal{O}\}.$$

If $P = (x, y) \in E_n(\mathbb{K})$ with $n > 0$ then $\operatorname{ord}(y) \leq -3n$. This can be seen from the equation

$$y^2 + a_1xy + a_3y = x^3 + a_2x^2 + a_4x + a_6.$$

Because $\operatorname{ord}(a_i) \geq 0$, the right side of the equation has

$$\operatorname{ord}(x^3 + a_2x^2 + a_4x + a_6) = \operatorname{ord}(x^3) \leq -6n.$$

If $\operatorname{ord}(y) > -3n$, then we must have

$$\operatorname{ord}(a_1xy) = \operatorname{ord}(x^3) \Leftrightarrow \operatorname{ord}(a_1) + \operatorname{ord}(y) = \operatorname{ord}(x^2) \leq -4n.$$

This is a contradiction, as

$$\operatorname{ord}(y) > -3n \quad \text{and} \quad \operatorname{ord}(a_1) \geq 0.$$

Therefore we get $\operatorname{ord}(y) \leq -3n$.

Definition 4.8. Let $E|\mathbb{K}$ be an elliptic curve in long Weierstraß normal form over the local field $\mathbb{K}$. Define the quantity

$$\mu = \mu(E) := \min\left\{\operatorname{ord}(b_2), \frac{1}{2}\operatorname{ord}(b_4), \frac{1}{3}\operatorname{ord}(b_6), \frac{1}{4}\operatorname{ord}(b_8)\right\}.$$

Observe that, if the equation of E has integral coefficients, then $\mu \leq 0$.

Proposition 4.9 (Lemma of Lutz). *Let $n \in \mathbb{Z}$ with $-2n < \mu$. Then $E_n(\mathbb{K})$ is a group. In particular we have for $P = (x_P, y_P), Q = (x_Q, y_Q) \in E(\mathbb{K})$:*

a) $P \in E_n(\mathbb{K}) \Rightarrow 2P \in E_n(\mathbb{K})$.

b) $P, Q \in E_n(\mathbb{K}) \Rightarrow \operatorname{ord}(x_{P\pm Q}) \leq \max\{\operatorname{ord}(x_P), \operatorname{ord}(x_Q)\}$.
If $\operatorname{ord}(x_P) \neq \operatorname{ord}(x_Q)$ then equality holds.

Proof. We shall prove Part b) for $P \neq \pm Q$, to show that $E_n(\mathbb{K})$ is a group. Part a) which we pose as an exercise (see Exercise 4) requires an extra consideration.

If P or Q is the zero point $\mathcal{O} = (\infty, \infty)$, nothing is to be shown. We suppose therefore that neither P nor Q is the zero point.

We need the following two formulas valid for $P \neq \pm Q$, which can be easily verified (see also Zimmer [258]):

$$x_{P+Q}x_{P-Q} = \frac{x_P^2 x_Q^2 - b_4 x_P x_Q - b_6(x_P + x_Q) - b_8}{(x_P - x_Q)^2} \tag{4.1}$$

$$x_{P+Q} + x_{P-Q} = \frac{2x_P^2 x_Q + 2x_P x_Q^2 + b_2 x_P x_Q + b_4(x_P + x_Q) + b_6}{(x_P - x_Q)^2} \tag{4.2}$$

There are two cases to be considered.

(i) Suppose that $\operatorname{ord}(x_P) < \operatorname{ord}(x_Q) \le -2n < \mu$. We first consider Equation (4.1). As

$$\begin{aligned}
&\operatorname{ord}(-b_4 x_P x_Q - b_6(x_P + x_Q) - b_8)\\
&\quad \ge \min\{\operatorname{ord}(b_4 x_P x_Q), \operatorname{ord}(b_6(x_P + x_Q)), \operatorname{ord}(b_8)\}\\
&\quad \ge \min\{2\mu + \operatorname{ord}(x_P) + \operatorname{ord}(x_Q), 3\mu + \operatorname{ord}(x_P), 4\mu\}\\
&\quad > 2\operatorname{ord}(x_P) + 2\operatorname{ord}(x_Q) = \operatorname{ord}(x_P^2 x_Q^2),
\end{aligned}$$

we get

$$\operatorname{ord}(x_P^2 x_Q^2 - b_4 x_P x_Q - b_6(x_P + x_Q) - b_8) = \operatorname{ord}(x_P^2 x_Q^2)$$

and from Equation (4.1)

$$\begin{aligned}
\operatorname{ord}(x_{P+Q}) + \operatorname{ord}(x_{P-Q}) &= \operatorname{ord}(x_P^2 x_Q^2) - 2\operatorname{ord}(x_P - x_Q)\\
&= 2\operatorname{ord}(x_P) + 2\operatorname{ord}(x_Q) - 2\operatorname{ord}(x_P)\\
&= 2\operatorname{ord}(x_Q).
\end{aligned}$$

From Equation (4.2), we obtain with the same method as above,

$$\begin{aligned}
\operatorname{ord}(x_{P+Q} + x_{P-Q}) &\ge \operatorname{ord}(x_P^2 x_Q) - 2\operatorname{ord}(x_P)\\
&= \operatorname{ord}(x_Q).
\end{aligned}$$

Hence, if $\operatorname{ord}(x_{P+Q}) < \operatorname{ord}(x_{P-Q})$, the first equation gives

$$\operatorname{ord}(x_Q) > \operatorname{ord}(x_{P+Q})$$

and the second equation

$$\begin{aligned}
\operatorname{ord}(x_Q) &> \operatorname{ord}(x_{P+Q})\\
&= \operatorname{ord}(x_{P+Q} + x_{P-Q})\\
&\ge \operatorname{ord}(x_Q),
\end{aligned}$$

which is obviously wrong. The case $\operatorname{ord}(x_{P-Q}) < \operatorname{ord}(x_{P+Q})$ is symmetrical. Altogether, we see that

$$\operatorname{ord}(x_{P\pm Q}) = \operatorname{ord}(x_Q).$$

The case $\operatorname{ord}(x_Q) < \operatorname{ord}(x_P) \leq -2n < \mu$ is symmetrical to the case just treated.

(ii) If $\operatorname{ord}(x_P) = \operatorname{ord}(x_Q) \leq -2n < \mu$, we suppose first that

$$\operatorname{ord}(x_{P\pm Q}) > -2n.$$

We show, that then

$$\operatorname{ord}(x_P) = \operatorname{ord}(x_{(P\pm Q)\mp Q}) > -2n,$$

which is a contradiction.

From Equation (4.1) for $x_{(P\pm Q)\mp Q}x_{(P\pm Q)\pm Q}$, it follows that

$$\begin{aligned}\operatorname{ord}(x_{(P\pm Q)\mp Q}) + \operatorname{ord}(x_{(P\pm Q)\pm Q}) &= \min\{2\operatorname{ord}(x_{(P\pm Q)}) + 2\operatorname{ord}(x_{\mp Q}),\\ &\qquad 2\mu + \operatorname{ord}(x_{(P\pm Q)}) + \operatorname{ord}(x_{\mp Q}),\\ &\qquad 3\mu + \operatorname{ord}(x_{\mp Q}), 4\mu\} - 2\operatorname{ord}(x_{\mp Q})\\ &\geq 2\min\{\operatorname{ord}(x_{(P\pm Q)}), \mu\}.\end{aligned}$$

On the other hand, we get from Equation (4.2) for $x_{(P\pm Q)\mp Q} + x_{(P\pm Q)\pm Q}$:

$$\begin{aligned}\operatorname{ord}(x_{(P\pm Q)\mp Q} + x_{(P\pm Q)\pm Q}) &\geq \min\{\operatorname{ord}(2) + 2\operatorname{ord}(x_{(P\pm Q)}) + \operatorname{ord}(x_{\pm Q}),\\ &\qquad \operatorname{ord}(2) + \operatorname{ord}(x_{(P\pm Q)}) + 2\operatorname{ord}(x_{\pm Q}),\\ &\qquad \mu + \operatorname{ord}(x_{(P\pm Q)}) + \operatorname{ord}(x_{\pm Q}),\\ &\qquad 2\mu + \operatorname{ord}(x_{\pm Q}), 3\nu\} - 2\operatorname{ord}(x_{\pm Q})\\ &\geq \min\{\operatorname{ord}(x_{(P\pm Q)}), \nu\}.\end{aligned}$$

Altogether, we arrive at the contradiction

$$\operatorname{ord}(x_P) = \operatorname{ord}(x_{(P\pm Q)\mp Q}) \geq \min\{\operatorname{ord}(x_{(P\pm Q)}), \mu\} > -2n.$$

Hence we have for $\operatorname{ord}(x_P) = \operatorname{ord}(x_Q) \leq -2n < \mu$ that

$$\operatorname{ord}(x_{P\pm Q}) \leq -2n.$$

Suppose now that

$$\operatorname{ord}(x_{P\pm Q}) > \operatorname{ord}(x_P) = \operatorname{ord}(x_Q).$$

Then we have the situation

$$\operatorname{ord}(x_{\mp Q}) = \operatorname{ord}(x_Q) < \operatorname{ord}(x_{P\pm Q}) \leq -2n < \mu.$$

We have proved in Part (i) that in this case

$$\operatorname{ord}(x_P) = \operatorname{ord}(x_{(P\pm Q)\mp Q}) = \operatorname{ord}(x_{P\pm Q}).$$

This is a contradiction to

$$\operatorname{ord}(x_{P\pm Q}) > \operatorname{ord}(x_P).$$

Therefore $\operatorname{ord}(x_{P\pm Q}) \leq \operatorname{ord}(x_P)$. □

Proposition 4.10. *Let $E|\mathbb{K}$ be an elliptic curve over the local field $\mathbb{K}$.*

a) *We have*

$$E(\mathbb{K}) \geq E_0(\mathbb{K}) \geq E_1(\mathbb{K}) \geq \cdots \geq E_{n-1}(\mathbb{K}) \geq E_n(\mathbb{K}) \geq \cdots \geq \{\mathcal{O}\}.$$

(This means that $E_n(\mathbb{K})$ is a subgroup of $E_{n-1}(\mathbb{K})$ for every $n \in \mathbb{N}$.)

b) *There is an exact sequence of abelian groups*

$$0 \to E_1(\mathbb{K}) \to E_0(\mathbb{K}) \to \widetilde{E}_{ns}(\mathbb{F}_\mathfrak{p}) \to 0,$$

where $\widetilde{E}_{ns}(\mathbb{F}_\mathfrak{p})$ are the nonsingular points on the mod $\mathfrak{p}$ reduced curve. In other words,

$$\begin{aligned} E_0(\mathbb{K}) &= \{P \in E(\mathbb{K}) : \textit{The reduction of } P \bmod \mathfrak{p} \textit{ is nonsingular}\} \\ E_1(\mathbb{K}) &= \{P \in E(\mathbb{K}) : \textit{The reduction of } P \bmod \mathfrak{p} \textit{ is } \mathcal{O}\} \end{aligned}$$

Proof. These assertions follow directly from the definition of $E_n(\mathbb{K})$ and from the Lemma of Lutz. □

Theorem 4.11. *Let $E|\mathbb{K}$ be an elliptic curve over the local field $\mathbb{K}$.*

a)

$$E(\mathbb{K})/E_0(\mathbb{K}) \textit{ is } \left\{ \begin{array}{l} \{\mathcal{O}\} \\ \textit{cyclic of order } -\operatorname{ord}(j) \\ \textit{of order } \leq 4 \end{array} \right\}$$

$$\textit{if } E \textit{ has } \left\{ \begin{array}{l} \textit{good} \\ \textit{multiplicative} \\ \textit{additive} \end{array} \right\} \textit{reduction.}$$

(The number

$$c_\mathfrak{p} := [E(\mathbb{K}) : E_0(\mathbb{K})]$$

is called the *Tamagawa number at* $\mathfrak{p}$. It can be easily computed by means of the algorithm of Tate [220].)

b)

$$E_0(\mathbb{K})/E_1(\mathbb{K}) \cong \widetilde{E}_{ns}(\mathbb{F}_\mathfrak{p}).$$

c) *For $n \in \mathbb{N}$:*

$$E_n(\mathbb{K})/E_{n+1}(\mathbb{K}) \cong \mathbb{F}_\mathfrak{p}^+.$$

Proof. a) The structure of $E(\mathbb{K})/E_0(\mathbb{K})$ is determined in the algorithm of Tate [220], see also the algorithm of Laska [119].

b) This follows from Part b) of Proposition 4.10.

c) This is also in the article of Tate [220]. □

Theorem 4.12. *Let $E|\mathbb{K}$ be an elliptic curve over the local field $\mathbb{K}$.*

a) *If E has good reduction, then*

$$\sharp(E(\mathbb{K})/E_1(\mathbb{K})) = \sharp(\widetilde{E}(\mathbb{F}_\mathfrak{p})).$$

b) *If E has multiplicative reduction, then*

$$\sharp(E(\mathbb{K})/E_1(\mathbb{K})) \mid \operatorname{ord}(j)(p^{2f_{\mathfrak{p}|p}} - 1),$$

where $f_{\mathfrak{p}|p}$ is the residue degree of $\mathfrak{p} \mid p$.

c) *If E has additive reduction, then*

$$\sharp(E(\mathbb{K})/E_1(\mathbb{K})) \mid \sharp(E(\mathbb{K})/E_0(\mathbb{K}))p^2.$$

Proof. If E has good reduction then $E(\mathbb{K}) = E_0(\mathbb{K})$ and therefore

$$E(\mathbb{K})/E_1(\mathbb{K}) = E_0(\mathbb{K})/E_1(\mathbb{K}) \cong \widetilde{E}_{\mathrm{ns}}(\mathbb{F}_\mathfrak{p}) = \widetilde{E}(\mathbb{F}_\mathfrak{p}),$$

which proves assertion a).

If E has bad reduction then

$$(E(\mathbb{K})/E_1(\mathbb{K}))/(E_0(\mathbb{K})/E_1(\mathbb{K})) \cong (E(\mathbb{K})/E_0(\mathbb{K})),$$

hence

$$\begin{aligned}\sharp(E(\mathbb{K})/E_1(\mathbb{K})) &= \sharp(E(\mathbb{K})/E_0(\mathbb{K}))\sharp(E_0(\mathbb{K})/E_1(\mathbb{K})) \\ &= \sharp(E(\mathbb{K})/E_0(\mathbb{K}))\sharp\widetilde{E}_{\mathrm{ns}}(\mathbb{F}_\mathfrak{p})\end{aligned}$$

by Theorem 4.11 b). In accordance with Theorem 4.11 a), we only have to show that

$$\sharp\widetilde{E}_{\mathrm{ns}}(\mathbb{F}_\mathfrak{p}) \mid (p^{2f_{\mathfrak{p}|p}} - 1),$$

in the case of multiplicative reduction and that

$$\sharp\widetilde{E}_{\mathrm{ns}}(\mathbb{F}_\mathfrak{p}) \mid p^2$$

in the case of additive reduction.

If E has multiplicative reduction, it is possible that the slopes of the tangents do not lie in the field $\mathbb{F}_\mathfrak{p}$. Then we have to consider a quadratic extension $\mathbb{F}$ of $\mathbb{F}_\mathfrak{p}$. Else we set $\mathbb{F} = \mathbb{F}_\mathfrak{p}$. Then, as $\widetilde{E}_{\mathrm{ns}}(\mathbb{F}_\mathfrak{p})$ is a subgroup of $\widetilde{E}_{\mathrm{ns}}(\mathbb{F})$, we have

$$\sharp\widetilde{E}_{\mathrm{ns}}(\mathbb{F}_\mathfrak{p}) \mid \sharp\widetilde{E}_{\mathrm{ns}}(\mathbb{F}).$$

In the case of multiplicative reduction, we have therefore

$$\widetilde{E}_{\mathrm{ns}}(\mathbb{F}) \cong \mathbb{F}^*$$

(see for example Silverman [204] Chapter III, Proposition 2.5). If $\mathbb{F}|\mathbb{F}_\mathfrak{p}$ is a quadratic extension, then

$$\sharp(\mathbb{F}^*) = (p^{2f_{\mathfrak{p}|p}} - 1).$$

Else

$$\sharp(\mathbb{F}^*) = (p^{f_{\mathfrak{p}|p}} - 1) \mid (p^{2f_{\mathfrak{p}|p}} - 1).$$

If E has additive reduction then

$$\widetilde{E}_{\mathrm{ns}}(\mathbb{F}_\mathfrak{p}) \cong \mathbb{F}_\mathfrak{p}^+$$

(see for example Silverman [204] Chapter III, Proposition 2.5). Hence

$$\sharp\widetilde{E}_{\mathrm{ns}}(\mathbb{F}_\mathfrak{p}) = p^{f_{\mathfrak{p}|p}}.$$

In fact the exponent $f_{\mathfrak{p}|p}$ may be replaced by 2 since $\mathbb{F}_\mathfrak{p}^+$ is an elementary abelian p-group and $E(\mathbb{K})[p]$ is isomorphic to a subgroup of $\mathbb{Z}/p\mathbb{Z} \times \mathbb{Z}/p\mathbb{Z}$. □

4.3 The theorem of Nagell, Lutz, and Cassels

In this section we prove the local theorem of Nagell, Lutz, and Cassels (see the articles of Nagell [152], Lutz [132], and Cassels [25], [27]), which is applied in Chapter 6 to estimate the denominator of torsion points. We present here the proof of Folz [64], see also Zimmer [250], [251].

Theorem 4.13. *Let $P = (x, y)$ be a point of order $m > 1$ on the elliptic curve E over the local field $\mathbb{K}$ of characteristic $\neq 2, 3$. Let φ be the Euler φ-function, $e = e_\mathfrak{p}$ the absolute ramification index of $\mathfrak{p}$, $\mathfrak{p} \mid p$, and μ be defined as in Definition* 4.8. *Then we have*

a) $\mathrm{ord}(x) \geq \nu$ *and* $\mathrm{ord}(\psi_2^2(x)) \geq 3\nu$ *with*

$$\nu = \begin{cases} \mu & \text{if } m \neq p^n, \\ \mu - \dfrac{2e}{\varphi(p^n)} & \text{if } m = p^n. \end{cases}$$

b) $\psi_2(x, y) = 0$ *if* $m = 2$ *and else* $\operatorname{ord}(\psi_2^2(x)) \le \operatorname{ord}(\Delta) - 2\mu - \nu$ *with*

$$\nu = \begin{cases} \mu & \text{if } m \neq 2p^n, m > 2, \\ \mu - \dfrac{2e}{\varphi(p^n)} & \text{if } m = 2p^n. \end{cases}$$

Proof (see Folz [64]). a) If $\operatorname{ord}(x) \ge \nu$ then

$$\begin{aligned} \operatorname{ord}(\psi_2^2(x)) &= \operatorname{ord}(4x^3 + b_2x^2 + 2b_4x + b_6) \\ &\ge \min\{\operatorname{ord}(4x^3), \operatorname{ord}(b_2x^2), \operatorname{ord}(2b_4x), \operatorname{ord}(b_6)\} \\ &\ge \min\{\operatorname{ord}(4) + 3\nu, \mu + 2\nu, \operatorname{ord}(2) + 2\mu + \nu, 3\mu\} \\ &\ge 3\nu. \end{aligned}$$

Thus it suffices to show that $\operatorname{ord}(x) \ge \nu$.

First let $m \neq p^n$ so that $\nu = \mu$. As $P = (x, y)$ is a point of order m, the first coordinate x is a root of the polynomial ψ_m^2. Here we write ψ_m^2 as a polynomial in one variable X. We have

$$\psi_m^2(X) = \sum_{i=0}^{m^2-1} f_i X^{m^2-(i+1)},$$

where $f_0 = m^2$ and f_i is a homogenous polynomial in b_2, b_4, b_6, and b_8 with integral coefficients and weight $2i$ (see Proposition 1.21) so that $\operatorname{ord}(f_i) \ge i\mu$. Hence there exists an $i \in \{1, \dots, m^2 - 1\}$, such that

$$\operatorname{ord}(f_0 x^{m^2-1}) \ge \operatorname{ord}(f_i x^{m^2-(i+1)}).$$

This implies that

$$\begin{aligned} 2\operatorname{ord}(m) + (m^2 - 1)\operatorname{ord}(x) &\ge \operatorname{ord}(f_i) + (m^2 - (i + 1))\operatorname{ord}(x) \\ &\ge i\mu + (m^2 - (i + 1))\operatorname{ord}(x), \end{aligned}$$

therefore

$$\operatorname{ord}(x) \ge \mu - 2\frac{\operatorname{ord}(m)}{i}.$$

If $\operatorname{ord}(m) = 0$ we get

$$\operatorname{ord}(x) \ge \mu = \nu.$$

If $\operatorname{ord}(m) > 0$ then $m = p^n m_0$ with $n, m_0 \in \mathbb{N}$ and $\gcd(m_0, p) = 1$. Consider the point $p^n P = (x_{p^n}, y_{p^n})$. This is a torsion point of order m_0 with $\operatorname{ord}(m_0) = 0$. For this point, we have shown that $\operatorname{ord}(x_{p^n}) \ge \mu$.

If $\operatorname{ord}(x) < \mu$, we get with Proposition 4.9 that also $\operatorname{ord}(x_{p^n}) < \mu$, a contradiction. Hence it follows that in this case

$$\operatorname{ord}(x) \ge \mu = \nu.$$

Now we come to the case $m = p^n$. First let $m = p = 2$. Then

$$\psi_2^2(x) = 4x^3 + b_2x^2 + 2b_4x + b_6 = 0,$$

therefore

$$\begin{aligned} \operatorname{ord}(4x^3) &= 2\operatorname{ord}(2) + 3\operatorname{ord}(x) \\ &= \operatorname{ord}(b_2x^2 + 2b_4x + b_6) \\ &\geq \min\{\operatorname{ord}(b_2x^2), \operatorname{ord}(2b_4x), \operatorname{ord}(b_6)\} \\ &\geq \min\{\mu + 2\operatorname{ord}(x), 2\mu + \operatorname{ord}(x), 3\mu\} \\ &= \mu + 2\min\{\operatorname{ord}(x), \mu\}. \end{aligned}$$

Now if $\operatorname{ord}(x) \leq \mu$ we get

$$2\operatorname{ord}(2) + 3\operatorname{ord}(x) \geq \mu + 2\operatorname{ord}(x),$$

hence

$$\operatorname{ord}(x) \geq \mu - 2\operatorname{ord}(2) = \mu - \frac{2\operatorname{ord}(2)}{\varphi(2)} = \mu - \frac{2e}{\varphi(2)} = \nu.$$

If $\operatorname{ord}(x) > \mu$, then the estimate

$$\operatorname{ord}(x) > \mu \geq \mu - \frac{2e}{\varphi(2)} = \nu$$

is trivial.

If $m = p^n > 2$, it follows that

$$\frac{\psi_{p^n}^2(x)}{\psi_{p^{n-1}}^2(x)} = \sum_{i=0}^{(p^2-1)p^{2(n-1)}} h_i x^{(p^2-1)p^{2(n-1)}-i} = 0,$$

where $h_0 = p^2$ and h_i is a polynomial in b_2, b_4, b_6, and b_8 with integral coefficients and weight $2i$ (see Proposition 1.22). Hence there exists an $i \in \{1, \ldots, (p^2-1)p^{2(n-1)}\}$ with

$$\operatorname{ord}(h_0 x^{(p^2-1)p^{2(n-1)}}) \geq \operatorname{ord}(h_i x^{(p^2-1)p^{2(n-1)}-i}),$$

which entails that

$$2\operatorname{ord}(p) + (p^2-1)p^{2(n-1)}\operatorname{ord}(x) \geq \operatorname{ord}(h_i) + ((p^2-1)p^{2(n-1)} - i)\operatorname{ord}(x),$$

hence

$$\operatorname{ord}(x) \geq \frac{1}{i}(\operatorname{ord}(h_i) - 2\operatorname{ord}(p)). \tag{4.3}$$

If $i \in \{1, \ldots, \frac{p-1}{2}p^{n-1} - 1\}$, then with Proposition 1.22 we get in Equation (4.3)

$$\operatorname{ord}(h_i) \geq \operatorname{ord}(p^2) + i\mu.$$

Therefore,

$$\operatorname{ord}(x) \geq \mu \geq \mu - \frac{2e}{\varphi(p^n)} = \nu.$$

If $i \in \left\{\frac{p-1}{2}p^{n-1}, \dots, (p^2-1)p^{2(n-1)}\right\}$, then we get in Equation (4.3) (since $\frac{\psi_{p^n}^2(x)}{\psi_{p^n-1}^2(x)}$ is a square)

$$\operatorname{ord}(x) \geq \mu - \frac{e}{i} \geq \mu - \frac{2e}{\varphi(p^n)} = \nu.$$

b) The assertion for $m = 2$ is trivial. Let $m > 2$ and $2P = (x_2, y_2)$. If $\operatorname{ord}(x) \geq \mu$, then with Equation (1.7) of Section 1.3, we get

$$\begin{aligned}\operatorname{ord}(\Delta) &\geq \min\{2\mu + \operatorname{ord}(\phi_2(x)), 3\mu + \operatorname{ord}(\psi_2^2(x))\}\\ &= \min\{2\mu + \operatorname{ord}(x_2), 3\mu\} + \operatorname{ord}(\psi_2^2(x))\\ &= 2\mu + \min\{\operatorname{ord}(x_2), \mu\} + \operatorname{ord}(\psi_2^2(x)),\end{aligned}$$

hence

$$\operatorname{ord}(\psi_2^2(x)) \leq \operatorname{ord}(\Delta) - 2\mu - \min\{\operatorname{ord}(x_2), \mu\}. \tag{4.4}$$

We first consider $m \neq 2p^n$. In this case $2P$ is not a torsion point of order p^n and hence from Part a) $\operatorname{ord}(x_2) \geq \mu$. It follows that

$$\operatorname{ord}(\psi_2^2(x)) \leq \operatorname{ord}(\Delta) - 3\mu.$$

If secondly, $m = 2p^n$, then the point $2P$ has order p^n and therefore it follows with Part a) that

$$\operatorname{ord}(x_2) \geq \mu - \frac{2e}{\varphi(p^n)}.$$

Hence if $\operatorname{ord}(x) \geq \mu$, the assertion is proved with Equation (4.4).

If $\operatorname{ord}(x) < \mu$, it follows from Part a) that $m = p^k$. In the case of $m \neq 2p^n$ we conclude that $p \neq 2$. Then, as $\operatorname{ord}(\Delta) \geq 6\mu$, we get in this case (remember that $\operatorname{char}(\mathbb{K}) \neq 2, 3$)

$$\operatorname{ord}(\psi_2^2(x)) = 3\operatorname{ord}(x) < \operatorname{ord}(\Delta) - 3\mu.$$

The last case is $\operatorname{ord}(x) < \mu$ and $m = 2p^n$. As it follows from Part a) that $m = p^k$, we get $p = 2$ and $m = 2^{n+1}$. We further know from a) that

$$\operatorname{ord}(x) \geq \mu - \frac{2e}{\varphi(2^{n+1})} = \mu - \frac{2e}{2^n} = \nu.$$

From Equation (1.7) of Section 1.3, we get then

$$\begin{aligned}\operatorname{ord}(\Delta) - \operatorname{ord}(\psi_2^2(x)) &\geq \min\{4e + 2\operatorname{ord}(x) + \operatorname{ord}(x_2), 3e + \mu + \operatorname{ord}(x) + \operatorname{ord}(x_2), \\ &\qquad 2\mu + \operatorname{ord}(x_2), 5e + 2\mu + \operatorname{ord}(x_2), 2e + 3\operatorname{ord}(x), \\ &\qquad \mu + 2\operatorname{ord}(x), e + 2\mu + \operatorname{ord}(x), 3\mu\} \\ &\geq 3\mu - \frac{2e}{2^{n-1}} \\ &= 3\mu - \frac{2e}{\varphi(2^n)}.\end{aligned}$$

□

4.4 Exercises

1) Compute the reduction type for
   a) $E : Y^2 = X^3 + X + 3$ over $\mathbb{Q}_2$, $\mathbb{Q}_{13}$, and $\mathbb{Q}_{19}$.
   b) $E : Y^2 + XY + Y = X^3 - X^2 + X$ over $\mathbb{Q}_3$, $\mathbb{Q}_5$, and $\mathbb{Q}_{11}$.
2) Compute a minimal equation and the reduction type for
   a) $E : Y^2 = X^3 + 75X^2 + 4375X + 171875$ over $\mathbb{Q}_5$.
   b) $E : Y^2 + 14XY - 686Y = X^3 - 49X^2 + 4802X$ over $\mathbb{Q}_7$.
3) Let $E|\mathbb{K}$ be an elliptic curve in long Weierstraß normal form over the local field $\mathbb{K}$. The coefficients need not be integral. Find a transformation such that the transformed curve has integral coefficients.
4) Prove Part a) of Proposition 4.9.
5) Prove the Formulas (4.1) and (4.2).

Chapter 5
The Mordell–Weil theorem and heights

The set of points of an elliptic curve over a number field forms a finitely generated abelian group, which is called the *Mordell–Weil group* of the elliptic curve. In this chapter we first give the proof of this theorem. The corresponding theorem holds similarly for an elliptic curve over a (congruence) function field.

A crucial part of this proof is the use of a height function defined on the elliptic curve. We define three different height functions on elliptic curves over number fields. Then we show how to compute heights of points and how to compute points of bounded height. There is a close connection between these height functions, which is the topic of the last section.

In this chapter let $\mathbb{K}$ be a number field and $M_{\mathbb{K}}$ the set of valuations of $\mathbb{K}$. For $v \in M_{\mathbb{K}}$ let n_v be the local degree at v. For a number field element x we write

$$v(x) = -\log |x|_v.$$

If v is a non-archimedean absolute value corresponding to the prime ideal $\mathfrak{p}_v$ then we have the normalized discrete additive valuation $\operatorname{ord}_v(x)$ with

$$|x|_v = \mathcal{N}(\mathfrak{p}_v)^{\operatorname{ord}_v(x)/n_v},$$

where $\mathcal{N}(\mathfrak{p}_v)$ is the norm of the ideal $\mathfrak{p}_v$. We denote by $M_{\mathbb{K}}^0$ the set of discrete valuations of $\mathbb{K}$.

5.1 Theorem of Mordell and Weil

The main goal of this section is to state and prove the fundamental theorem for the theory of elliptic curves over number fields: the theorem of Mordell–Weil. The theorem was proved by Mordell for elliptic curves over $\mathbb{Q}$ in [148], and in a more general setting by Weil [234]. The proof given below follows the approach of Heuß [98] (see also Serre [198] and Silverman [204]).

Theorem 5.1 (Weak theorem of Mordell–Weil). *Let $E|\mathbb{K}$ be an elliptic curve over the number field $\mathbb{K}$ and $m \in \mathbb{N}$, $m \geq 2$. Then*

$$E(\mathbb{K})/mE(\mathbb{K})$$

is finite.

For the proof of this theorem (which we only prove for $m = 2$) we need the following lemma:

Lemma 5.2. *Let $\mathbb{L}|\mathbb{K}$ be a finite Galois extension, $E|\mathbb{K}$ an elliptic curve and $m \in \mathbb{N}, m \geq 2$. If $E(\mathbb{L})/mE(\mathbb{L})$ is finite, then $E(\mathbb{K})/mE(\mathbb{K})$ is finite.*

Proof. Let

$$\Phi = (E(\mathbb{K}) \cap mE(\mathbb{L}))/mE(\mathbb{K})$$

denote the kernel of the natural map

$$f\colon E(\mathbb{K})/mE(\mathbb{K}) \to E(\mathbb{L})/mE(\mathbb{L}).$$

We show that Φ is finite. Then, because $E(\mathbb{L})/mE(\mathbb{L})$ is finite, it follows that $E(\mathbb{K})/mE(\mathbb{K})$ is finite.

Now we define for every point $P \bmod mE(\mathbb{K})$ in Φ a map

$$\lambda_P\colon \mathrm{Gal}(\mathbb{L}|\mathbb{K}) \to E[m].$$

To this end we choose a point $Q_P \in E(\mathbb{L})$ with $mQ_P = P$ and define

$$\lambda_P(\sigma) = Q_P^\sigma - Q_P$$

for $\sigma \in \mathrm{Gal}(\mathbb{L}|\mathbb{K})$. (As there are different possible choices of the point Q_P, it is possible that P is related to different maps λ_P.)

Let $P, P' \in E(\mathbb{K}) \cap mE(\mathbb{L})$. Then

$$\begin{aligned} \lambda_P = \lambda_{P'} &\Rightarrow (Q_P - Q_{P'})^\sigma = Q_P - Q_{P'} \quad \forall \sigma \in \mathrm{Gal}(\mathbb{L}|\mathbb{K}) \\ &\Rightarrow Q_P - Q_{P'} \in E(\mathbb{K}) \\ &\Rightarrow P - P' = m(Q_P - Q_{P'}) \in mE(\mathbb{K}) \\ &\Rightarrow P \equiv P' \bmod mE(\mathbb{K}). \end{aligned}$$

Therefore the map

$$\lambda\colon \Phi \to \mathrm{Map}(\mathrm{Gal}(\mathbb{L}|\mathbb{K}) \to E[m]), \quad P \mapsto \lambda_P,$$

sends different points to different maps. As the sets $\mathrm{Gal}(\mathbb{L}|\mathbb{K})$ and $E[m]$ are finite, there are only finitely many maps between those sets. Therefore Φ is finite. □

Using this lemma, we can assume that the 2-torsion group $E[2]$ of E is defined already over the number field $\mathbb{K}$. Indeed, if this is not the case, we replace $\mathbb{K}$ by $\mathbb{K}(E[2])$, i.e. by the Galois extension of $\mathbb{K}$ generated by the coordinates of the 2-torsion points.

Definition 5.3. Let $E|\mathbb{K}$ be an elliptic curve over the number field $\mathbb{K}$ given in the form

$$E : Y^2 = (X - e_1)(X - e_2)(X - e_3)$$

with $e_1, e_2, e_3 \in \mathbb{K}$. Define for $i = 1, 2, 3$ the map

$$\begin{aligned} \varphi_i : E(\mathbb{K}) &\to \mathbb{K}^*/\mathbb{K}^{*2}, \\ \mathcal{O} &\mapsto 1 \cdot \mathbb{K}^{*2}, \\ (e_i, 0) &\mapsto (e_i - e_j)(e_i - e_k)\mathbb{K}^{*2}, \quad \{i, j, k\} = \{1, 2, 3\} \\ (x, y) &\mapsto (x - e_i)\mathbb{K}^{*2}. \end{aligned}$$

Further define

$$\begin{aligned} \varphi : E(\mathbb{K}) &\to (\mathbb{K}^*/\mathbb{K}^{*2})^{\oplus 3}, \\ P &\mapsto (\varphi_1(P), \varphi_2(P), \varphi_3(P)). \end{aligned}$$

Remark. The two definitions in 5.3 for the points $\neq \mathcal{O}$ are essentially the same. For if $P = (x, y) \neq (e_i, 0)$, one has

$$\varphi_i(P) = (x - e_i)\mathbb{K}^{*2} = (x - e_j)(x - e_k)\mathbb{K}^{*2},$$

which follows from the equation

$$Y^2 = (X - e_i)(X - e_j)(X - e_k).$$

Lemma 5.4. *Let*

$$E : Y^2 = X^3 + a_2X^2 + a_4X + a_6$$

be an elliptic curve with integral coefficients a_2, a_4, a_6 over the number field $\mathbb{K}$ and $P = (x, y) \in E(\mathbb{K})$ with $y \neq 0$. Then there exist algebraic integers $r, s, t \in \mathbb{K}$ such that

$$x = \frac{r}{t^2}, \quad y = \frac{s}{t^3}$$

with

$$\gcd(r, t^2) = \mathfrak{c}^2, \quad \gcd(s, t^3) = \mathfrak{c}^3$$

for some integral divisor $\mathfrak{c}$.

Proof. We consider the divisors generated by x and y and write them as

$$(x) = \mathfrak{m}\mathfrak{t}^{-2}, \quad (y) = \mathfrak{n}\mathfrak{t}^{-3},$$

where $\mathfrak{m}, \mathfrak{n}, \mathfrak{t}$ are integral divisors of $\mathbb{K}$ with $\gcd(\mathfrak{m}, \mathfrak{t}) = \gcd(\mathfrak{n}, \mathfrak{t}) = 1$. This is possible, because the coefficients of the elliptic curve are integral.

Let $\mathfrak{c}$ be an integral divisor such that $\mathfrak{t}\mathfrak{c}$ becomes principal:

$$\mathfrak{t}\mathfrak{c} = (t).$$

We can write

$$(x) = (\mathfrak{m}\mathfrak{c}^2)(\mathfrak{t}^2\mathfrak{c}^2)^{-1}, \quad (y) = (\mathfrak{n}\mathfrak{c}^3)(\mathfrak{t}^3\mathfrak{c}^3)^{-1}.$$

It follows that both divisors $\mathfrak{m}\mathfrak{c}^2$ and $\mathfrak{n}\mathfrak{c}^3$ are generated by elements r' and s' of $\mathbb{K}$ respectively. As all divisors are integral, the elements r', s', t are also integral.

We can then write

$$x = \frac{r'\varepsilon_1}{t^2}, \qquad y = \frac{s'\varepsilon_2}{t^3}$$

with units $\varepsilon_1, \varepsilon_2 \in \mathbb{K}$. The lemma thus follows with $r = r'\varepsilon_1$ and $s = s'\varepsilon_2$. □

For the proof of the weak theorem of Mordell–Weil we have to show the following theorem.

Theorem 5.5. *Let $E|\mathbb{K}$ be an elliptic curve over the number field $\mathbb{K}$ in the form*

$$E : Y^2 = (X - e_1)(X - e_2)(X - e_3)$$

with $e_1, e_2, e_2 \in \mathbb{K}$. We can assume that e_1, e_2, e_3 are integral. Furthermore, let $\varphi\colon E(\mathbb{K}) \to (\mathbb{K}^*/\mathbb{K}^{*2})^{\oplus 3}$ be as above. Then

a) *φ is a group homomorphism.*

b) $\operatorname{Ker}(\varphi) = 2E(\mathbb{K})$.

c) *$\varphi(E(\mathbb{K}))$ is finite.*

Proof. If e_1, e_2, e_3 are not integral, let $d \in \mathbb{K}$ be integral such that de_i is integral for all $i = 1, 2, 3$. Multiply the equation for E with d^3 and set $X' := dX$ to get the equation

$$E : d^3Y^2 = (X' - de_1)(X' - de_2)(X' - de_3).$$

If d is a square in $\mathbb{K}$ we are done by defining $Y' := d\sqrt{d}Y$. If d is not a square in $\mathbb{K}$ we have shown in Lemma 5.2 that we can work in the field extension $\mathbb{K}(\sqrt{d})$ instead of $\mathbb{K}$.

Hence we can assume that the e_i are integral.

a) Let $P_1, P_2 \in E(\mathbb{K})$. We have to show that

$$\varphi_i(P_1 + P_2) = \varphi_i(P_1)\varphi_i(P_2)$$

for $i = 1, 2, 3$.

(i) If $P_1 = \mathcal{O}$, then

$$\varphi_i(P_1 + P_2) = \varphi_i(P_2) = \varphi_i(P_1)\varphi_i(P_2).$$

Analogous for $P_2 = \mathcal{O}$.

(ii) Let $P_1 + P_2 = \mathcal{O}$, that means $P_1 = -P_2$. Then the X-coordinates of the points P_1 and P_2 are equal: $x(P_1) = x(P_2)$. It follows that $\varphi_i(P_1) = \varphi_i(P_2)$ and thus

$$\varphi_i(P_1 + P_2) = 1 \cdot \mathbb{K}^{*2} = \varphi_i(P_1)\varphi_i(P_2).$$

(iii) Let $P_1 = (x_1, y_1)$, $P_2 = (x_2, y_2)$, $P_1 \neq -P_2$ and $P_1 + P_2 = P_3 = (x_3, y_3)$. Further let $x_l \neq e_i, l = 1, 2, 3$.

The line through P_1 and P_2 has the equation

$$Y = \lambda X + \nu,$$

where λ and ν depend on P_1 and P_2. As before, the intersection points of this line with the elliptic curve are given by the equation

$$(\lambda X + \nu)^2 = (X - e_1)(X - e_2)(X - e_3).$$

This is a cubic equation in X with the three solutions $x_1, x_2, x_3 \in \mathbb{K}$.

Let $\{i, j, k\} = \{1, 2, 3\}$ and set $X' := X - e_i, x'_l := x_l - e_i$ for $l = 1, 2, 3$. Then the above equation has the form

$$(\lambda X' + \lambda e_i + \nu)^2 = X'(X' - (e_j - e_i))(X' - (e_k - e_i))$$

$$\Leftrightarrow \; X'(X' - (e_j - e_i))(X' - (e_k - e_i)) - (\lambda X' + \lambda e_i + \nu)^2 = 0.$$

This cubic equation has three solutions $x'_1, x'_2, x'_3 \in \mathbb{K}$. Therefore we can write the equation as

$$(X' - x'_1)(X' - x'_2)(X' - x'_3) = 0.$$

For the constant term of this equation one has

$$-x'_1 x'_2 x'_3 = -(x_1 - e_i)(x_2 - e_i)(x_3 - e_i) = -(\lambda e_i + \nu)^2.$$

It follows that

$$\begin{aligned}(x_3 - e_i) &= \frac{1}{(x_1 - e_i)(x_2 - e_i)}(\lambda e_i + \nu)^2 \\ &= (x_1 - e_i)(x_2 - e_i)\left(\frac{\lambda e_i + \nu}{(x_1 - e_i)(x_2 - e_i)}\right)^2.\end{aligned}$$

Therefore

$$\varphi_i(P_1 + P_2) = (x_3 - e_i)\mathbb{K}^{*2} = (x_1 - e_i)(x_2 - e_i)\mathbb{K}^{*2} = \varphi_i(P_1)\varphi_i(P_2).$$

(iv) Let $P_1 = (x_1, y_1)$, $P_2 = (x_2, y_2) = (e_i, 0)$, $P_1 + P_2 = (x_3, y_3)$ with $P_1 \neq P_2, x_1 \neq e_j, e_k$ for $\{i, j, k\} = \{1, 2, 3\}$ (then $x_3 \neq e_1, e_2, e_3$). From

$$y_3^2 = (x_3 - e_1)(x_3 - e_2)(x_3 - e_3)$$

it follows that

$$\varphi_i(P_1 + P_2) = (x_3 - e_i)\mathbb{K}^{*2} = (x_3 - e_j)(x_3 - e_k)\mathbb{K}^{*2}.$$

Using Part (iii) with e_i replaced by e_j and e_k respectively, one gets

$$\begin{aligned}\varphi_i(P_1+P_2) &= (x_3-e_j)(x_3-e_k)\mathbb{K}^{*2}\\ &= (x_1-e_j)(x_2-e_j)(x_1-e_k)(x_2-e_k)\mathbb{K}^{*2}\\ &= (x_1-e_j)(x_1-e_k)(e_i-e_j)(e_i-e_k)\mathbb{K}^{*2}\\ &= (x_1-e_i)(e_i-e_j)(e_i-e_k)\mathbb{K}^{*2}\\ &= \varphi_i(P_1)\varphi_i(P_2).\end{aligned}$$

Analogous for $P_1=(e_i,0)$, $P_2=(x_2,y_2)$ with $P_1\neq P_2$, $x_2\neq e_j,e_k$.

(v) Let $P_1=(e_i,0)$, $P_2=(e_j,0)$ with $i\neq j$. Then $P_1+P_2=(e_k,0)$ with $\{j,k,i\}=\{1,2,3\}$. It follows that

$$\begin{aligned}\varphi_i(P_1+P_2) &= (e_k-e_i)\mathbb{K}^{*2}\\ &= (e_k-e_i)(e_j-e_i)(e_j-e_i)\mathbb{K}^{*2}\\ &= (e_i-e_j)(e_i-e_k)(e_j-e_i)\mathbb{K}^{*2}\\ &= \varphi_i(P_1)\varphi_i(P_2).\end{aligned}$$

Analogous for $P_1=(e_j,0)$, $P_2=(e_i,0)$, $i\neq j$.

(vi) Let $P_1+P_2=(e_i,0)$. If $P_2=(e_i,0)$ then $P_1=\mathcal{O}$ and the result follows directly from Part (i). Suppose now that $P_2\neq(e_i,0)$ and consider the equation $P_1+(P_2+(e_i,0))=\mathcal{O}$. From Part (ii) it follows that

$$\varphi_i(P_1)\varphi_i(P_2+(e_i,0))=1\cdot\mathbb{K}^{*2}.$$

With Part (iv) (or Part (v) if $P_2=(e_j,0)$) we get

$$\varphi_i(P_1)\varphi_i(P_2)(e_i-e_j)(e_i-e_k)=1\cdot\mathbb{K}^{*2}$$

with $\{i,j,k\}=\{1,2,3\}$. Hence

$$\varphi_i(P_1)\varphi_i(P_2)=(e_i-e_j)(e_i-e_k)\mathbb{K}^{*2}=\varphi_i((e_i,0))=\varphi_i(P_1+P_2).$$

b) As φ is a group homomorphism, one can easily see that

$$\varphi(2P)=\varphi(P)\varphi(P)=(1\mathbb{K}^{*2},1\mathbb{K}^{*2},1\mathbb{K}^{*2}).$$

We now show that $\mathrm{Ker}(\varphi)\subset 2E(\mathbb{K})$. Consider the equation

$$E: Y^2=(X-e_1)(X-e_2)(X-e_3)$$

with $e_i\in\mathbb{K}$. It is trivial that $\mathcal{O}\in\mathrm{Ker}(\varphi)$ and $\mathcal{O}\in 2E(\mathbb{K})$.

(i) Let $P=(x,y)\in\mathrm{Ker}(\varphi)$. We show that we may assume that $x=0$.

We consider the transformation $Y = Y'$, $X = X' + x$. The transformed equation is

$$E : (Y')^2 = (X' + x - e_1)(X' + x - e_2)(X' + x - e_3).$$

Then the point P maps to the point $P' = (0, y)$.

(ii) Now let $P = (0, y) \in E(\mathbb{K})$, where E is defined by the equation

$$E : Y^2 = (X - e_1)(X - e_2)(X - e_3),$$

with $P \in \operatorname{Ker}(\varphi)$. We show that $(-e_i) \in \mathbb{K}^2$ for $i = 1, 2, 3$.

If $y = 0$, then $e_i = 0$ for one $i = 1, 2, 3$. We may assume that $e_3 = 0$. Then

$$\begin{aligned} \varphi_1(P) &= (-e_1)\mathbb{K}^{*2} = 1 \cdot \mathbb{K}^{*2}, \\ \varphi_2(P) &= (-e_2)\mathbb{K}^{*2} = 1 \cdot \mathbb{K}^{*2}, \\ \varphi_3(P) &= (-e_1)(-e_2)\mathbb{K}^{*2} = 1 \cdot \mathbb{K}^{*2}, \end{aligned}$$

as $P \in \operatorname{Ker}(\varphi)$. It follows that

$$(-e_1) \in \mathbb{K}^2, \quad (-e_2) \in \mathbb{K}^2, \quad (-e_3) = 0 \in \mathbb{K}^2.$$

For $P = (0, y)$ with $y \neq 0$, this follows directly from the fact that $P \in \operatorname{Ker}(\varphi)$. We thus have in both cases

$$(-e_1) \in \mathbb{K}^2, \quad (-e_2) \in \mathbb{K}^2, \quad (-e_3) \in \mathbb{K}^2.$$

(iii) We take a point $Q \in E$ with $2Q = P$. To prove the claim it suffices to show that

$$(-e_1), (-e_2), (-e_3) \in \mathbb{K}^2 \Rightarrow Q \in E(\mathbb{K}).$$

Let $Q = (x_1, y_1) \in E$. For computing $2Q$, we consider the tangent line

$$Y = \lambda X + \nu$$

in the point Q at E, where λ and ν are dependent on Q. The point $-2Q = (0, -y)$ lies on this tangent line, hence

$$\nu = -y \in \mathbb{K}.$$

(iv) We exhibit now a quadratic equation for x_1.

The intersection points of the tangent with the elliptic curve are given by the equation

$$(\lambda X + \nu)^2 = (X - e_1)(X - e_2)(X - e_3) = X^3 + a_2X^2 + a_4X + a_6$$

with

$$\begin{aligned} a_2 &= -(e_1 + e_2 + e_3), \\ a_4 &= e_1e_2 + e_1e_3 + e_2e_3, \\ a_6 &= -e_1e_2e_3 = y^2 = \nu^2. \end{aligned}$$

Taking $(\lambda X + \nu)^2$ to the other side one gets the cubic equation

$$\begin{aligned} 0 &= X^3 + (a_2 - \lambda^2)X^2 + (a_4 - 2\nu\lambda)X + (a_6 - \nu^2) \\ &= X^3 + (a_2 - \lambda^2)X^2 + (a_4 - 2\nu\lambda)X. \end{aligned}$$

The solutions of this equation are $X = x_1$ (double root) and $X = 0$. Dividing by X results in the quadratic equation

$$X^2 + (a_2 - \lambda^2)X + (a_4 - 2\nu\lambda) = 0 \tag{5.1}$$

with the solutions

$$X_{1,2} = \frac{1}{2}\big((\lambda^2 - a_2) \pm \sqrt{(\lambda^2 - a_2)^2 - 4(a_4 - 2\nu\lambda)}\big).$$

(v) We finally show that $x_1 \in \mathbb{K}$.
As x_1 is a double root of the above equation, we get

$$(\lambda^2 - a_2)^2 = 4(a_4 - 2\nu\lambda).$$

This is equivalent to the fact that for an arbitrary $u \in \mathbb{K}$

$$\begin{aligned} (\lambda^2 - a_2 + u)^2 &= (\lambda^2 - a_2)^2 + 2(\lambda^2 - a_2)u + u^2 \\ &= 2u\lambda^2 + 4(a_4 - 2\nu\lambda) - 2a_2u + u^2 \\ &= 2u\lambda^2 - 8\nu\lambda + (u^2 - 2a_2u + 4a_4) \qquad (5.2) \\ &= (\sqrt{2u}\lambda)^2 - 2\frac{4\nu}{\sqrt{2u}}(\sqrt{2u}\lambda) + (u^2 - 2a_2u + 4a_4). \end{aligned}$$

The left hand side of this equation is a square. The right hand side is a square if and only if its discriminant (divided by 4)

$$\left(\frac{4\nu}{\sqrt{2u}}\right)^2 - (u^2 - 2a_2u + 4a_4) = \frac{8\nu^2}{u} - (u^2 - 2a_2u + 4a_4) = 0.$$

Multiplying by u, and using $\nu^2 = a_6$ ($= 0$, if $P = (0,0)$), we get the equation

$$0 = 8a_6 - u^3 + 2a_2u^2 - 4a_4u = -u^3 + 2a_2u^2 - 4a_4u + 8a_6.$$

With $u = -2u'$ this equation reads

$$0 = 8(u'^3 + a_2u'^2 + a_4u' + a_6).$$

The roots of this equation are $u' = e_1, e_2$ and e_3. Hence one gets for u itself the roots $u = -2e_1, -2e_2$ and $-2e_3$. (The computation needs some extra attention if $P = (0,0)$, but leads to the same result.)

Substitution of the root $u = -2e_1$ into equation (5.2) yields

$$\left((\lambda^2 - a_2) - 2e_1\right)^2 = -4e_1\lambda^2 - 8\nu\lambda + (4e_1^2 + 4a_2e_1 + 4a_4).$$

With the definitions of a_2 and a_4, it follows that

$$\begin{aligned}
&(\lambda^2 - e_1 + e_2 + e_3)^2 \\
&\quad = -4e_1\lambda^2 - 8\nu\lambda + 4(e_1^2 - (e_1 + e_2 + e_3)e_1 + e_1e_2 + e_1e_3 + e_2e_3) \\
&\quad = -4e_1\lambda^2 - 8\nu\lambda + 4e_2e_3 \\
&\quad = 4(-e_1\lambda^2 - 2\nu\lambda + e_2e_3) \\
&\quad = 4(e_1'\lambda \pm e_2'e_3')^2
\end{aligned}$$

with $(e_i')^2 := -e_i$ for $i = 1, 2, 3$. This follows from the fact that $\nu^2 = -e_1e_2e_3 = (e_1'e_2'e_3')^2$.

Taking roots on both sides leads to

$$\begin{aligned}
&\lambda^2 - e_1 + e_2 + e_3 = \pm 2(e_1'\lambda \pm e_2'e_3') \\
\Leftrightarrow\ &\lambda^2 \mp 2e_1'\lambda - e_1 = -e_2 \pm 2e_2'e_3' - e_3 \\
\Leftrightarrow\ &(\lambda \mp e_1')^2 = (e_2' \pm e_3')^2.
\end{aligned}$$

Again taking roots on both sides yields

$$\lambda = \pm e_1' \pm (e_2' \pm e_3').$$

From our assumption $(-e_i) \in \mathbb{K}^2$ it follows that $e_i' \in \mathbb{K}$ for $i = 1, 2, 3$ and hence $\lambda \in \mathbb{K}$. Then also by (5.1)

$$x_1 = \frac{\lambda^2 - a_2}{2} \in \mathbb{K}.$$

As $\nu = \pm e_1'e_2'e_3' = -y \in \mathbb{K}$ it follows that

$$y_1 = \lambda x_1 + \nu \in \mathbb{K}.$$

Therefore, $Q = (x_1, y_2) \in E(\mathbb{K})$.

c) We first define a group homomorphism

$$\begin{aligned}
\eta :\ \mathbb{K}^*/\mathbb{K}^{*2} &\to \mathbb{H}_\mathbb{K}/\mathbb{H}_\mathbb{K}^2 \\
x\mathbb{K}^{*2} &\mapsto (x)\mathbb{H}_\mathbb{K}^2,
\end{aligned}$$

where $\mathbb{H}_\mathbb{K}$ is the group of principal divisors of $\mathbb{K}$. The kernel of this homomorphism is

$$U_\mathbb{K}\mathbb{K}^*/\mathbb{K}^{*2} \cong U_\mathbb{K}/(U_\mathbb{K} \cap \mathbb{K}^{*2}) = U_\mathbb{K}/U_\mathbb{K}^2,$$

where $U_\mathbb{K}$ is the unit group of $\mathbb{K}$. The group $U_\mathbb{K}/U_\mathbb{K}^2$ is finite (Dirichlet, see for example Neukirch [157] Chapter I, Section 7 or Hasse [92], Chapter 28), hence the kernel of η is finite.

We now consider the homomorphism

$$\begin{aligned}\varphi' \colon E(\mathbb{K}) &\to (\mathbb{H}_{\mathbb{K}}/\mathbb{H}_{\mathbb{K}}^2)^{\oplus 3},\\ P &\mapsto (\eta(\varphi_1(P)), \eta(\varphi_2(P)), \eta(\varphi_3(P)))\end{aligned}$$

so that $\varphi' = \eta \circ \varphi$. It suffices to show that $\varphi'(E(\mathbb{K}))$ is finite in order to conclude that $\varphi(E(\mathbb{K}))$ is finite.

There are only finitely many points of order 2, so we consider only points $P = (x, y) \in E(\mathbb{K})$ with $y \neq 0$.

From Lemma 5.4 we know that we can write the coordinates as

$$x = \frac{r}{t^2}, \quad y = \frac{s}{t^3}$$

with algebraic integers $r, s, t \in \mathbb{K}$ such that

$$\gcd(r, t^2) = \mathfrak{c}^2, \quad \gcd(s, t^3) = \mathfrak{c}^3,$$

where $\mathfrak{c}$ is an integral divisor.

Hence,

$$\varphi_i(P) = (x - e_i)\mathbb{K}^{*2} = (r - e_i t^2)\mathbb{K}^{*2}.$$

The integral divisor $(r - e_i t^2)$ can be factored in a squarefree integral divisor $\mathfrak{a}$ and an integral divisor $\mathfrak{b}^2$:

$$(r - e_i t^2) = \mathfrak{a}\mathfrak{b}^2.$$

We take an integral divisor $\mathfrak{w}$ from the class of $\mathfrak{b}$ mod $\mathbb{H}_{\mathbb{K}}$. Then there exists an element $b \in \mathbb{K}^*$ such that

$$\mathfrak{b} = \mathfrak{w}(b).$$

We can further choose $c \in \mathbb{K}^*$ with

$$(c) = \mathfrak{a}\mathfrak{w}^2,$$

that is, up to elements of $\mathbb{H}_{\mathbb{K}}^2$, $(c) = \eta(\varphi_i(P))$. The element c is integral, as $\mathfrak{a}\mathfrak{w}^2$ is integral.

We show that there are only finitely many prime divisors which divide (c). As the class group of algebraic number fields is finite, it follows that $\eta(\varphi_i(E(\mathbb{K})))$ and hence $\varphi'(E(\mathbb{K}))$ is finite.

Let $\mathfrak{p}$ be prime divisor with $\operatorname{ord}_{\mathfrak{p}}(c) \neq 0$.

(i) Let $\operatorname{ord}_{\mathfrak{p}}(c) = 1$. We have for $P = (x, y)$

$$y^2 = (x - e_1)(x - e_2)(x - e_3).$$

Hence there must be a $j \in \{1, 2, 3\}$, $j \neq i$, with $\operatorname{ord}_{\mathfrak{p}}(c^{(j)}) \equiv 1 \bmod 2$, where $c^{(j)}$ is constructed in the same way as c above, for φ_j instead of φ_i.

It follows that

$$\mathfrak{p} \mid (r - e_i t^2) \quad \text{and} \quad \mathfrak{p} \mid (r - e_j t^2),$$

hence

$$\mathfrak{p} \mid t^2(e_i - e_j) \quad \text{and} \quad \mathfrak{p} \mid r(e_i - e_j).$$

It follows that $\mathfrak{p}$ divides the product of the discriminant of the polynomial $(X - e_1)(X - e_2)(X - e_3)$ and $\gcd(r, t^2)$. This product can be made independent of the point P and of i.

(ii) If $|\mathrm{ord}_{\mathfrak{p}}(c)| \geq 2$ then $\mathfrak{p}$ is a divisor of $\mathfrak{w}$ and hence is in a finite set. □

Proof (Weak theorem of Mordell–Weil). We only prove this theorem for $m = 2$. Let $E|\mathbb{K}$ be given in short Weierstraß normal form

$$Y^2 = X^3 + AX + B.$$

With Lemma 5.2 we can assume that $E[2] \subset E(\mathbb{K})$. If this is not the case, consider the finite Galois extension

$$\mathbb{L} = \mathbb{K}(e_1, e_2, e_3)$$

with the roots $e_1, e_2, e_3 \in \overline{\mathbb{K}}$ of the polynomial

$$X^3 + AX + B.$$

Consider the function φ defined above. As we have seen, φ is a group homomorphism with

$$\mathrm{Ker}(\varphi) = 2E(\mathbb{K}), \quad \varphi(E(\mathbb{K})) \text{ finite}.$$

Then the theorem follows from

$$E(\mathbb{K})/\mathrm{Ker}(\varphi) = E(\mathbb{K})/2E(\mathbb{K}) \cong \varphi(E(\mathbb{K})).$$ □

The next step towards the proof of the Mordell–Weil theorem is the descent theorem which is independent of the context of elliptic curves.

Theorem 5.6 (Descent theorem). *Let A be an abelian group. Suppose that there exists a height function*

$$h : A \to \mathbb{R}$$

which has the following properties:

(i) *For every $Q \in A$ exists a constant $C_1 = C_1(A, Q) \in \mathbb{R}$, such that for all $P \in A$*

$$h(P + Q) \leq 2h(P) + C_1.$$

(ii) *There exists an integer $m \geq 2$ and a constant $C_2 = C_2(A, m) \in \mathbb{R}_{\geq 0}$, such that for all $P \in A$*

$$h(mP) \geq m^2 h(P) - C_2.$$

(iii) *For every constant* $C_3 \in \mathbb{R}$ *the set*

$$\{P \in A : h(P) \leq C_3\}$$

is finite.

Furthermore, suppose that A/mA *is finite for the integer* m *in Part* (ii). *Then the group* A *is finitely generated.*

Proof (see also Appendix II written by Manin in Mumford's book [150]). Let

$$\{Q_1, \ldots, Q_r\}$$

be a system of representatives of A/mA in A and $P \in A$. We represent P as a linear combination of $Q_1, \ldots, Q_r$ and one further element $Q \in A$, which has height smaller than a constant independent of P. To this end, we write

$$P = mP_1 + Q_{i_1}$$

for some $1 \leq i_1 \leq r$ and $P_1 \in A$. Analogously, for $j \in \mathbb{N}$, setting $P_0 := P$, we write

$$P_{j-1} = mP_j + Q_{i_j}.$$

Here, we have $1 \leq i_j \leq r$ and $P_j \in A$ for all $j = 1, \ldots, n$, the number n being any positive integer. Then, by Parts (ii) and (i),

$$\begin{aligned} h(P_j) &\leq \frac{1}{m^2}(h(mP_j) + C_2) \\ &= \frac{1}{m^2}(h(P_{j-1} - Q_{i_j}) + C_2) \\ &\leq \frac{1}{m^2}(2h(P_{j-1}) + (C_1' + C_2)) \end{aligned}$$

where we have put

$$C_1' = \max\{C_1(A, -Q_i) : i = 1, \ldots, r\}.$$

These two constants C_1' and C_2 are independent of the point P. Using this inequality n times, it follows that

$$\begin{aligned} h(P_n) &\leq \left(\frac{2}{m^2}\right)^n h(P) + \left(\frac{1}{m^2} + \frac{2}{m^4} + \cdots + \frac{2^{n-1}}{m^{2n}}\right)(C_1' + C_2) \\ &< \left(\frac{2}{m^2}\right)^n h(P) + \frac{C_1' + C_2}{m^2 - 2} \\ &\leq 2^{-n}h(P) + \frac{C_1' + C_2}{2}, \end{aligned}$$

because $m \geq 2$. If n is large enough, we get

$$h(P_n) \leq 1 + \frac{C_1' + C_2}{2}.$$

Hence

$$P = m^n P_n + \sum_{j=1}^{n} m^{j-1} Q_{i_j},$$

and the element P is then a linear combination of the set

$$\{Q_1, \ldots, Q_r\} \cup \left\{Q \in A : h(Q) \leq 1 + \frac{C_1' + C_2}{2}\right\}.$$

Because this set is finite, A is finitely generated. □

We now come to the theorem of Mordell–Weil itself. (See also Appendix II in Mumford's book [150].)

Theorem 5.7 (Mordell–Weil). *Let $\mathbb{K}$ be a number field and $E|\mathbb{K}$ an elliptic curve. The group $E(\mathbb{K})$ is finitely generated.*

We have seen in the above discussion that to prove the Mordell–Weil theorem, we need a height function on the Mordell–Weil group which satisfies the conditions of the descent theorem. In the next section we shall give definitions of height functions on elliptic curves and then deliver a proof of the Mordell–Weil theorem.

From the above theorem it follows that the Mordell–Weil group $E(\mathbb{K})$ of an elliptic curve can be represented as

$$E(\mathbb{K}) \cong E(\mathbb{K})_{\text{tors}} \times \mathbb{Z}^r.$$

The *torsion group* $E(\mathbb{K})_{\text{tors}}$ is known to be finite (see Merel [143]). The natural number $r \in \mathbb{N}$ is the *rank* $\text{rk}(E(\mathbb{K}))$.

A basis for the infinite part is called *basis* of $E(\mathbb{K})$.

To determine the Mordell–Weil group $E(\mathbb{K})$ of an elliptic curve E over a number field $\mathbb{K}$ is a difficult problem. This is so since the proof of the Mordell–Weil theorem is not constructive. It is not difficult to compute the torsion group $E(\mathbb{K})_{\text{tors}}$ (Chapter 6), however, the trouble is to determine the free part $E(\mathbb{K})_{\text{fr}}$ of the group $E(\mathbb{K})$. We know that

$$E(\mathbb{K})_{\text{fr}} \cong \mathbb{Z}^r$$

but there does not even exist a general method for determining the rank r of the curve E over the field $\mathbb{K}$ (see Chapter 7). And if r independent points $P_1, \ldots, P_r$ of E over $\mathbb{K}$ are known, it is not easy to decide if they constitute a basis of $E(\mathbb{K})_{\text{fr}}$.

5.2 Heights

To finish the proof of the theorem of Mordell–Weil, a height function is needed. In this section we define several height functions for elliptic curves over number fields. We also show the properties of the heights and sketch the corresponding proof of the Mordell–Weil theorem. In the next section we show the close connection between these heights.

Definition 5.8. Let $\mathbb{K}$ be a number field.

a) Let $P = [x_0 : \ldots : x_N] \in \mathbb{P}^N(\mathbb{K})$ be a projective point over $\mathbb{K}$. For $v \in M_{\mathbb{K}}$, the *local* $\mathbb{K}$*-height* of P at v is

$$H_{\mathbb{K},v}(P) := \max\{|x_0|_v, \ldots, |x_N|_v\}.$$

The *global* $\mathbb{K}$*-height* of P is then

$$H_{\mathbb{K}}(P) := \prod_{v \in M_{\mathbb{K}}} H_{\mathbb{K},v}(P)^{n_v},$$

n_v, as before, being the local degree. The *global absolute height* of P is

$$H(P) := H_{\mathbb{K}}(P)^{1/[\mathbb{K}:\mathbb{Q}]}.$$

b) If $x \in \mathbb{K}$ is an element of $\mathbb{K}$, the *local* $\mathbb{K}$*-height* of x at $v \in M_{\mathbb{K}}$ is

$$H_{\mathbb{K},v}(x) := H_{\mathbb{K},v}([1 : x]) = \max\{1, |x|_v\}$$

and the *global* $\mathbb{K}$*-height* of x is then

$$H_{\mathbb{K}}(x) := \prod_{v \in M_{\mathbb{K}}} H_{\mathbb{K},v}(x)^{n_v}.$$

The *global absolute height* of x is

$$H(x) := H_{\mathbb{K}}(x)^{1/[\mathbb{K}:\mathbb{Q}]}.$$

c) Let $x \in \mathbb{K}$ be an element of $\mathbb{K}$. For $v \in M_{\mathbb{K}}$, the *ordinary logarithmic local* $\mathbb{K}$*-height* of x at v is

$$h_{\mathbb{K},v}(x) := -\min\{0, v(x)\} = \log H_{\mathbb{K},v}(x).$$

The (*global*) *absolute* or *ordinary logarithmic height* of x is

$$h(x) := \frac{1}{[\mathbb{K} : \mathbb{Q}]} \sum_{v \in M_{\mathbb{K}}} n_v h_{\mathbb{K},v}(x).$$

The (*global*) *absolute* or *ordinary logarithmic height at infinity* of x is

$$h_\infty(x) := \frac{1}{[\mathbb{K} : \mathbb{Q}]} \sum_{v \in M_{\mathbb{K}}^\infty} n_v h_{\mathbb{K},v}(x).$$

Definition 5.9. Let $E|\mathbb{K}$ be an elliptic curve in long Weierstraß normal form over the number field $\mathbb{K}$, $P = (x, y) \in E(\mathbb{K})$, and $v \in M_{\mathbb{K}}$.

a) We introduce the quantity

$$\lambda_v = \lambda_v(E) := \min\left\{v(b_2), \frac{1}{2}v(b_4), \frac{1}{3}v(b_6), \frac{1}{4}v(b_8)\right\}.$$

(This definition is similar to that of μ_v in Definition 4.8 in Chapter 4. Instead of the normed additive absolute value $\operatorname{ord}_v(x)$ we use here the absolute value $v(x) = -\log |x|_v$.)

b) The *ordinary logarithmic local height*[1] of P at v is

$$h_v(P) := -\frac{1}{2}\min\{0, v(x)\} = \frac{1}{2}\log(H_{\mathbb{K},v}(x)) = \frac{1}{2}h_{\mathbb{K},v}(x).$$

c) The *modified local height* of P at v is

$$d_v(P) := -\frac{1}{2}\min\{\lambda_v, v(x)\}.$$

Definition 5.10. Let $E|\mathbb{K}$ be an elliptic curve in long Weierstraß normal form over the number field $\mathbb{K}$ and $P = (x, y) \in E(\mathbb{K})$.

a) We put

$$\lambda = \lambda(E) := \frac{-1}{[\mathbb{K} : \mathbb{Q}]} \sum_{v \in M_{\mathbb{K}}} n_v \lambda_v.$$

(Then $\lambda \geq 0$.)

b) The *ordinary (logarithmic) height* of P is

$$h(P) := \log(H(x)) = \frac{1}{[\mathbb{K} : \mathbb{Q}]} \sum_{v \in M_{\mathbb{K}}} n_v h_v(P).$$

One defines $h(\mathcal{O}) := 0$.

c) The *modified height* of P is

$$d(P) := \frac{1}{[\mathbb{K} : \mathbb{Q}]} \sum_{v \in M_{\mathbb{K}}} n_v d_v(P).$$

One defines $d(\mathcal{O}) := \frac{1}{2}\frac{\lambda}{[\mathbb{K}:\mathbb{Q}]}$.

---

[1] The factor $\frac{1}{2}$ comes from the degree of the function $P = (x, y) \mapsto x$, which is equal to 2 (see for example the definition of the heights using arbitrary non constant functions in Silverman [204], Chapter VIII, Section 6). In the literature, the factor $\frac{1}{2}$ is often omitted and reintroduced at the definition of the canonical height (see Definition 5.16). Note that the factor $\frac{1}{2}$ is omitted in SIMATH for the ordinary logarithmic height as well as for the canonical height.

Proposition 5.11. *Let $P \in \mathbb{P}^N(\mathbb{K})$.*

a) *The $\mathbb{K}$-height $H_{\mathbb{K}}(P)$ does not depend on the choice of homogeneous coordinates for P.*

b) *Let $\mathbb{L}|\mathbb{K}$ be a finite extension. Then we have for the heights:*

$$H_{\mathbb{L}}(P) = H_{\mathbb{K}}(P)^{[\mathbb{L}:\mathbb{K}]}.$$

Proof. See Exercise 1) □

For the proof of the Mordell–Weil theorem, we need a height function on the elliptic curve which satisfies the properties in the descent theorem. We prove these properties for the ordinary height. It is equally possible to use the modified height for the proof of the Mordell–Weil theorem (see Exercise 5.6.4).

The following lemma relates the height of the coefficients of a polynomial to the heights of its roots.

Lemma 5.12. *Let*

$$f(T) = a_0T^d + a_1T^{d-1} + \cdots + a_d = a_0(T-\alpha_1)\ldots(T-\alpha_d) \in \overline{\mathbb{Q}}[T]$$

be a polynomial of degree d. Then

$$2^{-d}\prod_{j=1}^{d} H(\alpha_j) \le H([a_0 : \ldots : a_d]) \le 2^{d-1}\prod_{j=1}^{d} H(\alpha_j).$$

Proof. See for example Silverman [204] Chapter VIII, Theorem 5.9, or Lang [117]. □

To prove the properties of the ordinary height we also need the following theorem.

Theorem 5.13. *Let $E|\mathbb{K}$ be an elliptic curve over the number field $\mathbb{K}$. There are constants $c_1, c_2 \in \mathbb{R}$, depending only on E, such that for all $P, Q \in E$*

$$2h(P) + 2h(Q) - c_1 \le h(P+Q) + h(P-Q) \le 2h(P) + 2h(Q) + c_2.$$

Proof. The proof of this theorem can be found in Silverman [204] Chapter VIII, Theorem 6.2. The constants c_1, c_2 can be made explicit (see Zimmer [258] or Zimmer [252] for the modified height). They depend on the curve E, but are independent of the points P and Q. □

Proposition 5.14. *Let $E|\mathbb{K}$ be an elliptic curve over the number field $\mathbb{K}$.*

a) *Let $Q \in E(\mathbb{K})$. There exists a constant $C_1 = C_1(E, Q)$, such that for all $P \in E(\mathbb{K})$*

$$h(P+Q) \le 2h(P) + C_1.$$

b) *For all* $m \in \mathbb{Z}$ *there exists a constant* $C_2 = C_2(E, m) \geq 0$, *such that for all* $P \in E(\mathbb{K})$
$$h(mP) \geq m^2 h(P) - C_2.$$

c) *For every constant* $C_3 \in \mathbb{R}$ *the set*
$$\{P \in E(\mathbb{K}) : h(P) \leq C_3\}$$
is finite.

Proof. a) With the notation of Theorem 5.13, let $C_1 = 2h(Q) + c_2$. Then, since
$$h(P - Q) \geq 0,$$
it follows that
$$h(P + Q) \leq h(P + Q) + h(P - Q) \leq 2h(P) + C_1.$$

b) We show a stronger result:

For all $m \in \mathbb{Z}$ there exists a constant $C_2 = C_2(E, m) \geq 0$, such that for all $P \in E(\mathbb{K})$
$$m^2 h(P) - C_2 \leq h(mP) \leq m^2 h(P) + C_2.$$
We establish the estimates for $m \in \mathbb{N}_0$ by induction.

It is trivial for $m = 0$ and $m = 1$. Let $m \geq 1$. Then, by Theorem 5.13,
$$\begin{aligned} -h((m-1)P) &+ 2h(mP) + 2h(P) - c_1 \\ &\leq h((m+1)P) \\ &\leq -h((m-1)P) + 2h(mP) + 2h(P) + c_2. \end{aligned}$$
With the induction hypothesis we get
$$\begin{aligned} -(m-1)^2 h(P) &+ 2m^2 h(P) + 2h(P) - C_2(E, m-1) - 2C_2(E, m) - c_1 \\ &\leq h((m+1)P) \\ &\leq -(m-1)^2 h(P) + 2m^2 h(P) + 2h(P) \\ &\qquad + C_2(E, m-1) + 2C_2(E, m) + c_2. \end{aligned}$$
Taking
$$\begin{aligned} C_2 &= C_2(E, m+1) \\ &\geq \max\{0, C_2(E, m-1) + 2C_2(E, m) + c_1, \\ &\qquad C_2(E, m-1) + 2C_2(E, m) + c_2\} \end{aligned}$$
we get
$$(m+1)^2 h(P) - C_2 \leq h((m+1)P) \leq (m+1)^2 h(P) + C_2.$$

If $m < 0$ the result follows from $h(mP) = h(-mP)$.

c) Let $C_3 \in \mathbb{R}$. Then

$$\{x \in \mathbb{K} : \exists y \in \mathbb{K} \text{ with } P = (x, y) \in E(\mathbb{K}) \text{ and } h(P) \le C_3\} \subset \{x \in \mathbb{K} : H(x) \le e^{C_3}\}.$$

Let $x \in \mathbb{K}$ be an element of the above set. Then it satisfies an equation

$$f_x(T) = T^d + a_1 T^{d-1} + \cdots + a_d$$

where $a_j \in \mathbb{Q}$ and $d \le [\mathbb{K} : \mathbb{Q}]$. From Lemma 5.12 it follows that

$$H([1 : a_1 : \ldots : a_d]) \le 2^{d-1} \prod_{j=1}^{d} H(x^{(j)}),$$

where the $x^{(j)}$ are the conjugates of x. Since conjugates have the same height (see Exercise 5.6.3), we get

$$H([1 : a_1 : \ldots : a_d]) \le 2^{d-1} H(x)^d \le (2H(x))^{[\mathbb{K}:\mathbb{Q}]}.$$

Now we write $a_i = \frac{b_i}{b_0}$ with $b_0 \in \mathbb{N},\ b_i \in \mathbb{Z}$ and $\gcd(b_0, \ldots, b_d) = 1$. This leads to

$$H([1 : a_1 : \ldots : a_d]) = H([b_0 : b_1 : \ldots : b_d]) = \max\{|b_0|, \ldots, |b_d|\}$$

(see Exercise 5.6.1). Consequently we get that $\{x \in \mathbb{K} : H(x) \le e^{C_3}\}$ is a subset of the set of roots of the polynomials

$$b_0 T^d + b_1 T^{d-1} + \cdots + b_d \in \mathbb{Z}[T]$$

of degree $d \le [\mathbb{K} : \mathbb{Q}]$ with coefficients

$$|b_i| \le \left(2e^{C_3}\right)^{[\mathbb{K}:\mathbb{Q}]}, \quad 0 \le i \le d.$$

This set is finite. □

Proof (Theorem 5.7 of Mordell–Weil). From the weak theorem of Mordell–Weil (Theorem 5.1) it follows that $E(\mathbb{K})/mE(\mathbb{K})$ is finite for $m \in \mathbb{N}, m \ge 2$.

Further we have defined the ordinary height on the group $E(\mathbb{K})$, which satisfies the conditions of the descent Theorem 5.6. Employing this theorem, we conclude that the group $E(\mathbb{K})$ is finitely generated. □

Proposition 5.15. *Let $E|\mathbb{K}$ be an elliptic curve over the number field $\mathbb{K}$ and $P \in E(\mathbb{K})$. The limit*

$$\lim_{n\to\infty} \frac{h(2^n P)}{2^{2n}}$$

exists.

Proof. By Theorem 5.13 we know that there exist $c_1, c_2 \in \mathbb{R}$, such that for $P, Q \in E(\mathbb{K})$

$$2h(P) + 2h(Q) - c_1 \le h(P + Q) + h(P - Q) \le 2h(P) + 2h(Q) + c_2.$$

Setting $Q = P$, it follows that, for all $P \in E(\mathbb{K})$,

$$4h(P) - c_1 \le h(2P) \le 4h(P) + c_2.$$

Hence there exists a constant $C = C(E)$ with

$$|h(2P) - 4h(P)| \le C.$$

Let $N > M \ge 0$. Then

$$\begin{aligned}\left|\frac{h(2^N P)}{4^N} - \frac{h(2^M P)}{4^M}\right| &= \left|\sum_{n=M}^{N-1}\left(\frac{h(2^{n+1}P)}{4^{n+1}} - \frac{h(2^n P)}{4^n}\right)\right| \\ &\le \sum_{n=M}^{N-1} \frac{1}{4^{n+1}} |h(2^{n+1}P) - 4h(2^n P)| \\ &\le \sum_{n=M}^{N-1} \frac{C}{4^{n+1}} \\ &= \frac{C}{3}\left(\frac{1}{4^M} - \frac{1}{4^N}\right) \\ &\le \frac{C}{4^M}.\end{aligned}$$

Hence, this sequence is a Cauchy sequence and therefore converges. □

Definition 5.16. The (global) *Néron–Tate height* (or *canonical height*) $\hat{h}(P)$ is

$$\hat{h}(P) := \lim_{n\to\infty} \frac{h(2^n P)}{2^{2n}}.$$

Remark 5.17. The canonical height can be also defined as

$$\hat{h}(P) := \lim_{n\to\infty} \frac{h(m^n P)}{m^{2n}}$$

with any fixed $m \in \mathbb{N}, m \ge 2$.

There is a close connection between the different heights.

Proposition 5.18. *Let $E|\mathbb{K}$ be an elliptic curve over the number field $\mathbb{K}$. Let $d, h, \hat{h}$ be the modified height, the ordinary height and the canonical height on E respectively and λ_v the constant defined in Definition 5.9.*

a) *For all points* $P \in E(\mathbb{K})$ *we have the estimate*

$$\frac{1}{2[\mathbb{K}:\mathbb{Q}]} \sum_{v \in M_{\mathbb{K}}} n_v \min\{0, \lambda_v\} \le h(P) - d(P)$$

$$\le \frac{1}{2[\mathbb{K}:\mathbb{Q}]} \sum_{v \in M_{\mathbb{K}}} n_v \max\{0, \lambda_v\}.$$

b) *There exists a constant* C*, which only depends on* E *and* $\mathbb{K}$*, such that for all* $P \in E(\mathbb{K})$

$$|\hat{h}(P) - h(P)| \le C.$$

c) *There exists a constant* D*, which only depends on* E *and* $\mathbb{K}$*, such that for all* $P \in E(\mathbb{K})$

$$|\hat{h}(P) - d(P)| \le D.$$

d) *The canonical height can also be defined via the modified height:*

$$\hat{h}(P) = \lim_{n\to\infty} \frac{d(2^n P)}{2^{2n}}.$$

The constants C *and* D *can be made explicit (see Section* 5.5*).*

Proof. a) For the proof we assume that the point $P = (x, y) \in E(\mathbb{K})$ is not the point at infinity. Let $v \in M_{\mathbb{K}}$.

If $v(x) \ge 0$ we have $h_v(P) = 0$ and we get from $d_v(P) \ge -\frac{1}{2}\lambda_v$ the estimate

$$h_v(P) - d_v(P) \le 0 + \frac{1}{2}\lambda_v = \frac{1}{2}\lambda_v.$$

If $v(x) < 0$, then $h_v(P) = -\frac{1}{2}v(x)$ and from $d_v(P) \ge -\frac{1}{2}v(x)$ it follows that

$$h_v(P) - d_v(P) \le -\frac{1}{2}v(x) + \frac{1}{2}v(x) = 0.$$

If $v(x) \ge \lambda_v$ we have that $d_v(P) = -\frac{1}{2}\lambda_v$ and from $h_v(P) \ge 0$ it follows that

$$h_v(P) - d_v(P) \ge 0 + \frac{1}{2}\lambda_v = +\frac{1}{2}\lambda_v.$$

If $v(x) < \lambda_v$ then $d_v(P) = -\frac{1}{2}v(x)$ and we get from $h_v(P) \ge -\frac{1}{2}v(x)$ the estimate

$$h_v(P) - d_v(P) \ge -\frac{1}{2}v(x) + \frac{1}{2}v(x) = 0.$$

Combining these inequalities, we get the local estimate

$$\frac{1}{2}\min\{0, \lambda_v\} \le h_v(P) - d_v(P) \le \frac{1}{2}\max\{0, \lambda_v\}.$$

Multiplying this estimate with $\frac{1}{[\mathbb{K}:\mathbb{Q}]}n_v$ and summing up we obtain

$$\frac{1}{2[\mathbb{K}:\mathbb{Q}]}\sum_{v\in M_{\mathbb{K}}} n_v \min\{0,\lambda_v\} \le h(P)-d(P) \le \frac{1}{2[\mathbb{K}:\mathbb{Q}]}\sum_{v\in M_{\mathbb{K}}} n_v \max\{0,\lambda_v\}.$$

For $P=\mathcal{O}$, we get these inequalities too.

b) In the proof of Proposition 5.15 we have seen that for all $N > M \ge 0$ and, for all points $P \in E(\mathbb{K})$, one has

$$\left|\frac{h(2^N P)}{4^N}-\frac{h(2^M P)}{4^M}\right| \le \frac{C}{4^M}.$$

With $M=0$ and $N\to\infty$ it follows that

$$|\hat{h}(P)-h(P)| \le C.$$

c) From Part a) we obtain that there are constants $M_1, M_2 \in \mathbb{R}$, depending only on $\mathbb{K}$ and E, such that, for all points $P \in E(\mathbb{K})$,

$$M_1 \le h(P)-d(P) \le M_2.$$

From Part b) it follows that

$$-C \le \hat{h}(P)-h(P) \le C.$$

Taking the sum over these two estimates we get

$$M_1 - C \le \hat{h}(P)-d(P) \le M_2 + C,$$

hence $D=\max\{-M_1+C, M_2+C\}$, and the constant D depends only on E and $\mathbb{K}$.

d) From Part a) it follows that there is a constant $M\in\mathbb{R}$, depending only on E and $\mathbb{K}$, such that for all points $Q \in E(\mathbb{K})$

$$|h(Q)-d(Q)| \le M.$$

Consider for $n\in\mathbb{N}$

$$\left|\hat{h}(P)-\frac{d(2^n P)}{2^{2n}}\right| \le \left|\hat{h}(P)-\frac{h(2^n P)}{2^{2n}}\right| + \left|\frac{h(2^n P)}{2^{2n}}-\frac{d(2^n P)}{2^{2n}}\right| \le \left|\hat{h}(P)-\frac{h(2^n P)}{2^{2n}}\right| + \frac{M}{2^{2n}}.$$

The proof follows from $n\to\infty$. □

Proposition 5.19. *Let $E|\mathbb{K}$ be an elliptic curve and $\overline{\mathbb{K}}$ be an algebraic closure of the number field $\mathbb{K}$.*

a) *(Parallelogram law) Let $P, Q \in E$. Then*
$$\hat{h}(P+Q) + \hat{h}(P-Q) = 2\hat{h}(P) + 2\hat{h}(Q).$$

b) *For all $P \in E$ and $m \in \mathbb{Z}$ one has*
$$\hat{h}(mP) = m^2\hat{h}(P).$$

c) *Let $P \in E$. Then $\hat{h}(P) \geq 0$, and*
$$\hat{h}(P) = 0 \Leftrightarrow P \in E(\overline{\mathbb{K}})_{\text{tors}}.$$

d) *$\hat{h}$ is a quadratic form, that means $\hat{h}$ is even and the pairing*
$$\langle\ ,\ \rangle : E(\overline{\mathbb{K}}) \times E(\overline{\mathbb{K}}) \to \mathbb{R}$$
$$\langle P, Q\rangle = \hat{h}(P+Q) - \hat{h}(P) - \hat{h}(Q)$$
is symmetric and bilinear. This pairing is called Néron–Tate pairing.

(Sometimes a factor $\frac{1}{2}$ is placed in front of the right-hand side. This has the advantage that then $\hat{h}(P) = \langle P, P\rangle$.)

Proof. a) As we have seen in Theorem 5.13, one has
$$2h(P) + 2h(Q) - c_1 \leq h(P+Q) + h(P-Q) \leq 2h(P) + 2h(Q) + c_2$$
for all $P, Q \in E(\mathbb{K})$ with constants c_1 and c_2. Replacing P and Q by $2^n P$ and $2^n Q$, and dividing the equation by 2^{2n}, one gets the equation
$$\begin{aligned}2\frac{h(2^n P)}{2^{2n}} + 2\frac{h(2^n Q)}{2^{2n}} - \frac{c_1}{2^{2n}} &\leq \frac{h(2^n(P+Q))}{2^{2n}} + \frac{h(2^n(P-Q))}{2^{2n}}\\ &\leq 2\frac{h(2^n P)}{2^{2n}} + 2\frac{h(2^n Q)}{2^{2n}} + \frac{c_2}{2^{2n}}.\end{aligned}$$
Taking the limit $n \to \infty$, one ends up with the desired equation
$$\hat{h}(P+Q) + \hat{h}(P-Q) = 2\hat{h}(P) + 2\hat{h}(Q).$$

b) For the proof of the claim, we carry out induction for $m \in \mathbb{N}_0$. The claim is true for $m = 0$ and $m = 1$. Let $m \geq 1$. Using a) and the induction hypothesis, we get
$$\begin{aligned}\hat{h}((m+1)P) &= \hat{h}(mP + P)\\ &= -\hat{h}(mP - P) + 2\hat{h}(mP) + 2\hat{h}(P)\\ &= -\hat{h}((m-1)P) + 2m^2\hat{h}(P) + 2\hat{h}(P)\\ &= (-(m-1)^2 + 2m^2 + 2))\hat{h}(P)\\ &= (m+1)^2\hat{h}(P).\end{aligned}$$

Furthermore, $\hat{h}(-Q) = \hat{h}(Q)$ for all $Q \in E(\mathbb{K})$, hence it follows for $m < 0$ that

$$\hat{h}(mP) = \hat{h}(-mP) = (-m)^2\hat{h}(P) = m^2\hat{h}(P).$$

c) Because $h(P) \geq 0$ for all $P \in E$ we see that $\hat{h}(P) \geq 0$ for all $P \in E$. Let P be a torsion point of order m. From Part b) we obtain, considering the definitions of h and $\hat{h}$,

$$\hat{h}(P) = \frac{\hat{h}(mP)}{m^2} = 0.$$

If, conversely, $\hat{h}(P) = 0$, then it follows from Part b) that, for all $m \in \mathbb{N}$,

$$\hat{h}(mP) = m^2\hat{h}(P) = 0.$$

From the estimate of proposition 5.18 b), we conclude that, for all $m \in \mathbb{N}$,

$$h(mP) = |\hat{h}(mP) - h(mP)| \leq C.$$

Then the points $\{mP : m \in \mathbb{N}\}$ are a subset of the finite set

$$\{Q \in E(\mathbb{K}) : h(Q) \leq C\}.$$

So there can only be a finite set of points mP, $m \in \mathbb{N}$, meaning that P is a torsion point. Of course, P has an order dividing m.

d) From $\hat{h}(P) = \hat{h}(-P)$ we see that $\hat{h}$ is an even function. For proving the bilinearity of the Néron–Tate pairing, it suffices to show that, for $P, Q, R \in E(\mathbb{K})$,

$$\langle P + Q, R\rangle = \langle P, R\rangle + \langle Q, R\rangle.$$

because the pairing is symmetric. This is equivalent to the equation

$$\begin{aligned} &\hat{h}((P + Q) + R) - \hat{h}(P + Q) - \hat{h}(R) \\ &\quad = \hat{h}(P + R) - \hat{h}(P) - \hat{h}(R) + \hat{h}(Q + R) - \hat{h}(Q) - \hat{h}(R). \end{aligned}$$

This equation in turn is equivalent to the equation

$$\hat{h}(P + Q + R) - \hat{h}(P + Q) - \hat{h}(P + R) - \hat{h}(Q + R) + \hat{h}(P) + \hat{h}(Q) + \hat{h}(R) = 0.$$

Using the parallelogram law one gets the 4 equations

$$\begin{aligned} \hat{h}((P + Q) + R) + \hat{h}((P + Q) - R) - 2\hat{h}(P + Q) - 2\hat{h}(R) &= 0, \\ \hat{h}(P + (Q - R)) + \hat{h}(P - (Q - R)) - 2\hat{h}(P) - 2\hat{h}(Q - R) &= 0, \\ \hat{h}((P + R) + Q) + \hat{h}((P + R) - Q) - 2\hat{h}(P + R) - 2\hat{h}(Q) &= 0, \\ 2\hat{h}(Q + R) + 2\hat{h}(Q - R) - 4\hat{h}(Q) - 4\hat{h}(R) &= 0. \end{aligned}$$

Considering the alternating sum, we derive that

$$\begin{aligned} 0 &= \hat{h}((P+Q)+R) + \hat{h}((P+Q)-R) - 2\hat{h}(P+Q) - 2\hat{h}(R) \\ &\quad - \hat{h}(P+(Q-R)) - \hat{h}(P-(Q-R)) + 2\hat{h}(P) + 2\hat{h}(Q-R) \\ &\quad + \hat{h}((P+R)+Q) + \hat{h}((P+R)-Q) - 2\hat{h}(P+R) - 2\hat{h}(Q) \\ &\quad - 2\hat{h}(Q+R) - 2\hat{h}(Q-R) + 4\hat{h}(Q) + 4\hat{h}(R) \\ &= 2\hat{h}((P+Q)+R) - 2\hat{h}(P+Q) - 2\hat{h}(P+R) - 2\hat{h}(Q+R) \\ &\quad + 2\hat{h}(P) + 2\hat{h}(Q) + 2\hat{h}(R). \end{aligned}$$

□

Corollary 5.20. *Let $E|\mathbb{K}$ be an elliptic curve over the number field $\mathbb{K}$ and $P, Q \in E$ and $m, n \in \mathbb{Z}$. Then*

$$\hat{h}(mP + nQ) = m^2\hat{h}(P) + mn\langle P, Q\rangle + n^2\hat{h}(Q).$$

Proof. We have

$$\begin{aligned} mn\langle P, Q\rangle &= \langle mP, nQ\rangle \\ &= \hat{h}(mP+nQ) - \hat{h}(mP) - \hat{h}(nQ) \\ &= \hat{h}(mP+nQ) - m^2\hat{h}(P) - n^2\hat{h}(Q). \end{aligned}$$

□

Definition 5.21. Let $E|\mathbb{K}$ be an elliptic curve over the number field $\mathbb{K}$.

a) Let $P_1, \dots, P_n \in E(\mathbb{K})$, $n > 0$. The *regulator* of $P_1, \dots, P_n$ is the determinant

$$R_{P_1,\dots,P_n} := \det(\langle P_i, P_j\rangle)_{1\le i,j\le n}.$$

b) Let $\{P_1, \dots, P_r\}$ be a basis of $E(\mathbb{K})$. If $r > 0$ the *regulator of $E|\mathbb{K}$* is the regulator of the points $P_1, \dots, P_r \in E(\mathbb{K})$:

$$R_{E|\mathbb{K}} := R_{P_1,\dots,P_r}.$$

The *regulator matrix of $E|\mathbb{K}$* is the matrix

$$\mathcal{R}_{E|\mathbb{K}} = (\langle P_i, P_j\rangle)_{1\le i,j\le r}$$

where $\det(\mathcal{R}_{E|\mathbb{K}}) = R_{E|\mathbb{K}}$.

If $r = 0$, we put $R_{E|\mathbb{K}} := 1$.

Proposition 5.22. *Let $E|\mathbb{K}$ be an elliptic curve and $T \in E(\mathbb{K})_{\text{tors}}$.*

a) *If $P \in E(\mathbb{K})$ is an arbitrary point of $E(\mathbb{K})$, then*

$$\langle T, P\rangle = 0.$$

b) *Let* $\{P_1, \ldots, P_r\}$ *be a basis of* $E(\mathbb{K})$ *and* $Q = \sum_{i=1}^{r} m_i P_i + T$. *Then*

$$\hat{h}(Q) = \frac{1}{2} \sum_{i,j=1}^{r} m_i m_j \langle P_i, P_j \rangle.$$

Proof. a) Let $nT = \mathcal{O}, n \in \mathbb{N}$. Then

$$\begin{aligned}\langle T, P \rangle &= n^{-1} n \langle T, P \rangle \\ &= n^{-1} \langle nT, P \rangle \\ &= n^{-1} \langle \mathcal{O}, P \rangle \\ &= \frac{1}{n}(\hat{h}(P) - \hat{h}(P)) \\ &= 0.\end{aligned}$$

b) One has

$$\hat{h}(Q) = \frac{1}{2}\langle Q, Q \rangle = \frac{1}{2}\Big\langle \sum_{i=1}^{r} m_i P_i + T, \sum_{j=1}^{r} m_j P_j + T \Big\rangle.$$

Because of the bilinearity of $\langle\, ,\, \rangle$ it follows

$$\begin{aligned}\hat{h}(Q) = \; & \frac{1}{2} \sum_{i,j=1}^{r} m_i m_j \langle P_i, P_j \rangle + \frac{1}{2} \sum_{i=1}^{r} m_i \langle P_i, T \rangle \\ & + \frac{1}{2} \sum_{j=1}^{r} m_j \langle T, P_j \rangle + \frac{1}{2} \langle T, T \rangle.\end{aligned}$$

From Part a) and the symmetry of $\langle\, ,\, \rangle$ it follows that therein

$$\langle P_i, T \rangle = \langle T, P_j \rangle = \langle T, T \rangle = 0$$

for all $i, j = 1, \ldots, r$. □

Theorem 5.23. *Let* $E|\mathbb{K}$ *be an elliptic curve over the number field* $\mathbb{K}$ *of rank* $r = \operatorname{rk}(E(\mathbb{K})) > 0$. *Then the regulator matrix* $\mathcal{R}_{E|\mathbb{K}}$ *is symmetric and positive definite.*

Proof. It is easy to see that the regulator matrix is symmetric since the bilinear form is symmetric. It is further positive definite, because the Néron–Tate height is a positive definite quadratic form on $\mathbb{R} \otimes_{\mathbb{Q}} E(\mathbb{K})$. To show this, we consider $\mathbb{R} \otimes_{\mathbb{Q}} E(\mathbb{K})$ as a vector space of finite dimension $r = \operatorname{rk}(E(\mathbb{K}))$, and $E(\mathbb{K})/E(\mathbb{K})_{\text{tors}}$ as a lattice in this vector space. Then the Néron–Tate height extends to a quadratic form given by (cf. Proposition 5.22 b))

$$\hat{h}(\boldsymbol{x}) = \boldsymbol{x}^T \frac{1}{2} \mathcal{R}_{E|\mathbb{K}} \boldsymbol{x}.$$

This quadratic form is positive definite (Silverman [204] Chapter VIII, Proposition 9.6 or Lang [117]). □

5.3 Computation of the heights

For the computation of the ordinary height we factorize the ideal generated by the x-coordinate and compute $v_{\mathfrak{p}}(x)$ for all prime ideals such that the additive absolute value is negative. Then we add the portion of the non-archimedean absolute value.

Algorithm 5.24 (Computation of the ordinary height).

INPUT: A point P.
OUTPUT: $h(P)$.
1. If $P = \mathcal{O}$ then return(0).
2. $h \leftarrow 0$.
3. $x \leftarrow x$-coordinate of P.
4. $L \leftarrow$list of prime ideals $\mathfrak{p}$ with $v_{\mathfrak{p}}(x) < 0$.
5. For all prime ideals $\mathfrak{p}$ in L do: $h \leftarrow h - n_{v_{\mathfrak{p}}} v_{\mathfrak{p}}(x)$.
6. $h \leftarrow \frac{1}{2[\mathbb{K}:\mathbb{Q}]} h$.
7. $h \leftarrow h + \frac{1}{2} h_{\infty}(x)$.
8. return(h).

For the computation of the ordinary height, one can also apply a method which uses less factorization, so it should be faster than the usual method (see Heiser [93] and Silverman [205]). To explain this method let $P = (x, y)$ be a point on an elliptic curve over the number field $\mathbb{K}$. Then

$$\begin{aligned} h(P) &= \frac{-1}{2[\mathbb{K}:\mathbb{Q}]} \sum_{v \in M_{\mathbb{K}}} n_v \min\{0, v(x)\} \\ &= \frac{-1}{2[\mathbb{K}:\mathbb{Q}]} \sum_{v \in M_{\mathbb{K}}^0, v(x)<0} n_v v(x) + \frac{1}{2} h_{\infty}(x) \\ &= \frac{1}{2[\mathbb{K}:\mathbb{Q}]} \sum_{v \in M_{\mathbb{K}}^0, v(x)<0} n_v \log |x|_v + \frac{1}{2} h_{\infty}(x). \end{aligned}$$

Now, with $|x|_v = \mathcal{N}(\mathfrak{p}_v)^{-\mathrm{ord}_v(x)/n_v}$, we get

$$\begin{aligned} h(P) &= \frac{1}{2[\mathbb{K}:\mathbb{Q}]} \sum_{v \in M_{\mathbb{K}}^0, v(x)<0} \log\left(\mathcal{N}(\mathfrak{p}_v)^{-\mathrm{ord}_v(x)}\right) + \frac{1}{2} h_{\infty}(x) \\ &= \frac{1}{2[\mathbb{K}:\mathbb{Q}]} \log\Bigg(\prod_{v \in M_{\mathbb{K}}^0, v(x)<0} \mathcal{N}(\mathfrak{p}_v)^{-\mathrm{ord}_v(x)}\Bigg) + \frac{1}{2} h_{\infty}(x). \end{aligned}$$

If we put

$$(x)_\infty = \prod_{v \in M_{\mathbb{K}}^0, v(x)<0} \mathfrak{p}_v^{-\mathrm{ord}_v(x)}$$

considering only the denominator of the divisor (x), we get

$$h(P) = \frac{1}{2[\mathbb{K}:\mathbb{Q}]} \log(\mathcal{N}((x)_\infty)) + \frac{1}{2} h_\infty(x).$$

Now we write $x = \frac{\alpha}{m}$ with $\alpha \in \mathcal{O}_{\mathbb{K}}$ and $m \in \mathbb{N}$ and compute the integer

$$d := \prod_{p \in \mathbb{P},\ p | N_{\mathbb{Q}}^{\mathbb{K}}(\alpha),\ p|m} p^{\mathrm{ord}_p(N_{\mathbb{Q}}^{\mathbb{K}}(\alpha))}$$

and the divisor

$$\mathfrak{a} := \prod_{v \in M_{\mathbb{K}}^0,\ v(d)>0,\ v(x)<0} \mathfrak{p}_v^{-\mathrm{ord}_v(x)}.$$

Then we obtain the divisor norm (see Heiser [93] and Silverman [208])

$$\mathcal{N}((x)_\infty) = \frac{m^{[\mathbb{K}:\mathbb{Q}]}}{d} \cdot \mathcal{N}(\mathfrak{a}).$$

For the algorithm for the computation of the modified height and some special algorithms over $\mathbb{Q}$, we refer to the exercises.

The computation of the canonical height is more complicated. The usual definition of $\hat{h}$ as a limit is not practical for computations. Instead, one uses the fact that the canonical height can be written as a sum of local height functions. The proof of the existence of those local height functions can be found in Silverman [207], Chapter VI and in the papers of the second author [252], [253]. (See also Tschöpe–Zimmer [225].) There are also formulas for these functions. Here we present only the results.

Theorem 5.25 (Néron, Tate). *Let $\mathbb{K}$ be a field which is complete with respect to an absolute value $|\cdot|_v$ corresponding to the additive absolute value v of $\mathbb{K}$. Let $E|\mathbb{K}$ be an elliptic curve in long Weierstraß normal form.*

a) *There exists a unique function*

$$\hat{h}_v : E(\mathbb{K}) \backslash \{\mathcal{O}\} \to \mathbb{R}$$

with the following three properties:

(i) *$\hat{h}_v$ is continuous on $E(\mathbb{K})\backslash\{\mathcal{O}\}$ and is bounded on the complement of any v-adic neighborhood of $\mathcal{O}$.*

(ii) *The limit*

$$\lim_{P\ v\text{-adic}\to \mathcal{O}} (2\hat{h}_v(P) + v(x(P)))$$

exists.

(iii) *For all* $P = (x, y) \in E(\mathbb{K})$ *with* $2P \neq \mathcal{O}$,

$$\hat{h}_v(2P) = 4\hat{h}_v(P) + \frac{1}{2}v(\psi_2^2(x)) - \frac{1}{4}v(\Delta).$$

b) $\hat{h}_v$ *is independent of the choice of the Weierstraß equation for* $E|\mathbb{K}$.

The function $\hat{h}_v$ is called the *local height function* at v.

Proof. See Silverman [207] Chapter VI and the papers of the second author [252], [253]. □

We now consider the local height functions for the canonical height.

Theorem 5.26. a) *Let* $\mathbb{K}$ *be a local field with archimedean absolute value* v *and* $E|\mathbb{K}$ *be an elliptic curve given in Weierstraß form. Further let* $P = (x, y) \in E(\mathbb{K})$. *We define*

$$\begin{aligned} t = t(P) &:= \frac{1}{x}, \\ w = w(P) &:= 4t + b_2t^2 + 2b_4t^3 + b_6t^4 = \frac{\psi_2^2(x)}{x^4}, \\ z = z(P) &:= 1 - b_4t^2 - 2b_6t^3 - b_8t^4 = \frac{\phi_2(x)}{x^4}, \end{aligned}$$

and

$$\begin{aligned} t' = t'(P) &:= \frac{t}{1+t}, \\ b_2' &:= b_2 - 12, \\ b_4' &:= b_4 - b_2 + 6, \\ b_6' &:= b_6 - 2b_4 + b_2 - 4, \\ b_8' &:= b_8 - 3b_6 + 3b_4 - b_2 + 3, \\ w' &:= 4t' + b_2't'^2 + 2b_4't'^3 + b_6't'^4, \\ z' &:= 1 - b_4't'^2 - 2b_6't'^3 - b_8't'^4. \end{aligned}$$

Furthermore, we define

$$U := \{Q \in E(\mathbb{K}) : |t(Q)|_v \leq 2\}$$

and

$$U' := \{Q \in E(\mathbb{K}) : |t'(Q)|_v \leq 2\}.$$

For $P \in E(\mathbb{K})$, let us define the two series $(c_n)_{n\geq -1}$ and $(\beta_n)_{n\geq -1}$ as

$$(c_{-1}, \beta_{-1}) := \begin{cases} (-\log|t(P)|_v, 1) & \text{if } P \in U, \\ (-\log|t'(P)|_v, 0) & \text{if } P \notin U. \end{cases}$$

and for $n \geq 0$

$$(c_n, \beta_n) := \begin{cases} (\log|z(2^n P)|_v, 1) & \text{if } \beta_{n-1} = 1 \text{ and } 2^{n+1}P \in U, \\ (\log|z(2^n P) + w(2^n P)|_v, 0) & \text{if } \beta_{n-1} = 1 \text{ and } 2^{n+1}P \notin U, \\ (\log|z'(2^n P)|_v, 1) & \text{if } \beta_{n-1} = 0 \text{ and } 2^{n+1}P \in U', \\ (\log|z'(2^n P) - w'(2^n P)|_v, 0) & \text{if } \beta_{n-1} = 0 \text{ and } 2^{n+1}P \notin U'. \end{cases}$$

Then

$$2\hat{h}_v(P) = c_{-1} + \frac{1}{2^2}\sum_{n=0}^{\infty} \frac{1}{2^{2n}} c_n.$$

b) *Let $\mathbb{K}$ be a local field with non-archimedean absolute value v and $E|\mathbb{K}$ be an elliptic curve given by a minimal Weierstraß equation. Further let $P = (x, y) \in E(\mathbb{K})\backslash\{\mathcal{O}\}$.*

(i) *If $P \in E_0(\mathbb{K})$, the local height function is defined as*

$$\hat{h}_v(P) = h_v(P) + \frac{1}{12}v(\Delta).$$

Especially, if the curve has good reduction at v, then, for all points $P \in E(\mathbb{K})$, we have

$$\hat{h}_v(P) = h_v(P).$$

(ii) *If the curve has multiplicative reduction and $P = (x, y) \notin E_0(\mathbb{K})$, we define*

$$\alpha(P) := \min\left\{\frac{v(\psi_2(x, y))}{v(\Delta)}, \frac{1}{2}\right\}.$$

Then,

$$\hat{h}_v(P) = \frac{1}{2}B_2(\alpha(P))v(\Delta),$$

where $B_2(T) = T^2 - T + \frac{1}{6}$ is the second Bernoulli polynomial.

(iii) *If the curve has additive reduction and $P = (x, y) \notin E_0(\mathbb{K})$, then*

$$\hat{h}_v(P) = \begin{cases} -\frac{1}{6}v(\psi_2^2(x)) & \text{if } v(\psi_3^2(x)) \geq 3v(\psi_2^2(x)), \\ -\frac{1}{16}v(\psi_3^2(x)) & \text{else.} \end{cases}$$

Proof. See the article of Silverman [205], the diploma thesis [93] of Heiser, or the book of Silverman [207], Chapter VI and the papers of the second author [252], [253]. □

The canonical height can be decomposed as a sum of those local height functions.

Theorem 5.27. *Let $\mathbb{K}$ be a number field and $E|\mathbb{K}$ an elliptic curve over $\mathbb{K}$. For $v \in M_{\mathbb{K}}$ let $\hat{h}_v$ be the local height function given above. Then for $P \in E(\mathbb{K})$ the canonical height is*

$$\hat{h}(P) = \frac{1}{[\mathbb{K} : \mathbb{Q}]} \sum_{v \in M_{\mathbb{K}}} n_v \hat{h}_v(P).$$

Proof. Silverman [207], Chapter VI, theorem 2.1. □

A similar relation holds for the modified height

$$\hat{h}(P) = \frac{1}{[\mathbb{K} : \mathbb{Q}]} \sum_{v \in M_{\mathbb{K}}} n_v \hat{d}_v(P)$$

(see the paper of the second author [256]).

Hence we have the following algorithm for the computation of the canonical height of points on elliptic curves over number field.

Algorithm 5.28 (Computation of the canonical height).

INPUT: A point P on the curve $E|\mathbb{K}$.
OUTPUT: $\hat{h}(P)$.
1. If $P = \mathcal{O}$ then return(0).
2. $\hat{h} \leftarrow 0$.
3. $x \leftarrow x$-coordinate of P.
4. For all archimedean absolute values v do:
5. Compute $\hat{h} \leftarrow \hat{h} + n_v \hat{h}_v(P)$ with Theorem 5.26 a).
6. For all non-archimedean absolute values v do:
7. If E has good reduction at v then do: $\hat{h} \leftarrow \hat{h} + n_v h_v(P)$.
8. If E has multiplicative reduction at v then do:
9. If P reduces to a singular point then:
$\hat{h} \leftarrow \hat{h} + n_v \hat{h}_v(P)$ with Theorem 5.26 b) (ii).
10. If P reduces to a nonsingular point then:
$\hat{h} \leftarrow \hat{h} + n_v \hat{h}_v(P)$ with Theorem 5.26 b) (i).
11. If E has additive reduction at v then do:
12. If P reduces to a singular point then:
$\hat{h} \leftarrow \hat{h} + n_v \hat{h}_v(P)$ with Theorem 5.26 b) (iii).
13. If P reduces to a nonsingular point then:

$\hat{h} \leftarrow \hat{h} + n_v \hat{h}_v(P)$ with Theorem 5.26 b) (i).

14. $\hat{h} \leftarrow \frac{1}{[\mathbb{K}:\mathbb{Q}]} \hat{h}$.

15. return($\hat{h}$).

Note that the loop for non-archimedean valuations is finite, because we only have to consider a finite number of valuations. The valuations v where the curve has good reduction and $h_v(P) = 0$ do not contribute anything to the result.

5.4 Points of bounded height

Let $\mathbb{K}$ be a number field of degree $n = [\mathbb{K} : \mathbb{Q}]$. In this section we fix a positive number $A \in \mathbb{R}$, $A \geq 1$, and consider the set

$$S_A := \{x \in \mathbb{K} : H(x) \leq A\}.$$

This section is essentially the paper of Pethő and Schmitt [166].

During the computation of a basis of an elliptic curve over $\mathbb{K}$ (for details see Chapter 8 or the second author Schmitt [189]), one ends up with a maximal set of linearly independent points on the curve. The basis can then be computed using an index estimate of Siksek [203]. For this estimate one has to check, for a fixed A, and for all elements $x \in S_A$, whether there exists a $y \in \mathbb{K}$ such that the point (x, y) is lying on the curve. (See also Chapter 8.) A similar problem arises in the 2-descent algorithm, where one has to test an equation of the form $y^2 = g(x)$, where $g(x)$ is a quartic polynomial, whether such an equation has a solution (x, y) in $\mathbb{K}^2$ (see Chapter 7 Section 7.6 or, e.g., the papers of Cremona and Serf [43], and of Simon [210]). To solve such problems one has to provide an efficient method to enumerate all elements of bounded height.

In the case where $\mathbb{K} = \mathbb{Q}$, every $x \in \mathbb{Q}$ can be written as $x = a/b$, with integers a, b such that $b > 0$ and $\gcd(a, b) = 1$. For elements of number fields, where $\alpha \in \mathcal{O}_\mathbb{K}$ and $\gcd(\alpha, m)$ is a divisor of $\mathbb{K}$, we have an analogous representation. For every $x \in \mathbb{K}$ there exist an algebraic integer $\alpha \in \mathcal{O}_\mathbb{K}$ and a positive rational integer m such that $x = \alpha/m$. On choosing an integral basis $\omega_1, \ldots, \omega_n$ of $\mathcal{O}_\mathbb{K}$, the element $x \in \mathbb{K}$ can thus be written as

$$x = \frac{a_1\omega_1 + \cdots + a_n\omega_n}{m}$$

with $a_1, \ldots, a_n, m \in \mathbb{Z}$. These parameters are obviously not at all unique. For $x \in S_A$ the size of $a_1, \ldots, a_n$, and m depends not only on A, but also on the choice of the integral basis.

For $x \in \mathbb{K}$, we denote by $x^{(k)}$ the conjugates of x, $1 \leq k \leq n$. Adapting the notation of Pohst [169] let

$$T_2(x) = \sum_{k=1}^{n} |x^{(k)}|^2.$$

Theorem 5.29. *Let $A \in \mathbb{R}$, $A \geq 1$. Further, let $\Omega = \{\omega_1, \dots, \omega_n\}$ be an integral basis of $\mathcal{O}_{\mathbb{K}}$ and $d_{\mathbb{K}}$ denote the discriminant of $\mathbb{K}$. Then any $x \in S_A$ can be represented in the form*

$$x = \frac{a_1\omega_1 + \cdots + a_n\omega_n}{m}$$

with $a_1, \dots, a_n, m \in \mathbb{Z}$, satisfying

$$0 < m \leq A^n$$

and

$$|a_i| \leq \frac{A^n m n^{1/2}}{|d_{\mathbb{K}}|^{1/2}} \prod_{\substack{1 \leq j \leq n \\ j \neq i}} T_2(\omega_j)^{1/2} =: B_i(\mathbb{K}, \Omega, A, m)$$

for $i = 1, \dots, n$.

In the article [166] it is also proved that it is possible to choose the integral basis $\omega_1, \dots, \omega_n$ in such a way that for $x \in S_A$, the bound for the size of $a_1, \dots, a_n$ and m depends only on A and $\mathbb{K}$.

For the proof of the theorem we need some auxiliary result. To state it we fix some notation.

For a prime divisor $\mathfrak{p}$ of $\mathbb{K}$ let $v_{\mathfrak{p}} \in M_{\mathbb{K}}^0$ as before be the non-archimedean absolute value corresponding to $\mathfrak{p}$ and let $n_{\mathfrak{p}} = n_{v_{\mathfrak{p}}}$ be the corresponding local degree. For simplicity in this section, we write $\operatorname{ord}_{\mathfrak{p}}(x)$ instead of $\operatorname{ord}_{v_{\mathfrak{p}}}(x)$ and $|x|_{\mathfrak{p}}$ instead of $|x|_{v_{\mathfrak{p}}}$ so that $v_{\mathfrak{p}}(x) = -\log |x|_{\mathfrak{p}}$.

Lemma 5.30. *Let $x \in S_A$, i.e. $x \in \mathbb{K}$ such that $H(x) \leq A$. Then there exist an $\alpha \in \mathcal{O}_{\mathbb{K}}$ and an $m \in \mathbb{Z}$ such that $x = \frac{\alpha}{m}$ and $0 < m \leq A^n$.*

Proof. By the theorem on the unique prime ideal (or prime divisor) decomposition, we can write the fractional ideal (x) in the form $(x) = \mathfrak{z}\mathfrak{n}^{-1}$, where

$$\mathfrak{z} = \prod_{\substack{\mathfrak{p} \in M_{\mathbb{K}}^0 \\ \operatorname{ord}_{\mathfrak{p}}(x) > 0}} \mathfrak{p}^{\operatorname{ord}_{\mathfrak{p}}(x)}$$

and

$$\mathfrak{n} = \prod_{\substack{\mathfrak{p} \in M_{\mathbb{K}}^0 \\ \operatorname{ord}_{\mathfrak{p}}(x) < 0}} \mathfrak{p}^{-\operatorname{ord}_{\mathfrak{p}}(x)}.$$

Then $\mathfrak{n}$ is an integral ideal. Denote by m its norm: $\mathrm{N}(\mathfrak{n}) = m$. Then m is a positive integer and the principal ideal (m) is divisible by $\mathfrak{n}$. There exists an integral ideal $\mathfrak{m}$ such that $(m) = \mathfrak{n}\mathfrak{m}$. Hence

$$(x) = \mathfrak{z}\mathfrak{m}(m)^{-1},$$

which implies that $\mathfrak{z}\mathfrak{m}$ is a principal ideal, too. Denoting its generator by β we obtain

$$(x) = (\beta)(m)^{-1},$$

which proves the first assertion of the lemma by taking $\alpha = \beta\varepsilon$ for a suitable unit ε.

Now we estimate the size of m. We have

$$m = \mathcal{N}(\mathfrak{n}) = \prod_{\substack{\mathfrak{p}\in M_{\mathbb{K}}^0 \\ \operatorname{ord}_{\mathfrak{p}}(x)<0}} \mathcal{N}(\mathfrak{p})^{-\operatorname{ord}_{\mathfrak{p}}(x)} = \prod_{\substack{\mathfrak{p}\in M_{\mathbb{K}}^0 \\ \operatorname{ord}_{\mathfrak{p}}(x)<0}} |x|_{\mathfrak{p}}^{n_{\mathfrak{p}}} \le H(x)^n.$$

The lemma is proved. □

Proof of Theorem 5.29. If $H(x) \le A$, then there exist, by Lemma 5.30, an algebraic integer $\alpha \in \mathcal{O}_{\mathbb{K}}$ and a rational integer $m \in \mathbb{Z}$, $m > 0$ such that $x = \alpha/m$ and $m \le A^n$. The assumption $H(x) \le A$ implies that, for all $v \in M_{\mathbb{K}}$

$$|x|_{\mathfrak{p}} \le A^{n/n_{\mathfrak{p}}} \le A^n,$$

hence

$$|\alpha|_{\mathfrak{p}} \le A^n |m|_{\mathfrak{p}}.$$

If we consider archimedean absolute values and interpret the $\alpha^{(k)}$ as complex numbers, we see that all conjugates of α satisfy the inequality

$$|\alpha|_{\mathfrak{p}} = |\alpha^{(k)}| \le A^n m.$$

We choose an integral basis $\omega_1, \dots, \omega_n$ of $\mathcal{O}_{\mathbb{K}}$. Then α can be written in the form

$$\alpha = a_1\omega_1 + \cdots + a_n\omega_n$$

with $a_i \in \mathbb{Z}$. Taking conjugates and using Cramer's rule we see that

$$a_i = \frac{d_i}{d_{\mathbb{K}}^{1/2}},$$

where

$$d_i = \det\begin{pmatrix} \omega_1^{(1)} & \dots & \alpha^{(1)} & \dots & \omega_n^{(1)} \\ \cdot & & \cdot & & \cdot \\ \cdot & & \cdot & & \cdot \\ \cdot & & \cdot & & \cdot \\ \omega_1^{(n)} & \dots & \alpha^{(n)} & \dots & \omega_n^{(n)} \end{pmatrix}$$

is the determinant of the matrix $(\omega_j^{(k)})_{1\le j,k\le n}$, where the i-th column is replaced by the vector $(\alpha^{(k)})_{1\le k\le n}$.

By means of the Hadamard inequality for d_i the estimate

$$|d_i| \le \Big(\sum_{k=1}^{n} |\alpha^{(k)}|^2\Big)^{1/2} \prod_{\substack{1\le j\le n\\ j\ne i}} \Big(\sum_{k=1}^{n} |\omega_j^{(k)}|^2\Big)^{1/2}$$
$$= \Big(\sum_{k=1}^{n} |\alpha^{(k)}|^2\Big)^{1/2} \prod_{\substack{1\le j\le n\\ j\ne i}} T_2(\omega_j)^{1/2}$$

holds. The inequality implies for the k-th conjugate of α

$$\sum_{k=1}^{n} |\alpha^{(k)}|^2 \le \sum_{k=1}^{n} A^{2n} m^2 = A^{2n} m^2 n,$$

and this proves the theorem. □

5.5 The differences between the heights

It is important to find good estimates for the difference between the canonical height and the ordinary (or the modified) height. Such estimates can be computed, for example, with results of Siksek, Silverman or Zimmer. We use the notation given above. Actually there are little differences in the notation of the theorems here and the original literature.

These estimates are obtained by first estimating the local heights. There are two different methods for this. Silverman uses the local height functions for the canonical height. Employing these functions he obtains estimates for $\hat{h}_v(P) - h_v(P)$. Summing up the local estimates he results in estimates for $\hat{h}(P) - h(P)$.

Siksek and Zimmer compute estimates for $h_v(2P) - 4h_v(P)$ or $d_v(2P) - 4d_v(P)$ respectively and refer to the definition of the canonical height as a limit to obtain the global estimates.

We first state the estimates for the local height functions due to Tate and used and proved by Silverman.

Theorem 5.31. a) (Tate) *Let $\mathbb{K}$ be complete with respect to a non-archimedian absolute value v, and let $E|\mathbb{K}$ be an elliptic curve in long Weierstraß normal form with v-integral coefficients. For all $P \in E(\mathbb{K})$,*

$$-\frac{1}{24} h_{\mathbb{K},v}(j) \le \hat{h}_v(P) - h_v(P) \le \frac{1}{12} v(\Delta).$$

b) *Let $E|\mathbb{K}$ be an elliptic curve in long Weierstraß normal form. We define*

$$2^* := \begin{cases} 2 & \text{if } b_2 \ne 0, \\ 1 & \text{if } b_2 = 0. \end{cases}$$

Let v be an archimedean absolute value. Then, for all $P \in E(\mathbb{K})$,

$$\begin{aligned}
&-\frac{1}{12}h_{\mathbb{K},v}(\Delta) - \frac{1}{8}h_{\mathbb{K},v}(j) - \frac{1}{2}h_{\mathbb{K},v}\left(\frac{b_2}{12}\right) - \frac{1}{2}\log(2^*) - 0.973 \\
&\quad \leq \hat{h}_v(P) - h_v(P) \\
&\quad \leq \frac{1}{12}h_{\mathbb{K},v}(\Delta^{-1}) + \frac{1}{12}h_{\mathbb{K},v}(j) + \frac{1}{2}h_{\mathbb{K},v}\left(\frac{b_2}{12}\right) + \frac{1}{2}\log(2^*) + 1.07.
\end{aligned}$$

Proof. See Silverman [206] and Lang [115]. □

For the estimate of Zimmer (see [249]) we need the following lemma.

Lemma 5.32. *Let $\mathbb{K}$ be a number field, $a_1, \ldots, a_n \in \mathbb{K}$, $n \in \mathbb{N}$, and $v \in M_{\mathbb{K}}$. Define*

$$\alpha_v := \begin{cases} \log(2) & \text{if } v \text{ is archimedean,} \\ 0 & \text{if } v \text{ is non-archimedean.} \end{cases}$$

Then

$$v(a_1 + \cdots + a_n) \geq \min\{v(a_1), \ldots, v(a_n)\} - r\alpha_v$$

with $r \in \mathbb{N}_0$ such that

$$2^{r-1} < n \leq 2^r.$$

Proof. We first show that it suffices to prove the lemma for $n = 2^r$. If $n < 2^r$ we take the sum $a_1 + \cdots + a_n + 0 + \cdots + 0$ with 2^r summands. As $v(0) = \infty$ it follows that

$$\min\{v(a_1), \ldots, v(a_n), v(0), \ldots, v(0)\} = \min\{v(a_1), \ldots, v(a_n)\},$$

so that we may assume that $n = 2^r$. For this we perform induction over $r \in \mathbb{N}_0$. If $r = 0$, then the proof is trivial. The inequality for $r = 1$ follows from the triangle equation. Now we assume that the inequality has been shown for all $r' \leq r$. First using this inequality for $r = 1$ we get

$$\begin{aligned}
&v((a_1 + \cdots + a_{2^r}) + (a_{2^r+1} + \cdots + a_{2^{r+1}})) \\
&\quad \geq \min\{v(a_1 + \cdots + a_{2^r}), v(a_{2^r+1} + \cdots + a_{2^{r+1}})\} - \alpha_v.
\end{aligned}$$

Now the inequality for r yields

$$v(a_1 + \cdots + a_{2^{r+1}}) \geq \min\{v(a_1), \ldots, v(a_{2^{r+1}})\} - (r+1)\alpha_v.$$ □

Of course, Lemma 5.32 supplies only a coarse estimate.

Now we have the following local estimates.

Theorem 5.33. *Let $E|\mathbb{K}$ be an elliptic curve over the number field $\mathbb{K}$, $v \in M_{\mathbb{K}}$.*

a) (Siksek) *Suppose that E has integral coefficients. We define*

$$\begin{aligned}\sigma_v &:= \frac{1}{2}\inf_{P=(x,y)\in E(\mathbb{K}_v)}\{-\min\{v(\psi_2^2(x)), v(\phi_2(x))\} + 4\min\{0, v(x)\}\}\\ &= \inf_{P=(x,y)\in E(\mathbb{K}_v)}\left\{h_v(2P) - 4h_v(P) - \frac{1}{2}v(\psi_2^2(x))\right\}\end{aligned}$$

and for $P \in E(\mathbb{K})$

$$\sigma_v(P) := \begin{cases}0 & \text{if } v \in M_{\mathbb{K}}^0,\ E \text{ minimal at } v, \text{ and } P \in E_0(\mathbb{K}).\\ \sigma_v & \text{else.}\end{cases}$$

Then for all $P = (x, y) \in E(\mathbb{K})$

$$h_v(2P) - 4h_v(P) - \frac{1}{2}v(\psi_2^2(x)) \geq \sigma_v(P).$$

b) (Zimmer) *For all $P = (x, y) \in E(\mathbb{K})$*

$$-\frac{1}{2}v(\Delta) + 3\lambda_v - 4\alpha_v \leq d_v(2P) - 4d_v(P) - \frac{1}{2}v(\psi_2^2(x)) \leq \frac{3}{2}\alpha_v.$$

(Recall the definition of α_v in Lemma 5.32 *and of λ_v in Definition* 5.9.*)*

Proof. a) See Siksek [203]. Note that

$$\begin{aligned}&-\min\{v(\psi_2^2(x)), v(\phi_2(x))\}\\ &\quad = -\min\{0, v(\phi_2(x)) - v(\psi_2^2(x))\} - v(\psi_2^2(x))\\ &\quad = -\min\{0, v(x_{2P})\} - v(\psi_2^2(x))\\ &\quad = 2h_v(2P) - v(\psi_2^2(x)).\end{aligned}$$

It follows directly from the definition of σ_v that for all $v \in M_{\mathbb{K}}$

$$h_v(2P) - 4h_v(P) - \frac{1}{2}v(\psi_2^2(x)) \geq \sigma_v.$$

In the case that $v \in M_{\mathbb{K}}^0$, E minimal at v and $P \in E_0(\mathbb{K})$ it is sufficient to show that

$$\max\{|\psi_2^2(x)|_v, |\phi_2(x)|_v\} = \max\{1, |x|_v\}^4.$$

If $|x|_v > 1$ this claim follows from the fact that $|\psi_2^2(x)|_v \leq |x|_v^3$ and $|\phi_2(x)|_v = |x|_v^4$.

If $|x|_v \leq 1$ we have to show that

$$\max\{|\psi_2^2(x)|_v, |\phi_2(x)|_v\} = 1.$$

Let $\mathfrak{p}$ be the prime ideal (or prime divisor) corresponding to v. We show that $P = (x, y) \pmod{\mathfrak{p}}$ is singular on $E \pmod{\mathfrak{p}}$, if $\psi_2^2(x) \equiv 0 \bmod \mathfrak{p}$ and $\phi_2(x) \equiv 0 \bmod \mathfrak{p}$.

By a change of variable we may suppose that $P = (0,0)$ and hence $a_6 = 0$. Furthermore, since $\psi_2^2(x) \equiv 0 \bmod \mathfrak{p}$ and $\phi_2(x) \equiv 0 \bmod \mathfrak{p}$, we have $b_6 \equiv b_8 \equiv 0 \bmod \mathfrak{p}$ and hence $a_3 \equiv a_4 \equiv 0 \bmod \mathfrak{p}$. In this case the point $P = (0,0)$ is singular mod $\mathfrak{p}$.

b) If $v(x) \geq \lambda_v$, we consider Equation (1.7), Chapter 1. Applying Lemma 5.32 with $n = 2$ and the decomposition (1.7), Chapter 1, of Δ yields

$$\begin{aligned} v(\Delta) \geq \min\{&v((-48x^2 - 8b_2x + b_2^2 - 32b_4)\phi_2(x)),\\ &v((12x^3 - b_2x^2 - 10b_4x + b_2b_4 - 27b_6)\psi_2^2(x))\} - \alpha_v. \end{aligned}$$

Now with $n = 89$ in Lemma 5.32, we get

$$\begin{aligned} &v(-48x^2 - 8b_2x + b_2^2 - 32b_4)\\ &= v\Big(\sum_{j=1}^{48}(-x^2) + \sum_{j=1}^{8}(-b_2x) + b_2^2 + \sum_{j=1}^{32} b_4\Big)\\ &\geq \min\{v(-x^2), v(-b_2x), v(b_2^2), v(b_4)\} - 7\alpha_v\\ &= \min\{2v(x), v(b_2) + v(x), 2v(b_2), v(b_4)\} - 7\alpha_v\\ &\geq 2\lambda_v - 7\alpha_v \end{aligned}$$

Furthermore, Lemma 5.32 with $n = 51$ entails, according to the same method,

$$\begin{aligned} &v(12x^3 - b_2x^2 - 10b_4x + b_2b_4 - 27b_6)\\ &= v\Big(\sum_{j=1}^{12}(x^3) + (-b_2x^2) + \sum_{j=1}^{10}(-b_4x) + b_2b_4 + \sum_{j=1}^{27}(-b_6)\Big)\\ &\geq \min\{v(x^3), v(-b_2x^2), v(-b_4x), v(b_2b_4), v(-b_6)\} - 6\alpha_v\\ &= \min\{3v(x), v(b_2) + 2v(x), v(b_4) + v(x), v(b_2) + v(b_4), v(b_6)\} - 6\alpha_v\\ &\geq 3\lambda_v - 6\alpha_v. \end{aligned}$$

Altogether, we have for $v(x) \geq \lambda_v$

$$\begin{aligned} v(\Delta) &\geq \min\{2\lambda_v - 7\alpha_v + v(\phi_2(x)), 3\lambda_v - 6\alpha_v + v(\psi_2^2(x))\} - \alpha_v\\ &= 2\lambda_v - 7\alpha_v + \min\{-\alpha_v + v(\phi_2(x)), \lambda_v + v(\psi_2^2(x))\}\\ &\geq 2\lambda_v - 8\alpha_v + \min\{v(\phi_2(x)), \lambda_v + v(\psi_2^2(x))\}. \end{aligned}$$

The last estimate follows because $\alpha_v \geq 0$. The multiplication formulas yield

$$x_{2P} = \frac{\phi_2(x)}{\psi_2^2(x)},$$

so that for the valuations, $v(x_{2P}) = v(\phi_2(x)) - v(\psi_2^2(x))$. It follows thus from the decomposition (1.7) of Δ that

$$\begin{aligned} v(\Delta) &\geq 2\lambda_v - 8\alpha_v + \min\{v(x_{2P}) + v(\psi_2^2(x)), \lambda_v + v(\psi_2^2(x))\} \\ &= 2\lambda_v - 8\alpha_v + v(\psi_2^2(x)) + \min\{v(x_{2P}), \lambda_v\} \\ &= 6\lambda_v - 4\lambda_v - 8\alpha_v + v(\psi_2^2(x)) + \min\{v(x_{2P}), \lambda_v\}. \end{aligned}$$

Taking into account that $d_v(P) = -\frac{1}{2}\lambda_v$ and $d_v(2P) = -\frac{1}{2}\min\{v(x_{2P}), \lambda_v\}$, we obtain the relation

$$d_v(2P) - 4d_v(P) - \frac{1}{2}v(\psi_2^2(x)) \geq -\frac{1}{2}v(\Delta) + 3\lambda_v - 4\alpha_v.$$

For $v(x) \leq \mu_v$ we use similar methods and estimates as above and get from Equation (1.8) of Chapter 1 the estimate

$$\begin{aligned} v(\Delta) + 7v(x) &= v(\Delta) - 14d_v(P) \\ &\geq 6\lambda_v + 3v(x) - 8\alpha_v + v(\psi_2^2(x)) + \min\{v(x_{2P}), \lambda_v\} \\ &= 6\lambda_v - 6d_v(P) - 8\alpha_v + v(\psi_2^2(x)) - 2d_v(2P). \end{aligned}$$

In sum, we end up with the same estimate as in the former case:

$$d_v(2P) - 4d_v(P) - \frac{1}{2}v(\psi_2^2(x)) \geq -\frac{1}{2}v(\Delta) + 3\lambda_v - 4\alpha_v.$$

For the other side, we look first at the case $v(x_{2P}) < \lambda_v$, so that $d_v(2P) = -\frac{1}{2}v(x_{2P})$. We obtain the estimate

$$\begin{aligned} 2d_v(2P) &= -v(x_{2P}) = -v(\phi_2(x)) + v(\psi_2^2(x)) \\ &= -v(x^4 - b_4x^2 - 2b_6x - b_8) + v(\psi_2^2(x)) \\ &\leq -\min\{v(x^4), v(-b_4x^2), v(-b_6x), v(-b_8)\} + 3\alpha_v + v(\psi_2^2(x)) \\ &\leq -4\min\{v(x), \mu_v\} + 3\alpha_v + v(\psi_2^2(x)) \\ &= 8d_v(P) + 3\alpha_v + v(\psi_2^2(x)). \end{aligned}$$

Hence we have

$$d_v(2P) - 4d_v(P) - \frac{1}{2}v(\psi_2^2(x)) \leq \frac{3}{2}\alpha_v.$$

If $v(x_{2P}) \geq \lambda_v$, then $d_v(2P) = -\frac{1}{2}\lambda_v$, and

$$\begin{aligned} v(\psi_2^2(x)) &= v(4x^3 + b_2x^2 + 2b_4x + b_6) \\ &\geq \min\{v(x^3), v(b_2x^2), v(b_4x), v(b_6)\} - 3\alpha_v \\ &\geq 3\min\{v(x), \lambda_v\} + \lambda_v - \lambda_v - 3\alpha_v \\ &\geq 4\min\{v(x), \lambda_v\} - \lambda_v - 3\alpha_v \\ &= -8d_v(P) + 2d_v(2P) - 3\alpha_v. \end{aligned}$$

Together we get the same estimate

$$d_v(2P) - 4d_v(P) - \frac{1}{2}v(\psi_2^2(x)) \le \frac{3}{2}\alpha_v.$$

□

Summing up all local formulas, we obtain the desired global estimate.

Theorem 5.34. *Let $E|\mathbb{K}$ be an elliptic curve over the number field $\mathbb{K}$.*

a) (Siksek) *Define*

$$\sigma(P) := \frac{1}{[\mathbb{K}:\mathbb{Q}]} \sum_{v \in M_{\mathbb{K}}} n_v \sigma_v(P)$$

with $\sigma_v(P)$ as in Theorem 5.33. *Then, for all $P \in E(\mathbb{K})$,*

$$h(2P) - 4h(P) \ge \sigma(P).$$

b) (Zimmer) *Let μ be the constant defined in Definition* 5.10.
For all $P \in E(\mathbb{K})$

$$-3\lambda - 4\log(2) \le d(2P) - 4d(P) \le \frac{3}{2}\log(2).$$

Proof. a) From the sum formula for the absolute values of number fields we see that

$$\frac{1}{[\mathbb{K}:\mathbb{Q}]} \sum_{v \in M_{\mathbb{K}}} n_v v(\psi_2^2(x)) = 0 \quad \text{unless } \psi_2^2(x) = 0.$$

On multiplying the local estimate

$$h_v(2P) - 4h_v(P) - \frac{1}{2}v(\psi_2^2(x)) \ge \sigma_v(P)$$

by $\frac{1}{[\mathbb{K}:\mathbb{Q}]}n_v$ and summing over all $v \in M_{\mathbb{K}}$ we get

$$\frac{1}{[\mathbb{K}:\mathbb{Q}]} \sum_{v \in M_{\mathbb{K}}} n_v h_v(2P) - 4\frac{1}{[\mathbb{K}:\mathbb{Q}]} \sum_{v \in M_{\mathbb{K}}} n_v h_v(P) \ge \frac{1}{[\mathbb{K}:\mathbb{Q}]} \sum_{v \in M_{\mathbb{K}}} n_v \sigma_v(P).$$

With the definitions we arrive at the desired estimate.

b) The local estimates for $v \in M_{\mathbb{K}}$ are

$$-\frac{1}{2}v(\Delta) + 3\lambda_v - 4\alpha_v \le d_v(2P) - 4d_v(P) - \frac{1}{2}v(\psi_2^2(x)) \le \frac{3}{2}\alpha_v.$$

Multiplying with $\frac{1}{[\mathbb{K}:\mathbb{Q}]}n_v$, summing up over all $v \in M_{\mathbb{K}}$, and applying the sum formula for $\mathbb{K}$ we obtain

$$\begin{aligned} &3\frac{1}{[\mathbb{K}:\mathbb{Q}]}\sum_{v\in M_{\mathbb{K}}} n_v\lambda_v - 4\frac{1}{[\mathbb{K}:\mathbb{Q}]}\sum_{v\in M_{\mathbb{K}}} n_v\alpha_v \\ &\leq \frac{1}{[\mathbb{K}:\mathbb{Q}]}\sum_{v\in M_{\mathbb{K}}} n_v d_v(2P) - 4\frac{1}{[\mathbb{K}:\mathbb{Q}]}\sum_{v\in M_{\mathbb{K}}} n_v d_v(P) \\ &\leq \frac{3}{2}\frac{1}{[\mathbb{K}:\mathbb{Q}]}\sum_{v\in M_{\mathbb{K}}} n_v\alpha_v \end{aligned}$$

From the definition of α_v it follows that

$$\frac{1}{[\mathbb{K}:\mathbb{Q}]}\sum_{v\in M_{\mathbb{K}}} n_v\alpha_v = \frac{1}{[\mathbb{K}:\mathbb{Q}]}\sum_{v\in M_{\mathbb{K}}^{\infty}} n_v \log(2) = \log(2).$$

Further we use the definitions of λ and d and arrive at

$$-3\lambda - 4\log(2) \leq d(2P) - 4d(P) \leq \frac{3}{2}\log(2).$$

□

Theorem 5.35. *Let $E|\mathbb{K}$ be an elliptic curve in long Weierstraß normal form over the number field $\mathbb{K}$.*

a) (Silverman) *Define*

$$2^* := \begin{cases} 2 & \text{if } b_2 \neq 0, \\ 1 & \text{if } b_2 = 0 \end{cases}$$

and

$$H(E) := \frac{1}{12}h(\Delta) + \frac{1}{12}h_\infty(j) + \frac{1}{2}h_\infty\left(\frac{b_2}{12}\right) + \frac{1}{2}\log(2^*).$$

Then, for all $P \in E$,

$$-\frac{1}{24}h(j) - H(E) - 0.973 \leq \hat{h}(P) - h(P) \leq H(E) + 1.07.$$

b) (Siksek) *Let $E|\mathbb{K}$ be given with integral coefficients. For $v \in M_{\mathbb{K}}$ define k_v as follows:*

1. *If v is archimedean, then $k_v = \frac{1}{3}$.*
2. *If v is non-archimedean and E is not minimal at v, then $k_v = \frac{1}{3}$.*

3. *If v is non-archimedean and E is minimal at v, then*

$$k_v = \begin{cases} 0 & \text{if } [(E(\mathbb{K}_v) : E_0(\mathbb{K}_v)] = 1, \\ \frac{1}{4} & \text{if } E(\mathbb{K}_v)/E_0(\mathbb{K}_v) \cong (\mathbb{Z}/2\mathbb{Z})^2, \\ \left(1 - \frac{1}{4^\alpha}\right)/3 & \text{if } E(\mathbb{K}_v)/E_0(\mathbb{K}_v) \cong \mathbb{Z}/2^\alpha\mathbb{Z}, \alpha \geq 1, \\ \frac{1}{3} & \text{if } [(E(\mathbb{K}_v) : E_0(\mathbb{K}_v)] \text{ is not a power of } 2. \end{cases}$$

Then for all $P \in E$:

$$\hat{h}(P) - h(P) \leq \frac{1}{4[\mathbb{K} : \mathbb{Q}]} \sum_{v \in M_{\mathbb{K}}} k_v n_v \sigma_v(P).$$

c) (Zimmer) *Let μ be the constant defined in Definition* 5.10. *For all $P \in E$*

$$-\left(\lambda + \frac{4}{3}\log(2)\right) \leq \hat{h}(P) - d(P) \leq \frac{1}{2}\log(2).$$

Proof. a) See Silverman [206].

b) (See also Siksek [203].) We have

$$h(2P) - 4h(P) \geq \sigma \Leftrightarrow h(P) \leq \frac{h(2P)}{2^2} - \frac{1}{2^2}\sigma(P).$$

Hence we get, for $n \in \mathbb{N}$,

$$h(P) \leq \frac{h(2^n P)}{2^{2n}} - \frac{1}{2^2}\sum_{k=0}^{n-1}\frac{1}{2^{2k}}\sigma(2^k P).$$

Taking the limit we get

$$h(P) \leq \hat{h}(P) - \frac{1}{2^2}\sum_{k=0}^{\infty}\frac{1}{2^{2k}}\sigma(2^k P).$$

Now from the definition of σ

$$\sum_{k=0}^{\infty}\frac{1}{2^{2k}}\sigma(2^k P) = \frac{1}{[\mathbb{K} : \mathbb{Q}]}\sum_{v \in M_{\mathbb{K}}} n_v \sum_{k=0}^{\infty}\frac{1}{2^{2k}}\sigma_v(2^k P),$$

with

$$\sigma_v(P) := \begin{cases} 0 & \text{if } v \in M_{\mathbb{K}}^0, E \text{ minimal at } v, \text{ and } P \in E_0(\mathbb{K}). \\ \sigma_v & \text{else.} \end{cases}$$

With this definition one can easily show that

$$\sum_{k=0}^{\infty} \frac{1}{2^{2k}} \sigma_v(2^k P) \geq k_v \sigma_v(P)$$

for all $v \in M_{\mathbb{K}}$, so that

$$\sum_{k=0}^{\infty} \frac{1}{2^{2k}} \sigma(2^k P) \geq \frac{1}{[\mathbb{K} : \mathbb{Q}]} \sum_{v \in M_{\mathbb{K}}} n_v k_v \sigma_v(P).$$

c) We have, for $n \in \mathbb{N}$,

$$\begin{aligned} \frac{d(2^n P)}{2^{2n}} - d(P) &= \sum_{k=1}^{n} \left(\frac{d(2^k P)}{2^{2k}} - \frac{d(2^{k-1} P)}{2^{2(k-1)}} \right) \\ &= \sum_{k=1}^{n} \frac{d(2^k P) - 4d(2^{k-1} P)}{2^{2k}}. \end{aligned}$$

If we take the limit, we obtain

$$\hat{h}(P) - d(P) = \sum_{k=1}^{\infty} \frac{d(2^k P) - 4d(2^{k-1} P)}{2^{2k}}.$$

We know that $\sum_{k=1}^{\infty} \frac{1}{2^{2k}} = \frac{1}{3}$ and use the estimate

$$-3\lambda - 4\log(2) \leq d(2P) - 4d(P) \leq \frac{3}{2} \log(2)$$

to get

$$\begin{aligned} -(3\lambda + 4\log(2)) \sum_{k=1}^{\infty} \frac{1}{2^{2k}} = -\left(\lambda + \frac{4}{3}\log(2)\right) &\leq \hat{h}(P) - d(P) \\ &\leq \frac{3}{2}\log(2) \sum_{k=1}^{\infty} \frac{1}{2^{2k}} = \frac{1}{2}\log(2). \end{aligned}$$

□

With this theorem we have three different estimates for the difference between the canonical and the ordinary or the modified height. From the theoretical point of view, the definition of σ_v leads to the conclusion that in general the estimate of Siksek is the best possible estimate. One disadvantage of this method is that Siksek considers only one side of the estimate. Further the computation of the constants in the estimate is not at all straightforward. In [203] Siksek gives the algorithm to compute (or approximate) his estimates.

The estimates of Silverman and Zimmer can easily be computed. Zimmer uses the modified height, which is more general but not used in standard applications. One could combine the estimates of Zimmer with the estimate of $|d(P) - h(P)|$ to obtain an estimate of $|\hat{h}(P) - h(P)|$, but as this is a combination of two estimates, the error is too big in most applications.

In practice, one has to find good estimates for the difference of the canonical and the ordinary height. There is no easy way of deciding a priori which method should give the better estimate.

5.6 Exercises

1) (See also Silverman [204], Chapter VIII, Proposition 5.4.)

   a) Use the product formula to prove Part a) of Proposition 5.11.

   b) Prove Part b) of Proposition 5.11.

   c) Let $P \in \mathbb{P}^N(\mathbb{K})$ be a projective point over the number field $\mathbb{K}$ and $\mathbb{L}$ a finite extension of $\mathbb{K}$. Show that the height $H(P)$ is independent of the field $\mathbb{K}$, that means, show that

$$H(P) = H_{\mathbb{L}}(P)^{1/[\mathbb{L}:\mathbb{Q}]} = H_{\mathbb{K}}(P)^{1/[\mathbb{K}:\mathbb{Q}]}.$$

2) The ordinary height for points on elliptic curves over $\mathbb{Q}$ is defined as follows:

   Definition 5.36. a) Let $x = \frac{a}{b} \in \mathbb{Q}$ with $a, b \in \mathbb{Z}$ in lowest terms. Then the *height* of x is

$$H(x) := \max\{|a|, |b|\}.$$

   b) Let $P = [a_0 : \ldots : a_n] \in \mathbb{P}^n(\mathbb{Q})$ be a projective point which can be given with integral coefficients $a_i \in \mathbb{Z}$. Then the *height* of P is

$$H(P) := \max\{|a_0|, \ldots, |a_n|\}.$$

   Show that this definition is compatible with the definition for elliptic curves over arbitrary number fields.

3) Consider the formulas in Section 5.5 for elliptic curves in short (instead of long) Weierstraß normal form. Try to improve the estimate of the difference between the canonical height and the modified height in that special case.

4) Let $P \in \mathbb{P}^n(\overline{\mathbb{Q}})$ and $\sigma \in G_{\overline{\mathbb{Q}}|\mathbb{Q}}$. Show that

$$H(P^\sigma) = H(P).$$

5) Use Proposition 5.18 to prove the following as a corollary of Proposition 5.14.

   Let $E|\mathbb{K}$ be an elliptic curve in long Weierstraß normal form over the number field $\mathbb{K}$.

a) There exists a constant $C = C(E)$, such that for all points $P \in E(\mathbb{K})$

$$|h(P) - d(P)| \le C.$$

b) Let $Q \in E(\mathbb{K})$. There exists a constant $C_1 = C_1(Q, E)$, such that for all $P \in E(\mathbb{K})$

$$d(P + Q) \le 2d(P) + C_1.$$

c) For all $m \in \mathbb{Z}$ there exists a constant $C_2 = C_2(m, E)$, such that for all $P \in E(\mathbb{K})$

$$d(mP) \ge m^2 d(P) - C_2.$$

d) For every constant $C_3 \in \mathbb{R}$ the set

$$\{P \in E(\mathbb{K}) : d(P) \le C_3\}$$

is finite.

6) Let $E|\mathbb{K}$ be an elliptic curve over the number field $\mathbb{K}$ and $m \in \mathbb{N}$, $m \ge 2$.

a) Show that for all $P \in E(\mathbb{K})$ the limit

$$\hat{h}_m(P) := \lim_{n\to\infty} \frac{h(m^n P)}{m^{2n}}$$

exists.

b) Show that the function $\hat{h}_m(P)$ coincides with the usual canonical height in Definition 5.16.

7) Let $M_{\mathbb{K}}^{\infty}$ denote the archimedean places of a number field $\mathbb{K}$. Define, for an elliptic curve $E|\mathbb{K}$, the modified height

$$d_\infty(P) = \frac{-1}{2[\mathbb{K} : \mathbb{Q}]} \sum_{v \in M_{\mathbb{K}}^{\infty}} n_v \min\{\lambda_v + 2v(\zeta), v(\xi)\},$$

where n_v is the local degree at v and the point

$$P = \left(\frac{\xi}{\zeta^2}, \frac{\eta}{\zeta^3}\right) \in E(\mathbb{K})$$

is given with integers $\xi, \zeta, \eta \in \mathcal{O}_{\mathbb{K}}$ having the gcd's

$$(\xi, \zeta^2) = \mathfrak{c}^2, \quad (\eta, \zeta^3) = \mathfrak{c}^3;$$

with an integral divisor $\mathfrak{c}$ of $\mathbb{K}$. Show that the canonical height on $E(\mathbb{K})$ can also be defined by the limit relation

$$\hat{h}(P) = \lim_{m\to\infty} \frac{d_\infty(2^m P)}{2^{2m}}.$$

8) Find an elliptic curve $E|\mathbb{Q}$,

$$E : Y^2 + a_1XY + a_3Y = X^3 + a_2X^2 + a_4X + a_6 \quad (a_i \in \mathbb{Q})$$

such that, for a prime $p \in \mathbb{P}$, $\lambda_p > 0$, and in addition $\lambda > 0$.

9) a) Describe the algorithm for the computation of the modified height of points on an elliptic curve over a number field.

b) Describe the simplifications over $\mathbb{K} = \mathbb{Q}$ of the algorithms for the computation of the ordinary and the modified height for points on elliptic curves over $\mathbb{Q}$.

Chapter 6
Torsion group

In this chapter we consider the torsion group of elliptic curves. First we describe the structure of the torsion group. In Section 6.3 we explain the theorem of Nagell, Lutz, and Cassels, which gives bounds for the denominator of the coefficients of the torsion points of elliptic curves over number fields. For the computation of the torsion group, we need reduction modulo prime ideals, which is explained in Section 6.4. Then we show how to compute torsion points of elliptic curves over number fields.

The notation of this chapter is the same as in Chapter 5. Torsion groups are the subject of many investigations (see e.g. Frey [67]).

6.1 Structure of the torsion group

Definition 6.1. Let $E|\mathbb{K}$ be an elliptic curve. The set

$$E_{\text{tors}} := \{P \in E : \exists m \in \mathbb{N} \text{ with } mP = \mathcal{O}\} = \bigcup_{m\in\mathbb{N}} E[m]$$

is the *torsion group of E over* $\overline{\mathbb{K}}$. The set

$$E(\mathbb{K})_{\text{tors}} := \{P \in E(\mathbb{K}) : \exists m \in \mathbb{N} \text{ with } mP = \mathcal{O}\} = \bigcup_{m\in\mathbb{N}} E(\mathbb{K})[m]$$

is the *torsion group of E over* $\mathbb{K}$.

For a prime number p define

$$E[p^\infty] := \bigcup_{k\in\mathbb{N}} E[p^k]$$

and

$$E(\mathbb{K})[p^\infty] := \bigcup_{k\in\mathbb{N}} E(\mathbb{K})[p^k].$$

Theorem 6.2. *Let $E|\mathbb{K}$ be an elliptic curve over the field $\mathbb{K}$, $m \in \mathbb{Z}$, $m \neq 0$.*

a) *If* $\operatorname{char}(\mathbb{K}) = 0$ *or* $\operatorname{char}(\mathbb{K}) = p \nmid m$, *then*

$$E[m] \cong \mathbb{Z}/m\mathbb{Z} \times \mathbb{Z}/m\mathbb{Z}.$$

b) *If* char$(\mathbb{K}) = p > 0$, *then either*

$$E[p^k] = \{\mathcal{O}\} \quad \forall k \in \mathbb{N} \quad or \quad E[p^k] \cong \mathbb{Z}/p^k\mathbb{Z} \quad \forall k \in \mathbb{N}.$$

If $E[p^k] = \{\mathcal{O}\}$ *then* E *has Hasse invariant* 0, *otherwise* E *has Hasse invariant* 1.

Proof. a) In this case the multiplication with m is separable (see Proposition 1.34) with

$$\sharp E[m] = \deg(m) = m^2$$

and for all divisors d of m:

$$\sharp E[d] = \deg(d) = d^2.$$

The only possibility for the group structure is then

$$E[m] \cong \mathbb{Z}/m\mathbb{Z} \times \mathbb{Z}/m\mathbb{Z}.$$

b) Let φ_p be the p-th Frobenius-endomorphism. With Proposition 1.33 we get

$$\sharp E[p^k] = \deg_s(p^k) = \deg_s(\hat{\varphi_p} \circ \varphi_p)^k = \deg_s(\hat{\varphi_p})^k,$$

because φ_p is purely inseparable (see Chapter 3, Theorem 3.2). Now there are two possibilities for the degree of separability $\deg_s(\hat{\varphi_p})$. If $\hat{\varphi_p}$ is also purely inseparable, then $\deg_s(\hat{\varphi_p}) = 1$ and

$$E[p^k] = \{\mathcal{O}\}$$

for all $k \in \mathbb{N}$. If $\hat{\varphi_p}$ separable, then $\deg_s(\hat{\varphi_p}) = p$ and

$$E[p^k] \cong \mathbb{Z}/p^k\mathbb{Z}$$

for all $k \in \mathbb{N}$. □

There has been a longstanding conjecture, known as the strong uniform boundedness conjecture, which states that the number of torsion points of an elliptic curve over a number field is bounded by a constant which only depends on the degree of the number field. This conjecture has been proved by Merel [143]. We state the theorem in improved versions by Oesterlé (published by Merel [144]) and Parent [160].

We note that Manin [135] estimated the p-primary component of the torsion group of an elliptic curve over a number field with a constant which depends on p and on the number field.

Theorem 6.3. *Let* $\mathbb{K}$ *be a number field of degree* $n = [\mathbb{K} : \mathbb{Q}] > 1$ *and* $E|\mathbb{K}$ *an elliptic curve over* $\mathbb{K}$. *If* $E(\mathbb{K})$ *has a point of order* p, *where* p *is a prime number, then*

$$p \leq (1 + 3^{n/2})^2.$$

Proof. See Merel [144]. (The proof is beyond the scope of this book.) □

Theorem 6.4. *Let $E|\mathbb{K}$ be an elliptic curve over the number field $\mathbb{K}$ with degree $n = [\mathbb{K} : \mathbb{Q}]$. If $E(\mathbb{K})$ has a point P of order p^k, where $p \in \mathbb{P}$ and $k \in \mathbb{N}$, then*

1) *if $p \neq 2, 3$, then $p^k \leq 65(3^n - 1)(2n)^6$.*
2) *if $p = 3$, then $3^k \leq 65(5^n - 1)(2n)^6$.*
3) *if $p = 2$, then $2^k \leq 129(3^n - 1)(3n)^6$.*

Proof. See Parent [160] □

From the above theorems it follows that the torsion group of an elliptic curve over a number field is bounded by a bound which depends only on the degree of the number field.

Corollary 6.5. *Let $n \in \mathbb{N}$. There exists a real number $B(n) \in \mathbb{R}$, such that for every elliptic curve E defined over a number field $\mathbb{K}$ of degree n over $\mathbb{Q}$, every torsion point of $E(\mathbb{K})$ has order $\leq B(n)$. Equivalently, there is only a finite number of possible torsion group structures for elliptic curves over number fields of degree n.*

Proof. See for example Edixhoven [56] or Merel [144]. □

However, all this is of little help if the task is to actually compute torsion groups (see Section 6.2).

Before the proof of Merel, several other bounds for the torsion group have been found. For example Kubert [111] estimated the torsion for l-deficient elliptic curves with $l \geq 5$. This was extended to $l = 2, 3$ by Folz [64], [65] and Pfeifer [167].

Definition 6.6. Let $E|\mathbb{K}$ be an elliptic curve over a number field $\mathbb{K}$.

- For $l \in \mathbb{P}, l \geq 5$ the curve $E|\mathbb{K}$ is *l-deficient*, if the divisor generated by the denominator of the j-invariant of E is an l-th power of a divisor of $\mathbb{K}$.
- The curve E given by
$$E : Y^2 = X(X^2 + a_2X + a_4)$$
is *2-deficient*, if the j-invariant of E is not equal to 0, 1728 and if $a_4(a_2^2 - 4a_4)$ is a square divisor of $\mathbb{K}$.
- For $l \in \mathbb{P}, l \neq 2$, the curve E given by
$$E : Y^2 = X(X^2 + a_2X + a_4)$$
is *$(l, 2)$-deficient*, if the coefficients a_2, a_4 are integral such that a_4 is a 4th power divisor of $\mathbb{K}$ and $a_2^2 - 4a_4$ is a square divisor of $\mathbb{K}$.

We remark that the curve in the second and third case has discriminant $\Delta = 2^4 a_4^2(a_2^2 - 4a_4)$ and j-invariant

$$j = 2^8 \frac{(a_2^2 - 3a_4)^3}{a_4^2(a_2^2 - 4a_4)}.$$

Therefore 2-deficiency is almost l-deficiency for $l = 2$. On the other hand, 3-deficiency cannot be defined in the same way as 2-deficiency for the Weierstraß model $Y^2 = X(X^2 + a_2X + a_4)$. But $(l, 2)$-deficiency is defined for all primes $l \neq 2$

Theorem 6.7. *Let $E|\mathbb{K}$ be an elliptic curve being either*

- *l-deficient for some $l \in \mathbb{P}, l = 2$ or $l \geq 5$, or*
- *$(l, 2)$-deficient for some $l \in \mathbb{P}, l \neq 2$.*

There exists a constant $B(l, \mathbb{K})$ depending on l and on the number field $\mathbb{K}$, such that for a torsion point of $E(\mathbb{K})$ of order $p \in \mathbb{P}$:

$$p \leq B(l, \mathbb{K}).$$

Proof. This has been proved by Kubert [111] for l-deficient elliptic curves with primes $l \geq 5$. It was extended for $l = 2$ by Folz [64], [65] and for $(l, 2)$-deficient curves by Pfeifer [167]. Of course, Merel's proof of the strong boundedness conjecture makes these earlier proofs of special cases obsolete. □

For number fields of low degree there are precise statements about the possible structures making no use of Merel's (later) papers [143], [144]:

Theorem 6.8 (Mazur). *Let $E|\mathbb{Q}$ be an elliptic curve. Then*

$$E(\mathbb{Q})_{\text{tors}} \cong \begin{cases} \mathbb{Z}/m\mathbb{Z} & \text{for } m = 1, \dots, 10, 12 \\ \text{or } \mathbb{Z}/2\mathbb{Z} \times \mathbb{Z}/2m\mathbb{Z} & \text{for } m = 1, \dots, 4. \end{cases}$$

Proof. See Mazur ([140] and [141]). □

Theorem 6.9 (Kamienny, Kenku, Momose). *Let $\mathbb{K} = \mathbb{Q}(\sqrt{D})$ be a quadratic number field and $E|\mathbb{K}$ an elliptic curve over $\mathbb{K}$. Then*

$$E(\mathbb{K})_{\text{tors}} \cong \begin{cases} \mathbb{Z}/m\mathbb{Z} & \text{for } m = 1, \dots, 16, 18 & \\ \text{or } \mathbb{Z}/2\mathbb{Z} \times \mathbb{Z}/2m\mathbb{Z} & \text{for } m = 1, \dots, 6 & \\ \text{or } \mathbb{Z}/3m\mathbb{Z} \times \mathbb{Z}/3m\mathbb{Z} & \text{for } m = 1, 2 & \text{only if } \mathbb{K} = \mathbb{Q}(\sqrt{-3}) \\ \text{or } \mathbb{Z}/4\mathbb{Z} \times \mathbb{Z}/4\mathbb{Z} & & \text{only if } \mathbb{K} = \mathbb{Q}(\sqrt{-1}). \end{cases}$$

Proof. Combination of the articles of Kenku, Momose [106] and Kamienny [105]. □

We need later the following connection between the m-torsion points of an elliptic curve and the m-th roots of unity.

Theorem 6.10. *Let $E|\mathbb{K}$ be an elliptic curve over the number field $\mathbb{K}$ and $m \in \mathbb{N}$. If the m-torsion points of E are given over $\mathbb{K}$:*

$$E[m] = E(\mathbb{K})[m] \cong \mathbb{Z}/m\mathbb{Z} \times \mathbb{Z}/m\mathbb{Z},$$

then the m-th roots of unity lie in $\mathbb{K}$.

Proof. This is proven using the Weil pairing, which is not in the scope of this book. See for example Silverman [204], Chapter III, Corollary 8.1.1. □

6.2 Elliptic curves with integral j-invariant

Independently of the work of Kenku and Momose there has been work on the torsion group of elliptic curves with integral j-invariant over quadratic number fields. This is still interesting, because the work makes use of computers and for many torsion structures there are only a finite number of possible elliptic curves and number fields of small degree such as quadratic, general cubic and certain biquadratic number fields. The curves and fields can all be determined using the computer. However, the assumption that j be integral is essential!

Theorem 6.11. a) *Let E be an elliptic curve with integral j-invariant over a quadratic number field $\mathbb{K}$. Then, up to isomorphism, the torsion group of E over $\mathbb{K}$ is one of the following groups:*

$$E(\mathbb{K})_{\text{tors}} \cong \begin{cases} \mathbb{Z}/m\mathbb{Z} & \text{for } m = 1, \ldots, 8, 10 \\ \text{or } \mathbb{Z}/2\mathbb{Z} \times \mathbb{Z}/2m\mathbb{Z} & \text{for } m = 1, 2, 3 \\ \text{or } \mathbb{Z}/3\mathbb{Z} \times \mathbb{Z}/3\mathbb{Z} & \text{provided that } \mathbb{K} = \mathbb{Q}(\sqrt{-3}). \end{cases}$$

All these groups do occur. Furthermore, except for the cases where

$$E(\mathbb{K})_{\text{tors}} \cong \{0\}, \mathbb{Z}/2\mathbb{Z}, \mathbb{Z}/3\mathbb{Z}, \text{ or } \mathbb{Z}/2\mathbb{Z} \times \mathbb{Z}/2\mathbb{Z},$$

the number of isomorphism classes of elliptic curves and the number of the quadratic fields with one of the given torsion structures are both finite.

b) *Let E be an elliptic curve with integral j-invariant over a general cubic number field $\mathbb{K}$. Then, up to isomorphism, the torsion group of E over $\mathbb{K}$ is one of the following groups:*

$$E(\mathbb{K})_{\text{tors}} \cong \begin{cases} \mathbb{Z}/m\mathbb{Z} & \text{for } m = 1, \ldots, 10, 14 \\ \text{or } \mathbb{Z}/2\mathbb{Z} \times \mathbb{Z}/2m\mathbb{Z} & \text{for } m = 1, 2, 3. \end{cases}$$

All these groups do occur. Furthermore, except for the cases where

$$E(\mathbb{K})_{\mathrm{tors}} \cong \{0\}, \mathbb{Z}/2\mathbb{Z}, \mathbb{Z}/3\mathbb{Z}, \mathbb{Z}/4\mathbb{Z}, \mathbb{Z}/5\mathbb{Z}, \text{ or } \mathbb{Z}/2\mathbb{Z} \times \mathbb{Z}/2\mathbb{Z},$$

the number of isomorphism classes of elliptic curves and the number of the cubic fields with one of the given torsion structures are both finite.

In the cyclic cubic case, the torsion group isomorphic to $\mathbb{Z}/5\mathbb{Z}$ *is no exception* (*see Pethő, Zimmer* [163]).

c) *Let E be an elliptic curve with integral j-invariant over a biquadratic number field $\mathbb{K}$. Then, up to isomorphism, the torsion group of E over $\mathbb{K}$ is one of the following groups:*

$$E(\mathbb{K})_{\mathrm{tors}} \cong \begin{cases} \mathbb{Z}/m\mathbb{Z} & \text{for } m = 1, \dots, 8, 10, 14 \\ \text{or } \mathbb{Z}/2\mathbb{Z} \times \mathbb{Z}/2m\mathbb{Z} & \text{for } m = 1, 2, 3 \\ \text{or } \mathbb{Z}/3\mathbb{Z} \times \mathbb{Z}/3m\mathbb{Z} & \text{for } m = 1, 2, \text{provided that } \sqrt{-3} \in \mathbb{K} \\ \text{or } \mathbb{Z}/4\mathbb{Z} \times \mathbb{Z}/4\mathbb{Z} & \text{provided that } \sqrt{-1} \in \mathbb{K}. \end{cases}$$

The corresponding elliptic curves E and biquadratic fields $\mathbb{K}$, as long as they are totally real or totally complex, except for the cases in which again

$$E(\mathbb{K})_{\mathrm{tors}} \cong \{0\}, \mathbb{Z}/2\mathbb{Z}, \mathbb{Z}/3\mathbb{Z}, \mathbb{Z}/4\mathbb{Z}, \mathbb{Z}/5\mathbb{Z}, \text{ or } \mathbb{Z}/2\mathbb{Z} \times \mathbb{Z}/2\mathbb{Z},$$

are each finite in number.

The result of c) can be generalized to multiquadratic fields $\mathbb{K}$ of arbitrary degree. However, for higher degrees, it is difficult to explicitly determine the curves and fields.

T. Vessis [227] was able to find the possible torsion groups of elliptic curves E over Galois fields $\mathbb{K}$ of degree 6, but it turned out to be impossible to determine explicitly the corresponding curves E and fields $\mathbb{K}$ in the finiteness cases because the orders of possible torsion groups were again too high.

Proof. a) See the paper of Müller, Ströher, and Zimmer [149]. The finitely many elliptic curves and the finitely many quadratic fields found by means of the SIMATH package are listed there. Schmitt [189] has computed the exact torsion group for all curves and fields listed in this article.

b) This part is proved in the paper of Pethő, Weis, and Zimmer [162]. The finitely many elliptic curves and the finitely many cubic fields are listed there. We will give an outline of the proof below.

c) This is a combination of the diploma thesis of Hollinger ([100], see also [1]) for totally complex biquadratic number fields, and the diploma thesis of Stein [215] for totally real biquadratic number fields.

The general strategy behind the proofs is standard. It is the following. First the number of torsion points (and hence the possible structure of the torsion group) is estimated using mainly reduction theory (see Theorem 6.30). Then the possible torsion groups are restricted by employing a special parametrization for elliptic curves over number fields. In this way the parametrizations for elliptic curves with special torsion groups are found. In order to determine all elliptic curves with one of the torsion structures exhibited before and to determine the corresponding fields, norm equations have to be solved. These norm equations arise from the corresponding parametrization of the curves. (See also the article of the second author [257].) □

To explain the strategy in more detail, we give an overview of the proof of Part b).

We recall that a model of E over $\mathbb{K}$ or $\mathbb{K}_\mathfrak{p}$ is said to be $\mathfrak{p}$-*minimal* if the coefficients a_i are $\mathfrak{p}$-integral and the $\mathfrak{p}$-value of the discriminant Δ of E is minimal. The assumption about the absolute invariant j of E over $\mathbb{K}$ to be globally integral, i.e. $v_\mathfrak{p}(j) \geq 0$ for all finite places $\mathfrak{p}$ of $\mathbb{K}$, implies that E has *potentially good*, that is, either *good* or *additive* reduction modulo $\mathfrak{p}$. Under this assumption, reduction theory yields a bound for the order of the torsion group $E(\mathbb{K})_{\text{tors}}$ which depends only on the ground field $\mathbb{K}$ but not on the curve E (see Theorem 6.12).

We need the following theorem, which we shall prove in Section 6.5 (see Theorem 6.30).

Theorem 6.12. *Let $\mathbb{K}$ be a number field and $\text{ord}_\mathfrak{p}$ a non-archimedean normalized (additive) valuation of $\mathbb{K}$ with prime ideal or prime divisor $\mathfrak{p}$. The ideal $\mathfrak{p}$ divides the ideal (p) with ramification index $e_{\mathfrak{p}|p}$ and residue degree $f_{\mathfrak{p}|p}$. Let E be an elliptic curve over $\mathbb{K}$ given by a minimal equation at $\mathfrak{p}$. Then one has for the torsion group $E(\mathbb{K})_{\text{tors}}$ the following estimates:*

$$\sharp(E(\mathbb{K})_{\text{tors}}) \textit{ divides } \begin{cases} \sharp(\widetilde{E}(k(\mathfrak{p})))p^{2t} & \textit{at good reduction mod } \mathfrak{p}, \\ |\text{ord}_\mathfrak{p}(j)|(p^{2f_{\mathfrak{p}|p}}-1)p^{2t} & \textit{at multiplicative reduction mod } \mathfrak{p}, \\ \sharp(E(\mathbb{K})/E_0(\mathbb{K}))p^{2+2t} \leq 4p^{2+2t} & \textit{at additive reduction mod } \mathfrak{p}, \end{cases}$$

with

$$t = \begin{cases} 0, & \textit{if } \varphi(p) > e_{\mathfrak{p}|p}, \\ \max\{r \in \mathbb{N} : \varphi(p^r) \leq e_{\mathfrak{p}|p}\}, & \textit{else.} \end{cases}$$

Here $\sharp(\widetilde{E}(k(\mathfrak{p})))$ is the number of points on the reduced curve $\widetilde{E} = E$ modulo $\mathfrak{p}$ over the residue field $k(\mathfrak{p})$, j is the j-invariant of the curve E and $E_0(\mathbb{K})$ is the set of points on $E(\mathbb{K})$, which are nonsingular at reduction modulo $\mathfrak{p}$.

The number t can be strengthened (see Frey [66]).

Corollary 6.13. *Let $\mathbb{K}$ be a cubic field. Denote by $\mathfrak{p}$ a finite place of $\mathbb{K}$ lying above a rational prime p. Let $E|\mathbb{K}$ be an elliptic curve with integral j-invariant over $\mathbb{K}$ and $\widetilde{E}$ the modulo $\mathfrak{p}$ reduced curve over the finite field $k(\mathfrak{p})$. The order of the torsion group $E(\mathbb{K})_{\text{tors}}$ satisfies the following divisibility relations.*

(1) *If E has good reduction modulo $\mathfrak{p} \mid p$,*
   (a) $\#E(\mathbb{K})_{\text{tors}} \mid \#\widetilde{E}(k(\mathfrak{p})) \cdot 2^4$ *in case $p = 2$,*

(b) $\#E(\mathbb{K})_{\text{tors}} \mid \#\widetilde{E}(k(\mathfrak{p})) \cdot 3^2$ *in case* $p = 3$,
(c) $\#E(\mathbb{K})_{\text{tors}} \mid \#\widetilde{E}(k(\mathfrak{p}))$ *in case* $p > 3$.

(2) *If E has additive reduction modulo* $\mathfrak{p} \mid p$,
(a) $\#E(\mathbb{K})_{\text{tors}} \mid 2^8 \cdot 3$ *in case* $p = 2$,
(b) $\#E(\mathbb{K})_{\text{tors}} \mid 2^2 \cdot 3^5$ *in case* $p = 3$,
(c) $\#E(\mathbb{K})_{\text{tors}} \mid 2^2 \cdot 3 \cdot p^2$ *in case* $p > 3$.

Proof. For proving the corollary, one applies Theorem 6.12 and observes that, for cubic fields $\mathbb{K}$, the ramification index and the residue degree are bounded by 3:

$$e_{\mathfrak{p}|p},\ f_{\mathfrak{p}|p} \le 3 = [\mathbb{K} : \mathbb{Q}].$$

This fact leads to the following bounds for the number t defined in Theorem 6.12:

$$\begin{aligned} t &\le 2 \ \text{ if } \ p = 2,\\ t &\le 1 \ \text{ if } \ p = 3,\\ t &= 0 \ \text{ if } \ p \ge 5. \end{aligned}$$

The corollary then follows from Theorem 6.12. □

Remark 6.14. By Corollary 6.13 (1), if E has good reduction at $\mathfrak{p} \mid p$, then Hasse's theorem (Theorem 3.3) implies that, for cubic fields $\mathbb{K}$,

(i) for $p = 2$, $\#E(\mathbb{K})_{\text{tors}} \mid \#\widetilde{E}(k(\mathfrak{p})) \cdot 2^4$, where $\#\widetilde{E}(k(\mathfrak{p})) < 15$,

(ii) for $p = 3$, $\#E(\mathbb{K})_{\text{tors}} \mid \#\widetilde{E}(k(\mathfrak{p})) \cdot 3^2$, where $\#\widetilde{E}(k(\mathfrak{p})) < 39$,

(iii) for $p = 5$, $\#E(\mathbb{K})_{\text{tors}} \mid \#\widetilde{E}(k(\mathfrak{p}))$, where $\#\widetilde{E}(k(\mathfrak{p})) < 149$.

We are now in a position to determine the torsion structures that can occur for elliptic curves with $\mathfrak{p}$-integral j-invariant at the places $\mathfrak{p}$ lying over the primes $p = 2, 3$ or 5 of a cubic field $\mathbb{K}$.

Step 1*:* We first observe that the primes q dividing the order of the torsion group are the following:

$$q \mid \#E(\mathbb{K})_{\text{tors}} \Rightarrow q = 2, 3, 5, 7, 11, 13.$$

This is obvious from Part (i) of Remark 6.14, provided that E has good reduction at $\mathfrak{p}/2$.

Step 2*:* Next we derive upper estimates for the orders of the q-Sylow subgroups $E(\mathbb{K})[q^\infty]$ of $E(\mathbb{K})_{\text{tors}}$ with respect to the primes q which arose in Step 1.

$q = 2$: If E has good reduction at $\mathfrak{p}$ for $\mathfrak{p} \mid 3$, Part (ii) of Remark 6.14 shows that

$$\#E(\mathbb{K})[2^\infty] = 2^{n_2} < 39 \Rightarrow n_2 \le 5.$$

If E has additive reduction at $\mathfrak{p}$ for $\mathfrak{p} \mid 3$, Part (2)(b) of Corollary 6.13 shows that

$$\#E(\mathbb{K})[2^\infty] = 2^{n_2} \mid 2^2 \cdot 3^5 \;\Rightarrow\; n_2 \leq 2.$$

$q = 3$: If E has good reduction at $\mathfrak{p}$ for $\mathfrak{p} \mid 2$, Part (i) of Remark 6.14 yields

$$\#E(\mathbb{K})[3^\infty] = 3^{n_3} \mid \#\widetilde{E}(k(\mathfrak{p})) < 15 \quad\Rightarrow\quad n_3 \leq 2.$$

If E has additive reduction at $\mathfrak{p}$ for $\mathfrak{p} \mid 2$, Part (2)(a) of Corollary 6.13 yields

$$\#E(\mathbb{K})[3^\infty] = 3^{n_3} \mid 2^8 \cdot 3 \;\Rightarrow\; n_3 \leq 1.$$

$q \geq 5$: If $E(\mathbb{K})_{\text{tors}}$ contains a point of order $q \geq 5$, then, by Part (2)(a) of Corollary 6.13, E must have good reduction at $\mathfrak{p}$ for $\mathfrak{p} \mid 2$. Then Part (i) of Remark 6.14 shows that

$$\#E(\mathbb{K})[q^\infty] = q^{n_q} \mid \#\widetilde{E}(k(\mathfrak{p})) < 15 \;\Rightarrow\; n_q \leq 1.$$

In sum we have proved the following lemma. (We may assume w.l.o.g. that E has good reduction at $\mathfrak{p}/2$ provided that E has potentially good reduction at $\mathfrak{p}/2$. The same is true for any other finite place $\mathfrak{p}$.)

Lemma 6.15. *Let E be an elliptic curve over a cubic field $\mathbb{K}$. Suppose that E has $\mathfrak{p}$-integral j-invariant at the places $\mathfrak{p}$ of $\mathbb{K}$ lying over $p = 2$ or 3. Then $E(\mathbb{K})_{\text{tors}}$ has a non-trivial q-Sylow subgroup $E(\mathbb{K})[q^\infty]$ at most for the primes*

$$q \in \{2, 3, 5, 7, 11, 13\},$$

and their order is bounded as follows:

$$\#E(\mathbb{K})[q^\infty] = q^{n_q} \text{ with } n_q \leq \begin{cases} 5 & \text{if } q = 2, \\ 2 & \text{if } q = 3, \\ 1 & \text{if } q \geq 5. \end{cases}$$

However, a closer look at Corollary 6.13 and Remark 6.14 reveals some further restrictions on the torsion group $E(\mathbb{K})_{\text{tors}}$.

Step 3: We show that not all combinations of the q-Sylow subgroups of $E(\mathbb{K})_{\text{tors}}$ found in Step 2 really occur. In fact we shall prove

Lemma 6.16. *Let E be an elliptic curve over a cubic field $\mathbb{K}$. Suppose that E has $\mathfrak{p}$-integral j-invariant for the places $\mathfrak{p}$ of $\mathbb{K}$ lying over the primes $p = 2$ or 3.*

(i) *If $E(\mathbb{K})_{\text{tors}}$ does not contain a point of prime order $q \geq 5$, then $\#E(\mathbb{K})_{\text{tors}} \mid 2^{n_2} \cdot 3^{n_3}$ for $n_2 \leq 5, n_3 \leq 2$ subject to $n_3 \leq 1$ if $n_2 = 5$.*
(ii) *If $E(\mathbb{K})_{\text{tors}}$ contains a point of order $q = 5$ or 7, then $\#E(\mathbb{K})_{\text{tors}} \mid 2^2 \cdot q$.*
(iii) *If $E(\mathbb{K})_{\text{tors}}$ contains a point of order $q = 11$ or 13, then $\#E(\mathbb{K})_{\text{tors}} \mid 2 \cdot q$.*

Proof. Case (i): If E has good reduction at $\mathfrak{p} \mid 5$, we know, by Lemma 6.15 and Part (iii) of Remark 6.14 that

$$\#E(\mathbb{K})_{\text{tors}} = 2^{n_2} \cdot 3^{n_3} < 149 \quad \text{for } n_2 \leq 5, n_3 \leq 2,$$

which rules out the simultaneous occurrence of the exponents $n_2 = 5$ and $n_3 = 2$. If E has additive reduction at $\mathfrak{p} \mid 5$, Lemma 6.15 and Part (2)(c) of Corollary 6.13 show that

$$\#E(\mathbb{K})_{\text{tors}} = 2^{n_2} \cdot 3^{n_3} \mid 2^2 \cdot 3,$$

so that $n_2 \leq 2, n_3 \leq 1$.

Case (ii): Consider the case in which $E(\mathbb{K})_{\text{tors}}$ contains a point of order $q = 5$. Then, by (2)(a), (b) of Corollary 6.13, E must have good reduction at all places $\mathfrak{p} \mid p$ for $p = 2$ and 3. Part (i) of Remark 6.14 then implies that

$$5 \mid \#\widetilde{E}(k(\mathfrak{p})) \quad \text{and} \quad \#E(\mathbb{K})_{\text{tors}} \mid \#\widetilde{E}(k(\mathfrak{p})) \cdot 2^4 \quad \text{for } \#\widetilde{E}(k(\mathfrak{p})) < 15,$$

hence

$$\#E(\mathbb{K})_{\text{tors}} \mid 2^5 \cdot 5,$$

and Part (ii) of Remark 6.14 yields *a fortiori* (see also Zimmer [257], Proposition 3)

$$\#E(\mathbb{K})_{\text{tors}} \mid 2^2 \cdot 5 = 20.$$

In particular, there are no points of prime orders $q = 3, 7, 11$ or 13 in $E(\mathbb{K})_{\text{tors}}$.

The case of $q = 7$ is treated analogously.

Case (iii): As in Case (ii), E must have good reduction at all places $\mathfrak{p} \mid p$ for $p = 2$ and 3 if $E(\mathbb{K})_{\text{tors}}$ contains a point of order $q = 11$ or 13. Part (i) of Remark 6.14 then shows that

$$\#E(\mathbb{K})_{\text{tors}} \mid 2^4 \cdot q$$

and Part (ii) *a fortiori* leads to (see also [257], Proposition 3)

$$\#E(\mathbb{K})_{\text{tors}} \mid 2 \cdot q$$

for $q = 11$ and $q = 13$. This proves Lemma 6.16. □

Now we make use of the fact that, for any positive integer m, the group of m-torsion points of E over the algebraic closure $\overline{\mathbb{K}}$ of $\mathbb{K}$ is of isomorphism type (cf. Theorem 6.2)

$$E[m] \cong \mathbb{Z}/m\mathbb{Z} \times \mathbb{Z}/m\mathbb{Z} \ .$$

From this fact and Lemma 6.16, we derive

Theorem 6.17. *Let E be an elliptic curve over a cubic field $\mathbb{K}$, and suppose that E has $\mathfrak{p}$-integral j-invariant for the finite places $\mathfrak{p}$ of $\mathbb{K}$ lying over $p = 2, 3$ or 5. Then the torsion group of E over $\mathbb{K}$ is at most of one of the following isomorphism types:*

$$E(\mathbb{K})_{\text{tors}} \cong \begin{cases} \mathbb{Z}/m\mathbb{Z} & \textit{for } m = 1, 2, 3, 4, 5, 6, 7, 8, 9, 10, 11, 12, 13, 14 \\ & \quad 16, 18, 20, 22, 24, 26, 28, 32, 36, 48 \\ & \quad 72, 96, 144 \\ \mathbb{Z}/2\mathbb{Z} \times \mathbb{Z}/2m\mathbb{Z} & \textit{for } m = 1, 2, 3, 4, 5, 6, 7, 8, 9, 12, 18, 24, 36 \\ \mathbb{Z}/3\mathbb{Z} \times \mathbb{Z}/3m\mathbb{Z} & \textit{for } m = 1, 2, 4, 8, 16 \\ \mathbb{Z}/4\mathbb{Z} \times \mathbb{Z}/2m\mathbb{Z} & \textit{for } m = 2, 3, 4, 6, 9, 12, 18 \\ \mathbb{Z}/6\mathbb{Z} \times \mathbb{Z}/2m\mathbb{Z} & \textit{for } m = 3, 4, 6, 8, 12 \\ \mathbb{Z}/8\mathbb{Z} \times \mathbb{Z}/3m\mathbb{Z} & \textit{for } m = 4, 6 \\ \mathbb{Z}/12\mathbb{Z} \times \mathbb{Z}/12\mathbb{Z}. & \end{cases}$$

Proof. In the cyclic case, one writes down all positive integers from 1 up to 144 (see Remark 6.14) and removes those numbers which are not in compliance with Lemma 6.16.

In the non-cyclic cases, one observes the above-mentioned facts about m-torsion points and proceeds analogously. □

In order to obtain a complete overview over all torsion groups that really occur for elliptic curves E with (globally) integral j-invariant over cubic fields $\mathbb{K}$, we shall employ a parametrization theorem for torsion groups of low order and deal with the groups of higher order by referring back to those of low order.

As opposed to the weak hypothesis made in Theorem 6.17 about local $\mathfrak{p}$-integrality of the absolute invariante j for the places $\mathfrak{p}$ dividing 2, 3 or 5, we require now the invariant j to be a global integer in $\mathbb{K}$. This condition will lead to constraints on the parameters of the corresponding elliptic curves E over $\mathbb{K}$.

In fact the parametrization theorem to be stated is a generalization of a previous result proved in the article of Müller, Ströher, and Zimmer [149] and in the article of Fung, Ströher, Williams, and Zimmer [74]. It holds for an arbitrary algebraic number field $\mathbb{L}$ in place of the cubic field $\mathbb{K}$. In most cases, we shall use Tate's normal form (see [103]) of an elliptic curve which contains $P = (0, 0)$ as a rational torsion point of maximal order. Nagell [153] was one of the first number theorists who invented parametrizations of a similar type.

But our starting point was the paper [111] of Kubert. We denote by

$\mathcal{O}_{\mathbb{K}}$ the ring of integers of $\mathbb{K}$,
$U_{\mathbb{K}}$ the unit group of $\mathbb{K}$,
$\mathbb{P}_{\mathbb{K}}$ the set of finite places of $\mathbb{K}$,
$\text{ord}_{\mathfrak{p}}$ the normalized additive valuation of $\mathbb{K}$ corresponding to a place $\mathfrak{p}$.

Theorem 6.18. *An elliptic curve E over an algebraic number field $\mathbb{K}$ has torsion group $E(\mathbb{K})_{\text{tors}}$ containing one of the additive abelian groups listed below (up to isomorphism) if and only if E admits one of the following normal forms. Here the point $P = (0,0)$ is always a torsion point of maximal order.*

1. $\mathbb{Z}/2\mathbb{Z}$:
$$E: \ Y^2 = X^3 + a_2X^2 + a_4X \quad (a_2, a_4 \in \mathbb{K}; a_4 \neq 0)$$
On putting
$$c := \frac{a_2^2}{a_4}, \quad b := c - 4, \quad \alpha := 2^4 b$$
we obtain
$$j = \frac{(\alpha + 2^4)^3}{\alpha};$$
and hence
$$j \in \mathcal{O}_{\mathbb{K}} \Leftrightarrow 0 \leq \operatorname{ord}_{\mathfrak{p}}(\alpha) \leq 12\operatorname{ord}_{\mathfrak{p}}(2) \quad \textit{for } \mathfrak{p} \in \mathbb{P}_{\mathbb{K}}.$$

2. $\mathbb{Z}/2\mathbb{Z} \times \mathbb{Z}/2\mathbb{Z}$:
$$E: \ Y^2 = aX(X-1)(X-\lambda) \quad (a, \lambda \in \mathbb{K}; a \neq 0, \lambda \neq 0, 1)$$
On putting
$$\alpha := 2^4 \lambda$$
we obtain
$$j = \frac{(\alpha(\alpha - 2^4) + 2^8)^3}{\alpha^2(\alpha - 2^4)^2};$$
and hence
$$j \in \mathcal{O}_{\mathbb{K}} \Leftrightarrow \left\{ \begin{array}{c} \text{(i) } 0 \leq \operatorname{ord}_{\mathfrak{p}}(\alpha) \leq 8\operatorname{ord}_{\mathfrak{p}}(2) \\ \text{(ii) } 0 \leq \operatorname{ord}_{\mathfrak{p}}(\alpha - 2^4) \leq 8\operatorname{ord}_{\mathfrak{p}}(2) \\ \text{(iii) } 0 \leq \operatorname{ord}_{\mathfrak{p}}(\alpha(\alpha - 2^4)) \leq 12\operatorname{ord}_{\mathfrak{p}}(2) \end{array} \right\} \textit{for } \mathfrak{p} \in \mathbb{P}_{\mathbb{K}}.$$
Here we have $\sqrt{-1} \in \mathbb{K}$.

3. $\mathbb{Z}/4\mathbb{Z}$:
$$E: \ Y^2 + XY - bY = X^3 - bX^2 \quad \left(b \in \mathbb{K}; b \neq 0, -\tfrac{1}{16}\right)$$
On putting
$$\alpha = \frac{1}{b}$$
we obtain
$$j = \frac{(\alpha(\alpha + 2^4) + 2^4)^3}{\alpha(\alpha + 2^4)};$$

and hence

$$j \in \mathcal{O}_{\mathbb{K}} \Leftrightarrow \left\{ \begin{array}{l} \text{(i) } 0 \le \operatorname{ord}_{\mathfrak{p}}(\alpha) \le 8\operatorname{ord}_{\mathfrak{p}}(2) \\ \text{(ii) } 0 \le \operatorname{ord}_{\mathfrak{p}}(\alpha + 2^4) \le 8\operatorname{ord}_{\mathfrak{p}}(2) \\ \text{(iii) } 0 \le \operatorname{ord}_{\mathfrak{p}}(\alpha(\alpha + 2^4)) \le 12\operatorname{ord}_{\mathfrak{p}}(2) \end{array} \right\} \textit{ for } \mathfrak{p} \in \mathbb{P}_{\mathbb{K}}.$$

4. $\mathbb{Z}/8\mathbb{Z}$:

$$E: \quad Y^2 + (1-c)XY - bY = X^3 - bX^2 \quad (b, c \in \mathbb{K}; b \neq 0, c \neq 0)$$

Here b and c are parametrized by $\alpha \in \mathbb{K}$, $\alpha \neq 0, 1, 2, \alpha^2 - 2^3\alpha + 2^3 \neq 0$, according to

$$b = \frac{(\alpha-1)(\alpha-2)}{\alpha^2}, \quad c = \frac{(\alpha-1)(\alpha-2)}{\alpha},$$

so that

$$c = \alpha b\,.$$

We obtain

$$j = \frac{(\alpha^8 - 2^4\alpha^7 + 2^5\cdot 3\alpha^6 - 2^5\cdot 3^2\alpha^5 + 2^5\cdot 3\cdot 5\alpha^4 - 2^6\cdot 7\alpha^3 + 2^5\cdot 7\alpha^2 - 2^6\alpha + 2^4)^3}{\alpha^2(\alpha-2)^4(\alpha-1)^8(\alpha^2 - 2^3\alpha + 2^3)};$$

and hence

$$j \in \mathcal{O}_{\mathbb{K}} \Rightarrow \left\{ \begin{array}{l} \text{(i) } 0 \le \operatorname{ord}_{\mathfrak{p}}(\alpha) \le \frac{5}{2}\operatorname{ord}_{\mathfrak{p}}(2) \\ \text{(ii) } (\alpha - 1) \in U_{\mathbb{K}} \\ \text{(iii) } 0 \le \operatorname{ord}_{\mathfrak{p}}(\alpha - 2) \le 2\operatorname{ord}_{\mathfrak{p}}(2) \\ \text{(iv) } 0 \le \operatorname{ord}_{\mathfrak{p}}(\alpha^2 - 2^3\alpha + 2^3) \le 5\operatorname{ord}_{\mathfrak{p}}(2) \end{array} \right\} \textit{ for } \mathfrak{p} \in \mathbb{P}_{\mathbb{K}}.$$

5. $\mathbb{Z}/2\mathbb{Z} \times \mathbb{Z}/4\mathbb{Z}$:

$$E: \quad Y^2 = X(X^2 + 2(a^2+1)X + (a^2-1)^2) \quad (a \in \mathbb{K}; a \neq 0, \pm 1)$$

We put

$$\alpha := 4a$$

and obtain

$$j = \frac{((\alpha^2 - 2^4)^2 + 2^8\alpha^2)^3}{\alpha^2(\alpha^2 - 2^4)^4};$$

hence

$$j \in \mathcal{O}_{\mathbb{K}} \Leftrightarrow \left\{ \begin{array}{l} \text{(i) } 0 \le \operatorname{ord}_{\mathfrak{p}}(\alpha) \le 4\operatorname{ord}_{\mathfrak{p}}(2) \\ \text{(ii) } 0 \le \operatorname{ord}_{\mathfrak{p}}(\alpha^2 - 2^4) \le 8\operatorname{ord}_{\mathfrak{p}}(2) \end{array} \right\} \textit{ for } \mathfrak{p} \in \mathbb{P}_{\mathbb{K}}.$$

Here we have $\sqrt{-1} \in \mathbb{K}$.

6. $\mathbb{Z}/4\mathbb{Z} \times \mathbb{Z}/4\mathbb{Z}$:

$$E: \quad Y^2 = X(X^2 + 2(a^2+1)X + (a^2-1)^2) \quad (a \in \mathbb{K}; a \neq 0, \pm 1)$$

Here we must have $\sqrt{-1}, \sqrt{a} \in \mathbb{K}$. *On putting again*

$$\alpha := 4a$$

we obtain as before

$$j = \frac{((\alpha^2 - 2^4)^2 + 2^8\alpha^2)^3}{\alpha^2(\alpha^2 - 2^4)^4};$$

and hence as before

$$j \in \mathcal{O}_{\mathbb{K}} \iff \left\{ \begin{array}{l} \text{(i) } 0 \leq \mathrm{ord}_{\mathfrak{p}}(\alpha) \leq 4\mathrm{ord}_{\mathfrak{p}}(2) \\ \text{(ii) } 0 \leq \mathrm{ord}_{\mathfrak{p}}(\alpha^2 - 2^4) \leq 8\mathrm{ord}_{\mathfrak{p}}(2) \end{array} \right\} \text{ for } \mathfrak{p} \in \mathbb{P}_{\mathbb{K}}.$$

7. $\mathbb{Z}/2\mathbb{Z} \times \mathbb{Z}/8\mathbb{Z}$:

$$E: \quad Y^2 + (1-c)XY - bY = X^3 - bX^2 \quad (b, c \in \mathbb{K}; b \neq 0, c \neq 0)$$

Here b and c are parametrized by $\alpha \in \mathbb{K}, \alpha \neq 0, \frac{1}{2\sqrt{2}}, -\frac{1}{2}, -\frac{1}{4}, 2^3\alpha^2 + 2^2\alpha + 1 \neq 0, 2^3\alpha^2 + 2^3\alpha + 1 \neq 0$, *according to*

$$b = \frac{(\alpha(2^3\alpha + 2) + (2\alpha + 1))(2\alpha + 1)}{(2^3\alpha^2 - 1)^2},$$
$$c = \frac{(\alpha(2^3\alpha + 2) + (2\alpha + 1))(2\alpha + 1)}{\alpha(2^3\alpha + 2)(2^3\alpha^2 - 1)}$$

so that

$$c = \frac{2^3\alpha^2 - 1}{\alpha(2^3\alpha + 2)} b.$$

We obtain

$$j = \frac{jz(\alpha)^3}{2^8\,\alpha^8\,(2\alpha+1)^8(2^2\alpha+1)^8(2^3\alpha^2-1)^2(2^3\alpha^2+2^2\alpha+1)^4(2^3\alpha^2+2^3\alpha+1)^2}$$

for

$$\begin{aligned} jz(\alpha) = {} & 2^{24} \cdot \alpha^{16} + 2^{26} \cdot \alpha^{15} + 2^{24} \cdot 7 \cdot (\alpha^{14} + \alpha^{13}) + 2^{20} \cdot 3 \cdot 23 \cdot \alpha^{12} \\ & + 2^{20} \cdot 5^2 \cdot \alpha^{11} + 2^{17} \cdot 5^2 \cdot \alpha^{10} - 2^{16} \cdot 29 \cdot \alpha^9 - 2^{12} \cdot 17^2 \cdot \alpha^8 \\ & - 2^{13} \cdot 29 \cdot \alpha^7 + 2^{11} \cdot 5^2 \cdot \alpha^6 + 2^{11} \cdot 5^2 \cdot \alpha^5 + 2^8 \cdot 3 \cdot 23 \cdot \alpha^4 \\ & + 2^9 \cdot 7 \cdot \alpha^3 + 2^6 \cdot 7 \cdot \alpha^2 + 2^5 \cdot \alpha + 1; \end{aligned}$$

hence

$$j \in \mathcal{O}_{\mathbb{K}} \Leftrightarrow \left\{ \begin{array}{r} \text{(i)}\ -2\mathrm{ord}_{\mathfrak{p}}(2) \le \mathrm{ord}_{\mathfrak{p}}(\alpha) \le 0 \\ \text{(ii)}\ -\mathrm{ord}_{\mathfrak{p}}(2) \le \mathrm{ord}_{\mathfrak{p}}(2\alpha + 1) \le \mathrm{ord}_{\mathfrak{p}}(2) \\ \text{(iii)}\ -\mathrm{ord}_{\mathfrak{p}}(2) \le \mathrm{ord}_{\mathfrak{p}}(2^2\alpha + 1) = 0 \\ \text{(iv)}\ -\mathrm{ord}_{\mathfrak{p}}(2) \le \mathrm{ord}_{\mathfrak{p}}(2^3\alpha^2 - 1) \le \frac{7}{2}\mathrm{ord}_{\mathfrak{p}}(2) \\ \text{(v)}\ -\mathrm{ord}_{\mathfrak{p}}(2) \le \mathrm{ord}_{\mathfrak{p}}(2^3\alpha^2 + 2^3\alpha + 1) \le \frac{7}{2}\mathrm{ord}_{\mathfrak{p}}(2) \\ \text{(vi)}\ -\mathrm{ord}_{\mathfrak{p}}(2) \le \mathrm{ord}_{\mathfrak{p}}(2^3\alpha^2 + 2^2\alpha + 1) \le 3\mathrm{ord}_{\mathfrak{p}}(2) \end{array} \right\}$$

for $\mathfrak{p} \in \mathbb{P}_{\mathbb{K}}$. *Again we must have* $\sqrt{-1} \in \mathbb{K}$.

8. $\mathbb{Z}/3\mathbb{Z}$:

$$E: \ Y^2 + cXY + c^2Y = X^3 \quad (c \in \mathbb{K}, c \neq 0)$$

For

$$\alpha := c - 3^3 \neq 0$$

we obtain

$$j = \frac{(\alpha + 3)^3(\alpha + 3^3)}{\alpha};$$

and hence

$$j \in \mathcal{O}_{\mathbb{K}} \Leftrightarrow 0 \le \mathrm{ord}_{\mathfrak{p}}(\alpha) \le 6\mathrm{ord}_{\mathfrak{p}}(3) \quad \text{for } \mathfrak{p} \in \mathbb{P}_{\mathbb{K}}.$$

9. $\mathbb{Z}/9\mathbb{Z}$:

$$E: \ Y^2 + (1 - c)XY - bY = X^3 - bX^2 \quad (b, c \in \mathbb{K}; b \neq 0, c \neq 0)$$

Here b *and* c *are parametrized by* $\alpha \in \mathbb{K}$, $\alpha \neq 0, 1$, *according to*

$$b = \alpha^2(\alpha - 1)(\alpha^2 - \alpha + 1), \quad c = \alpha^2(\alpha - 1)$$

so that

$$c = \frac{b}{\alpha^2 - \alpha + 1},$$

and we obtain

$$j = \frac{jz(\alpha)^3}{\alpha^9(\alpha - 1)^9(\alpha^2 - \alpha + 1)^3(\alpha^3 - 2 \cdot 3\alpha^2 + 3\alpha + 1)}$$

for

$$\begin{aligned} jz(\alpha) = (&\alpha^9 - 3^2 \cdot \alpha^8 + 3^3 \cdot \alpha^7 - 2^4 \cdot 3 \cdot \alpha^6 + 2 \cdot 3^3 \cdot \alpha^5 - 3^2 \cdot 5 \cdot \alpha^4 \\ &+ 3^3 \cdot \alpha^3 - 3^2 \cdot \alpha^2 + 1) \cdot (\alpha^3 - 3 \cdot \alpha^2 + 1) \end{aligned}$$

hence

$$j \in \mathcal{O}_{\mathbb{K}} \Rightarrow \begin{cases} \text{(i) } \alpha \in U_{\mathbb{K}} \\ \text{(ii) } (\alpha - 1) \in U_{\mathbb{K}} \\ \left.\begin{array}{l} \text{(iii) } 0 \le \operatorname{ord}_{\mathfrak{p}}(\alpha^2 - \alpha + 1) \le 2\operatorname{ord}_{\mathfrak{p}}(3) \\ \text{(iv) } 0 \le \operatorname{ord}_{\mathfrak{p}}(\alpha^3 - 2\cdot 3\alpha^2 + 3\alpha + 1) \le 6\operatorname{ord}_{\mathfrak{p}}(3) \end{array}\right\} \textit{for } \mathfrak{p} \in \mathbb{P}_{\mathbb{K}}. \end{cases}$$

10. $\mathbb{Z}/3\mathbb{Z} \times \mathbb{Z}/3\mathbb{Z}$:

$$E: \quad Y^2 = X^3 + AX + B \quad (A, B \in \mathbb{K})$$

Here we must have $\sqrt{-3} \in \mathbb{K}$, *and* A *and* B *are parametrized by* $\alpha \in \mathbb{K}$, $\alpha \neq 3, -\frac{3}{2}(1 \pm \sqrt{-3})$, *according to*

$$\begin{aligned} A &= -\frac{1}{3^3}\alpha(\alpha + 2\cdot 3)(\alpha^2 - 2\cdot 3\alpha + 2^2\cdot 3^2), \\ B &= -\frac{2}{3^6}(\alpha^2 - 2\cdot 3\alpha - 2\cdot 3^2)(\alpha^4 + 2\cdot 3\alpha^3 + 2\cdot 3^3\alpha^2 - 2^2\cdot 3^3\alpha + 2^2\cdot 3^4). \end{aligned}$$

We obtain

$$j = \left(\frac{\alpha((\alpha - 3)(\alpha^2 + 3\alpha + 3^2) + 3^5)}{(\alpha - 3)(\alpha^2 + 3\alpha + 3^2)}\right)^3;$$

and hence

$$j \in \mathcal{O}_{\mathbb{K}} \Leftrightarrow \begin{cases} \text{(i) } \alpha \in \mathcal{O}_{\mathbb{K}} \\ \left.\begin{array}{l} \text{(ii) } 0 \le \operatorname{ord}_{\mathfrak{p}}(\alpha - 3) \le 3\operatorname{ord}_{\mathfrak{p}}(3) \\ \text{(iii) } 0 \le \operatorname{ord}_{\mathfrak{p}}(\alpha^3 - 3^3) \le 6\operatorname{ord}_{\mathfrak{p}}(3) \end{array}\right\} \textit{for } \mathfrak{p} \in \mathbb{P}_{\mathbb{K}}. \end{cases}$$

11. $\mathbb{Z}/6\mathbb{Z}$:

$$E: \quad Y^2 + (1 - c)XY - c(1 - c)Y = X^3 - c(1 - c)X^2$$

$$\left(c \in \mathbb{K};\, c \neq 0, -1, -\tfrac{1}{3^2}\right)$$

We put

$$\alpha := \frac{1}{c}$$

and obtain

$$j = \frac{((\alpha + 1)^3(\alpha + 3^2) - 2^4\alpha)^3}{\alpha^2(\alpha + 1)^3(\alpha + 3^2)};$$

hence

$$j \in \mathcal{O}_{\mathbb{K}} \Leftrightarrow \left\{ \begin{array}{ll} \text{(i) } 0 \leq \operatorname{ord}_{\mathfrak{p}}(\alpha) \leq 2\operatorname{ord}_{\mathfrak{p}}(3) & \\ \text{(ii) } 0 \leq \operatorname{ord}_{\mathfrak{p}}(\alpha+1) \leq 3\operatorname{ord}_{\mathfrak{p}}(2) & \\ \text{(iii) (a) } \operatorname{ord}_{\mathfrak{p}}(\alpha+3^2) = 0 & \textit{if } \mathfrak{p} \nmid 6 \\ \quad\text{(b) } \operatorname{ord}_{\mathfrak{p}}(\alpha+3^2) = \operatorname{ord}_{\mathfrak{p}}(\alpha+1) & \textit{if } \mathfrak{p} \mid 2 \\ \quad\text{(c) } \operatorname{ord}_{\mathfrak{p}}(\alpha+3^2) = \operatorname{ord}_{\mathfrak{p}}(\alpha) & \textit{if } \mathfrak{p} \mid 3 \end{array} \right\} \textit{ for } \mathfrak{p} \in \mathbb{P}_{\mathbb{K}}.$$

12. $\mathbb{Z}/12\mathbb{Z}$*:*

$$E: \quad Y^2 + (1-c)XY - bY = X^3 - bX^2 \quad (b, c \in \mathbb{K}; b \neq 0)$$

Here b, c *are parametrized by* α *according to*

$$b = -\frac{\alpha^5 - 7\alpha^4 + 3\cdot 7\alpha^3 - 2\cdot 17\alpha^2 + 2\cdot 3\cdot 5\alpha - 2^2\cdot 3}{\alpha^2(\alpha-1)^4},$$

$$c = -\frac{\alpha^3 - 5\alpha^2 + 3^2\alpha - 2\cdot 3}{\alpha(\alpha-1)^3}$$

so that

$$b = \frac{\alpha^2 - 2\alpha + 2}{\alpha(\alpha-1)} c$$

We obtain

$$j = \frac{jz(\alpha)^3}{\alpha^2(\alpha-1)^{12}(\alpha-2)^6(\alpha^2-2\alpha+2)^3(\alpha^2-3\alpha+3)^4(\alpha^2-6\alpha+6)}$$

for

$$\begin{aligned} jz(\alpha) = (&\alpha^{12} - 2\cdot 3^2\alpha^{11} + 2^4\cdot 3^2\alpha^{10} - 2^2\cdot 3^2\cdot 19\alpha^9 + 2\cdot 3\cdot 359\alpha^8 \\ &- 2^3\cdot 3\cdot 197\alpha^7 + 2^3\cdot 3\cdot 307\alpha^6 - 2^4\cdot 3\cdot 13^2\alpha^5 \\ &+ 2^2\cdot 3\cdot 7\cdot 73\alpha^4 - 2^3\cdot 3\cdot 5^3\alpha^3 + 2^5\cdot 3^3\alpha^2 \\ &- 2^4\cdot 3^2\alpha + 2^3\cdot 3)(\alpha^4 - 2\cdot 3\alpha^3 \\ &+ 2^2\cdot 3\alpha^2 - 2^2\cdot 3\alpha + 2\cdot 3); \end{aligned}$$

hence

$$j \in \mathcal{O}_{\mathbb{K}} \Rightarrow \left\{\begin{array}{ll} \text{(i) } 0 \leq \mathrm{ord}_{\mathfrak{p}}(\alpha) \leq \mathrm{ord}_{\mathfrak{p}}(2) + \frac{1}{2}\mathrm{ord}_{\mathfrak{p}}(3) & \\ \text{(ii) } (\alpha - 1) \in U_{\mathbb{K}} & \\ \text{(iii) (a) } \mathrm{ord}_{\mathfrak{p}}(\alpha - 2) = 0 & \textit{if } \mathfrak{p} \nmid 2 \\ \qquad \text{(b) } \mathrm{ord}_{\mathfrak{p}}(\alpha - 2) = \mathrm{ord}_{\mathfrak{p}}(\alpha) & \textit{if } \mathfrak{p} \mid 2 \\ \text{(iv) } 0 \leq \mathrm{ord}_{\mathfrak{p}}(\alpha^2 - 2\alpha + 2) \leq 4\mathrm{ord}_{\mathfrak{p}}(2) & \\ \text{(v) } 0 \leq \mathrm{ord}_{\mathfrak{p}}(\alpha^2 - 3\alpha + 3) \leq \frac{3}{2}\mathrm{ord}_{\mathfrak{p}}(3) & \\ \text{(vi) } 0 \leq \mathrm{ord}_{\mathfrak{p}}(\alpha^2 - 2\cdot 3\alpha + 2\cdot 3) \leq 12\mathrm{ord}_{\mathfrak{p}}(2) + 2\cdot 3\mathrm{ord}_{\mathfrak{p}}(3) & \end{array}\right\}$$

for $\mathfrak{p} \in \mathbb{P}_{\mathbb{K}}$.

13. $\mathbb{Z}/2\mathbb{Z} \times \mathbb{Z}/6\mathbb{Z}$:

$$E: \quad Y^2 + (1-c)XY - bY = X^3 - bX^2 \quad \left(b, c \in \mathbb{K};\, b \neq 0,\, c \neq 0, -1, -\tfrac{1}{3^2}\right)$$

Here b and c are parametrized by $\alpha \in \mathbb{K}$, $\alpha \neq 1, \pm 3, 3^2, 5$, *according to*

$$b = -2\frac{(\alpha-1)^2(\alpha-5)}{(\alpha-3)^2(\alpha+3)^2}, \quad c = -2\frac{\alpha-5}{(\alpha-3)(\alpha+3)}$$

so that

$$b = \frac{(\alpha-1)^2}{(\alpha-3)(\alpha+3)}c.$$

The j-invariant is

$$j = \frac{j_Z(\alpha)^3}{2^6\,(\alpha-1)^6(\alpha+3)^2(\alpha-3)^2(\alpha-5)^6(\alpha-3^2)^2}$$

for

$$j_Z(\alpha) = (\alpha^2 - 2\cdot 3\alpha + 3\cdot 7)\cdot(\alpha^6 - 2\cdot 3^2\alpha^5 + 3\cdot 5^2\alpha^4 + 2^2\cdot 3^2\cdot 5\alpha^3 - 3\cdot 5^2\cdot 11\alpha^2 - 2\cdot 3^2\cdot 11^2\alpha + 3\cdot 2287);$$

hence

$$j \in \mathcal{O}_{\mathbb{K}} \Leftrightarrow \left\{\begin{array}{l} \text{(i) } \mathrm{ord}_{\mathfrak{p}}(2) \leq \mathrm{ord}_{\mathfrak{p}}(\alpha - 1) \leq 3\mathrm{ord}_{\mathfrak{p}}(2) \\ \text{(ii) } \mathrm{ord}_{\mathfrak{p}}(2) \leq \mathrm{ord}_{\mathfrak{p}}(\alpha + 3) \leq 3\mathrm{ord}_{\mathfrak{p}}(2) + \mathrm{ord}_{\mathfrak{p}}(3) \\ \text{(iii) } \mathrm{ord}_{\mathfrak{p}}(2) \leq \mathrm{ord}_{\mathfrak{p}}(\alpha - 3) \leq \mathrm{ord}_{\mathfrak{p}}(2) + \mathrm{ord}_{\mathfrak{p}}(3) \\ \text{(iv) } \mathrm{ord}_{\mathfrak{p}}(2) \leq \mathrm{ord}_{\mathfrak{p}}(\alpha - 3^2) \leq 3\mathrm{ord}_{\mathfrak{p}}(2) + \mathrm{ord}_{\mathfrak{p}}(3) \\ \text{(v) } \mathrm{ord}_{\mathfrak{p}}(2) \leq \mathrm{ord}_{\mathfrak{p}}(\alpha - 5) \leq 3\mathrm{ord}_{\mathfrak{p}}(2) \end{array}\right\}$$

for $\mathfrak{p} \in \mathbb{P}_{\mathbb{K}}$. *Here* $\sqrt{-1} \in \mathbb{K}$.

14. $\mathbb{Z}/5\mathbb{Z}$:

$$E: \quad Y^2 + (1-b)XY - bY = X^3 - bX^2 \quad (b \in \mathbb{K}; b \neq 0)$$

We put

$$\alpha := b$$

and obtain

$$j = \frac{((\alpha^2 - 11\alpha - 1)^2 + 5\alpha(2(\alpha^2 - 11\alpha - 1) + \alpha))^3}{\alpha^5(\alpha^2 - 11\alpha - 1)};$$

hence

$$j \in \mathcal{O}_{\mathbb{K}} \Leftrightarrow \left\{ \begin{array}{ll} \text{(i) } \alpha \in U_{\mathbb{K}} & \\ \text{(ii) } 0 \leq \mathrm{ord}_{\mathfrak{p}}(\alpha^2 - 11\alpha - 1) \leq 3\mathrm{ord}_{\mathfrak{p}}(5) & \\ \text{(iii) } \mathrm{ord}_{\mathfrak{p}}(\alpha + 1) = 0 & \textit{if } \mathfrak{p} \mid 11 \end{array} \right\}$$

for $\mathfrak{p} \in \mathbb{P}_{\mathbb{K}}$.

15. $\mathbb{Z}/10\mathbb{Z}$:

$$E: \quad Y^2 + (1-c)XY - bY = X^3 - bX^2 \quad (b, c \in \mathbb{K}; b \neq 0)$$

Here b and c are parametrized by $\alpha \in \mathbb{K}$, $\alpha \neq 0, 1, 2, \frac{1}{2}(3 \pm \sqrt{5})$, according to

$$b = \frac{(\alpha-1)(\alpha-2)}{\alpha((\alpha-1)^2 - \alpha)^2}, \qquad c = \frac{(\alpha-1)(\alpha-2)}{\alpha((\alpha-1)^2 - \alpha)}$$

so that

$$b = \frac{1}{(\alpha-1)^2 - \alpha} c.$$

We obtain

$$j = \frac{jz(\alpha)^3}{\alpha^5(\alpha-1)^{10}(\alpha-2)^5((\alpha-1)^2-\alpha)^2(\alpha^2+2(\alpha-2))}$$

for

$$jz(\alpha) = \alpha^{12} - 2^3\alpha^{11} + 2^4\alpha^{10} + 2^3 \cdot 5\alpha^9 - 2^4 \cdot 3 \cdot 5\alpha^8 + 2^4 \cdot 3^3\alpha^7 - 2^8\alpha^6 \\ - 2^5 \cdot 3^2\alpha^5 + 2^4 \cdot 3^2 \cdot 5\alpha^4 - 2^4 \cdot 3^2 \cdot 5\alpha^3 + 2^5 \cdot 13\alpha^2 - 2^7\alpha + 2^4.$$

We obtain

$$j \in \mathcal{O}_{\mathbb{K}} \Rightarrow \left\{ \begin{array}{l} \text{(i) } 0 \leq \mathrm{ord}_{\mathfrak{p}}(\alpha) \leq \mathrm{ord}_{\mathfrak{p}}(2) \\ \text{(ii) } (\alpha - 1) \in U_{\mathbb{K}} \\ \text{(iii) } 0 \leq \mathrm{ord}_{\mathfrak{p}}(\alpha - 2) \leq \mathrm{ord}_{\mathfrak{p}}(2) \\ \text{(iv) } 0 \leq \mathrm{ord}_{\mathfrak{p}}((\alpha-1)^2 - \alpha) \leq \frac{3}{2}\mathrm{ord}_{\mathfrak{p}}(5) \\ \text{(v) } 0 \leq \mathrm{ord}_{\mathfrak{p}}(\alpha^2 + 2(\alpha - 2)) \leq 12\mathrm{ord}_{\mathfrak{p}}(2) + 3\mathrm{ord}_{\mathfrak{p}}(5) \end{array} \right\}$$

for $\mathfrak{p} \in \mathbb{P}_{\mathbb{K}}$.

16. $\mathbb{Z}/7\mathbb{Z}$:

$$E: \quad Y^2 + (1-c)XY - bY = X^3 - bX^2 \quad (b, c \in \mathbb{K}; b \neq 0)$$

Here b and c are parametrized by $\alpha \in \mathbb{K}$, $\alpha \neq 0, 1$, *according to*

$$b = \alpha^2(\alpha - 1), \quad c = \alpha(\alpha - 1)$$

so that

$$b = \alpha c.$$

We obtain

$$\begin{aligned} j &= \frac{(\alpha^8 - 2^2 \cdot 3\alpha^7 + 2 \cdot 3 \cdot 7\alpha^6 - 2^3 \cdot 7\alpha^5 + 5 \cdot 7\alpha^4 - 2 \cdot 7\alpha^2 + 2^2\alpha + 1)^3}{\alpha^7(\alpha - 1)^7(\alpha^3 - 2^3\alpha^2 + 5\alpha + 1)} \\ &= \frac{(\alpha^2 - \alpha + 1)^3(\alpha^6 - 11\alpha^5 + 2 \cdot 3 \cdot 5\alpha^4 - 3 \cdot 5\alpha^3 - 2 \cdot 5\alpha^2 + 5\alpha + 1)^3}{\alpha^7(\alpha - 1)^7(\alpha^3 - 2^3\alpha^2 + 5\alpha + 1)}; \end{aligned}$$

hence

$$j \in \mathcal{O}_{\mathbb{K}} \Rightarrow \left\{ \begin{array}{l} \text{(i)}\ \alpha \in U_{\mathbb{K}} \\ \text{(ii)}\ (\alpha - 1) \in U_{\mathbb{K}} \\ \text{(iii)}\ 0 \leq \mathrm{ord}_{\mathfrak{p}}(\alpha^3 - 2^3\alpha^2 + 5\alpha + 1) \leq 2\mathrm{ord}_{\mathfrak{p}}(7) \end{array} \right\} \text{ for } \mathfrak{p} \in \mathbb{P}_{\mathbb{K}}.$$

17. $\mathbb{Z}/11\mathbb{Z}$:

$$E: \quad Y^2 + (1-c)XY - bY = X^3 - bX^2 \quad (b, c \in \mathbb{K}; b \neq 0)$$

Here we have two parameters $\alpha, \beta \in \mathbb{K}$, $\alpha \neq 0$, $\beta \neq \pm 4$ *by which b and c are given according to*

$$\begin{aligned} b &= \tfrac{1}{2^7\alpha}(2\alpha + (\beta - 2^2))(\beta - 2^2)(\beta + 2^2), \\ c &= \tfrac{1}{2^4\alpha}(2\alpha + (\beta - 2^2))(\beta - 2^2), \end{aligned}$$

so that

$$b = \tfrac{1}{2^3}(\beta + 2^2)c.$$

In addition, the parameters constitute a rational point (α, β) *on the elliptic curve*

$$X_1(11): \quad V^2 = U^3 - 2^2U^2 + 2^4.$$

The j-invariant is

$$j = \frac{(((1-c)^2 - 2^2b)^2 + 2^3 \cdot 3b(1-c))^3}{b^3(((1-c)^2 - 2^2b)^2 + 2^3(1-c)^3 - 3^3b - 3^2(1-c)((1-c)^2 - 2^2b))}.$$

The integrality condition on j leads to

$$j \in \mathcal{O}_{\mathbb{K}} \Rightarrow \left\{ \begin{array}{l} \text{(i) } 0 \leq \operatorname{ord}_{\mathfrak{p}}(\alpha) \leq 2\operatorname{ord}_{\mathfrak{p}}(2) \\ \text{(ii) } 0 \leq \operatorname{ord}_{\mathfrak{p}}(\alpha - 2^2) \leq 2\operatorname{ord}_{\mathfrak{p}}(2) \\ \text{(iii) } 0 \leq \operatorname{ord}_{\mathfrak{p}}(\beta - 2^2) = \frac{3}{2}\operatorname{ord}_{\mathfrak{p}}(\alpha) \\ \text{(iv) } 0 \leq \operatorname{ord}_{\mathfrak{p}}(\beta + 2^2) = \frac{3}{2}\operatorname{ord}_{\mathfrak{p}}(\alpha) \\ \text{(v) } 0 \leq \operatorname{ord}_{\mathfrak{p}}(2\alpha + (\beta - 2^2)) = \frac{3}{2}\operatorname{ord}_{\mathfrak{p}}(\alpha) \end{array} \right\} \text{ for } \mathfrak{p} \in \mathbb{P}_{\mathbb{K}}.$$

18. $\mathbb{Z}/13\mathbb{Z}$*:*

$$E: \quad Y^2 + (1-c)XY - bY = X^3 - bX^2 \quad (b, c \in \mathbb{K}; b \neq 0)$$

Again b, c can be expressed in terms of two parameters $\alpha, \beta \in \mathbb{K}, \beta \neq 0$, according to

$$b = \frac{\alpha^2(\alpha-1)}{\beta^2} \cdot \frac{(\alpha^2(\alpha-1)+\beta)(\alpha(\alpha^2-1)+\beta)}{\alpha(\alpha-1)+\beta},$$

$$c = \frac{\alpha^2(\alpha-1)}{\beta} \cdot \frac{\alpha(\alpha^2-1)+\beta}{\alpha(\alpha-1)+\beta}$$

so that

$$b = \frac{\alpha^2(\alpha-1)+\beta}{\beta} c.$$

Similar to Case 17*, the parameters constitute a rational point (α, β) on the elliptic curve*

$$X_1(13): \quad V^2 + (U^3 - U^2 - 1)V = U^2 - U.$$

The j-invariant is the same as in Case 17 *and hence we have here*

$$j \in \mathcal{O}_{\mathbb{K}} \Rightarrow \left\{ \begin{array}{ll} \text{(i) } \alpha & \in U_{\mathbb{K}} \\ \text{(ii) } (\alpha - 1) & \in U_{\mathbb{K}} \\ \text{(iii) } \beta & \in U_{\mathbb{K}} \\ \text{(iv) } (\beta - 1) & \in U_{\mathbb{K}} \\ \text{(v) } (\alpha - \beta) & \in U_{\mathbb{K}} \\ \text{(vi) } ((\alpha - 1) + \beta) & \in U_{\mathbb{K}} \\ \text{(vii) } (\alpha(\alpha - 1) + \beta) & \in U_{\mathbb{K}} \\ \text{(viii) } (\alpha(\alpha^2 - 1) + \beta) & \in U_{\mathbb{K}} \\ \text{(ix) } (\alpha^2(\alpha - 1) + \beta) & \in U_{\mathbb{K}} \\ \text{(x) } (\alpha^2(\alpha - 1) + (\beta - 1)) & \in U_{\mathbb{K}} \end{array} \right\}.$$

Proof. The proofs of the most cases are essentially straightforward, and we refer to the articles of Müller, Ströher, and Zimmer [149], Fung, Ströher, Williams, and Zimmer [74] and the diploma thesis of Stein [215] for the details (see also the paper of Reichert [174]). The rather tedious proof of Case 18 is given in the diploma thesis of Weis [236]. We confine ourselves to carrying out the proof of Case 9 only.

We have $j = \frac{c_4^3}{\Delta}$ with

$$c_4 = (\alpha^9 - 3^2\alpha^8 + 3^3\alpha^7 - 2^4 \cdot 3\alpha^6 + 2 \cdot 3^3\alpha^5 - 3^2 \cdot 5\alpha^4 + 3^3\alpha^3 - 3^2\alpha^2 + 1) \\ \cdot (\alpha^3 - 3\alpha^2 + 1)$$

and

$$\Delta = \alpha^9(\alpha - 1)^9(\alpha^2 - \alpha + 1)^3(\alpha^3 - 2 \cdot 3\alpha^2 + 3\alpha + 1).$$

Suppose there exists a prime $\mathfrak{p} \in \mathbb{P}_{\mathbb{K}}$ with $\operatorname{ord}_{\mathfrak{p}}(\alpha) > 0$. It then follows easily that $\operatorname{ord}_{\mathfrak{p}}(\Delta) = 9\operatorname{ord}_{\mathfrak{p}}(\alpha) > 0$ and that $\operatorname{ord}_{\mathfrak{p}}(c_4) = 0$, hence $\operatorname{ord}_{\mathfrak{p}}(j) < 0$, which is a contradiction, as j has to be integral. If $\operatorname{ord}_{\mathfrak{p}}(\alpha) < 0$ for a prime $\mathfrak{p}$, then

$$\operatorname{ord}_{\mathfrak{p}}(j) = 36\operatorname{ord}_{\mathfrak{p}}(\alpha) - 27\operatorname{ord}_{\mathfrak{p}}(\alpha) = 9\operatorname{ord}_{\mathfrak{p}}(\alpha) < 0,$$

which is also a contradiction. It follows that $\operatorname{ord}_{\mathfrak{p}}(\alpha) = 0$ for all primes $\mathfrak{p} \in \mathbb{P}_{\mathbb{K}}$, hence $\alpha \in U_{\mathbb{K}}$ and Condition (i) of Case 9 is proved.

From this we see that $\operatorname{ord}_{\mathfrak{p}}(\alpha - 1) \geq 0$ for all primes $\mathfrak{p} \in \mathbb{P}_{\mathbb{K}}$. Suppose now that there exists a prime $\mathfrak{p} \in \mathbb{P}_{\mathbb{K}}$ with $\operatorname{ord}_{\mathfrak{p}}(\alpha - 1) > 0$. In this case one also easily derives that $\operatorname{ord}_{\mathfrak{p}}(j) < 0$. Hence Condition (ii) of Case 9 is proved.

Let us now prove (iii). Consider the polynomials

$$\begin{aligned} p(X) &= X^3 - 3X^2 + 1, \\ q(X) &= X^9 - 3^2X^8 + 3^3X^7 - 2^4 \cdot 3X^6 + 2 \cdot 3^3X^5 - 3^2 \cdot 5X^4 + 3^3X^3 - 3^2X^2 + 1, \\ r(X) &= X^2 - X + 1, \end{aligned}$$

which – for $X = \alpha$ – occur in the numerator or denominator of the j-invariant. They satisfy the equation

$$p(X)q(X)s(X) + r(X)t(X) = 3^2 \tag{6.1}$$

involving the polynomials

$$\begin{aligned} s(X) &= -X, \\ t(X) &= X^{11} - 11X^{10} + 2 \cdot 3 \cdot 7X^9 - 3 \cdot 5^2X^8 + 2^3 \cdot 3^2X^7 \\ &\quad - 3 \cdot 11X^6 + 3^2X^5 + 2 \cdot 3X^4 - 3 \cdot 7X^3 + X^2 + 2 \cdot 5X + 3^2. \end{aligned}$$

The hypothesis

$$\operatorname{ord}_{\mathfrak{p}}(j) \geq 0$$

implies the inequality

$$\mathrm{ord}_{\mathfrak{p}}(p(\alpha)) + \mathrm{ord}_{\mathfrak{p}}(q(\alpha)) \geq \mathrm{ord}_{\mathfrak{p}}(r(\alpha)) \geq 0 \tag{6.2}$$

since (i), (ii) of Case 9 tell us that $\alpha, \alpha - 1 \in U_{\mathbb{K}}$, and we know therefore that $\mathrm{ord}_{\mathfrak{p}}(\alpha^3 - 2 \cdot 3\alpha^2 + 3\alpha + 1) \geq 0$. Suppose now that

$$\mathrm{ord}_{\mathfrak{p}}(r(\alpha)) > 2\mathrm{ord}_{\mathfrak{p}}(3).$$

Then, also $\mathrm{ord}_{\mathfrak{p}}(p(\alpha)q(\alpha)) > 2\mathrm{ord}_{\mathfrak{p}}(3)$ by (6.2) and (6.1) – for $X = \alpha$ – leads to a contradiction. This shows that Condition (iii) of Case 9 must hold.

The last Condition (iv) of Case 9 is proved analogously. □

The integrality conditions for the parameters α and β in Theorem 6.18 can be transformed into norm equations as follows. The advantage of these norm equations is that they can be solved in the rational field $\mathbb{Q}$ instead of the number field $\mathbb{K}$. To explain this, let

$$p \cong \mathfrak{p}_1^{e_1} \ldots \mathfrak{p}_s^{e_s}$$

be the decomposition of a prime p of $\mathbb{Q}$ into primes $\mathfrak{p}_i$ of $\mathbb{K}$, where $e_i = e_{\mathfrak{p}_i}$ is the ramification index of $\mathfrak{p}_i$ over p. The residue degree $f_i = f_{\mathfrak{p}_i}$ is fixed by the norm equation

$$\mathcal{N}(\mathfrak{p}_i) = p^{f_i},$$

where $\mathcal{N} = \mathcal{N}_{\mathbb{K}|\mathbb{Q}}$ denotes the divisor norm for $\mathbb{K}|\mathbb{Q}$. We have the well known relation

$$\sum_{i=1}^{s} e_i f_i = n$$

with the field degree $n = [\mathbb{K} : \mathbb{Q}]$. In addition, for numbers in $\mathbb{Q}$,

$$\mathrm{ord}_{\mathfrak{p}_i} = e_i \mathrm{ord}_p.$$

The norm equations come from the relation

$$\mathrm{ord}_p(\mathcal{N}(\alpha)) = \sum_{i=1}^{s} f_i \mathrm{ord}_{\mathfrak{p}_i}(\alpha)$$

for a non-zero number $\alpha \in \mathbb{K}$, where w.l.o.g. $\mathcal{N} = \mathcal{N}_{\mathbb{K}|\mathbb{Q}}$ also denotes the element norm for $\mathbb{K}|\mathbb{Q}$ (see e.g. Hasse [92]).

Case 1 of Theorem 6.18 states, for instance,

$$j \in \mathcal{O}_{\mathbb{K}} \Leftrightarrow 0 \leq \mathrm{ord}_{\mathfrak{p}}(\alpha) \leq 12\mathrm{ord}_{\mathfrak{p}}(2) \quad \text{for } \mathfrak{p} \in \mathbb{P}_{\mathbb{K}}.$$

This condition yields

$$\begin{aligned} 0 &\leq \operatorname{ord}_p(\mathcal{N}(\alpha)) \\ &= \sum_{i=1}^{s} f_i \operatorname{ord}_{\mathfrak{p}_i}(\alpha) \\ &\leq 12 \sum_{i=1}^{s} f_i e_i \operatorname{ord}_p(2) \\ &= 12n\operatorname{ord}_p(2); \end{aligned}$$

hence the norm equation

$$\mathcal{N}(\alpha) = \pm 2^m \qquad \text{with } 0 \leq m \leq 12n,\ m \in \mathbb{Z}.$$

Of course, such norm equations can be solved only for number field $\mathbb{K}$ of small degree n over $\mathbb{Q}$.

Theorem 6.19. *We shall use the abbreviation $\mathcal{N} = \mathcal{N}_{\mathbb{K}/\mathbb{Q}}$. The condition $j \in \mathcal{O}_{\mathbb{K}}$ implies the norm equations listed below.*

1. $\mathbb{Z}/2\mathbb{Z}$:

$$\mathcal{N}(\alpha) = \pm 2^m \ ,\ 0 \leq m \leq 12n \ .$$

2. $\mathbb{Z}/2\mathbb{Z} \times \mathbb{Z}/2\mathbb{Z}$:

$$\begin{aligned} &\text{(i)} & \mathcal{N}(\alpha) &= \pm 2^l, & 0 &\leq l \leq 8n, \\ &\text{(ii)} & \mathcal{N}(\alpha - 2^4) &= \pm 2^l, & 0 &\leq l \leq 8n, \\ &\text{(iii)} & \mathcal{N}(\alpha(\alpha - 2^4)) &= \pm 2^m, & 0 &\leq m \leq 12n. \end{aligned}$$

3. $\mathbb{Z}/4\mathbb{Z}$:

$$\begin{aligned} &\text{(i)} & \mathcal{N}(\alpha) &= \pm 2^l, & 0 &\leq l \leq 8n, \\ &\text{(ii)} & \mathcal{N}(\alpha + 2^4) &= \pm 2^l, & 0 &\leq l \leq 8n, \\ &\text{(iii)} & \mathcal{N}(\alpha(\alpha + 2^4)) &= \pm 2^m, & 0 &\leq m \leq 12n. \end{aligned}$$

4. $\mathbb{Z}/8\mathbb{Z}$:

$$\begin{aligned} &\text{(i)} & \mathcal{N}(\alpha) &= \pm 2^l, & 0 &\leq l \leq [\tfrac{5}{2}n], \\ &\text{(ii)} & \mathcal{N}(\alpha - 1) &= \pm 1, & & \\ &\text{(iii)} & \mathcal{N}(\alpha - 2) &= \pm 2^m, & 0 &\leq m \leq 2n, \\ &\text{(iv)} & \mathcal{N}(\alpha^2 - 2^3\alpha + 2^3) &= \pm 2^\mu, & 0 &\leq \mu \leq 5n. \end{aligned}$$

5. $\mathbb{Z}/2\mathbb{Z} \times \mathbb{Z}/4\mathbb{Z}$, *or* 6. $\mathbb{Z}/4\mathbb{Z} \times \mathbb{Z}/4\mathbb{Z}$:

$$\begin{array}{lrl} \text{(i)} & \mathcal{N}(\alpha) = \pm 2^l, & 0 \le l \le 4n, \\ \text{(ii)} & \mathcal{N}(\alpha^2 - 2^4) = \pm 2^m, & 0 \le m \le 8n. \end{array}$$

In fact, a more detailed analysis shows that

$$\begin{array}{lrl} \text{(i)} & \mathcal{N}(\alpha) = \pm 2^l, & 0 \le l \le 4n, \\ \text{(ii)} & \mathcal{N}(\alpha \pm 2^2) = \pm 2^m, & 0 \le m \le 5n. \end{array}$$

7. $\mathbb{Z}/2\mathbb{Z} \times \mathbb{Z}/8\mathbb{Z}$: *For* $\beta = 4\alpha$ *we get*

$$\begin{array}{lrl} \text{(i)} & \mathcal{N}(\beta) = \pm 2^l, & 0 \le l \le 2n, \\ \text{(ii)} & \mathcal{N}(\beta + 1) = \pm 1, & \\ \text{(iii)} & \mathcal{N}(\beta + 2) = \pm 2^l, & 0 \le l \le 2n, \\ \text{(iv)} & \mathcal{N}(\beta^2 - 2) = \pm 2^m, & 0 \le m \le [\frac{9}{2}n], \\ \text{(v)} & \mathcal{N}(\beta^2 + 2^2\beta + 2) = \pm 2^m, & 0 \le m \le [\frac{9}{2}n], \\ \text{(vi)} & \mathcal{N}(\beta^2 + 2\beta + 2) = \pm 2^k, & 0 \le k \le 4n. \end{array}$$

8. $\mathbb{Z}/3\mathbb{Z}$:

$$\mathcal{N}(\alpha) = \pm 3^m, \quad 0 \le m \le 6n.$$

9. $\mathbb{Z}/9\mathbb{Z}$:

$$\begin{array}{lrl} \text{(i)} & \mathcal{N}(\alpha) = \pm 1, & \\ \text{(ii)} & \mathcal{N}(\alpha - 1) = \pm 1, & \\ \text{(iii)} & \mathcal{N}(\alpha^2 - \alpha + 1) = \pm 3^l, & 0 \le l \le 2n, \\ \text{(iv)} & \mathcal{N}(\alpha^3 - 2 \cdot 3\alpha^2 + 3\alpha + 1) = \pm 3^m, & 0 \le m \le 6n. \end{array}$$

10. $\mathbb{Z}/3\mathbb{Z} \times \mathbb{Z}/3\mathbb{Z}$:

$$\begin{array}{lrl} \text{(i)} & \mathcal{N}(\alpha) \in \mathbb{Z}, & \\ \text{(ii)} & \mathcal{N}(\alpha - 3) = \pm 3^l, & 0 \le l \le 3n, \\ \text{(iii)} & \mathcal{N}(\alpha^3 - 3^3) = \pm 3^m, & 0 \le m \le 6n. \end{array}$$

11. $\mathbb{Z}/6\mathbb{Z}$:

$$\begin{array}{lll} \text{(i)} & \mathcal{N}(\alpha) = \pm 3^l, & 0 \le l \le 2n, \\ \text{(ii)} & \mathcal{N}(\alpha + 1) = \pm 2^m, & 0 \le m \le 3n, \\ \text{(iii)} & \mathcal{N}(\alpha + 3^2) = \pm 2^m 3^l, & 0 \le m \le 3n, 0 \le l \le 2n. \end{array}$$

12. $\mathbb{Z}/12\mathbb{Z}$:

$$\begin{array}{lll} \text{(i)} & \mathcal{N}(\alpha) = \pm 2^l 3^m, & 0 \le l \le n, 0 \le m \le [\frac{n}{2}], \\ \text{(ii)} & \mathcal{N}(\alpha - 1) = \pm 1, & \\ \text{(iii)} & \mathcal{N}(\alpha - 2) = \pm 2^l, & 0 \le l \le n, \\ \text{(iv)} & \mathcal{N}(\alpha^2 - 2\alpha + 2) = \pm 2^k, & 0 \le k \le 4n, \\ \text{(v)} & \mathcal{N}(\alpha^2 - 3\alpha + 3) = \pm 3^h, & 0 \le h \le [\frac{3}{2}n], \\ \text{(vi)} & \mathcal{N}(\alpha^2 - 2 \cdot 3\alpha + 2 \cdot 3) = \pm 2^k 3^h, & 0 \le k \le 12n, 0 \le h \le 6n. \end{array}$$

13. $\mathbb{Z}/2\mathbb{Z} \times \mathbb{Z}/6\mathbb{Z}$: *For* $\beta = \frac{\alpha-1}{2}$ *we get*

$$\begin{array}{lll} \text{(i)} & \mathcal{N}(\beta) = \pm 2^l, & 0 \le l \le 2n, \\ \text{(ii)} & \mathcal{N}(\beta + 2) = \pm 2^l 3^m, & 0 \le l \le 2n, 0 \le m \le n, \\ \text{(iii)} & \mathcal{N}(\beta - 1) = \pm 3^m, & 0 \le m \le n, \\ \text{(iv)} & \mathcal{N}(\beta - 2) = \pm 2^l, & 0 \le l \le 2n, \\ \text{(v)} & \mathcal{N}(\beta - 2^2) = \pm 2^l 3^m, & 0 \le l \le 2n, 0 \le m \le n. \end{array}$$

14. $\mathbb{Z}/5\mathbb{Z}$:

$$\begin{array}{lll} \text{(i)} & \mathcal{N}(\alpha) = \pm 1, & \\ \text{(ii)} & \mathcal{N}(\alpha^2 - 11\alpha - 1) = \pm 5^m, & 0 \le m \le 3n. \end{array}$$

15. $\mathbb{Z}/10\mathbb{Z}$:

$$\begin{array}{lll} \text{(i)} & \mathcal{N}(\alpha) = \pm 2^l, & 0 \le l \le n, \\ \text{(ii)} & \mathcal{N}(\alpha - 1) = \pm 1, & \\ \text{(iii)} & \mathcal{N}(\alpha - 2) = \pm 2^l, & 0 \le l \le n\ , \\ \text{(iv)} & \mathcal{N}((\alpha - 1)^2 - \alpha) = \pm 5^m, & 0 \le m \le [\frac{3}{2}n], \\ \text{(v)} & \mathcal{N}(\alpha^2 + 2(\alpha - 2)) = \pm 2^l 5^m, & 0 \le l \le 12n, 0 \le m \le 3n. \end{array}$$

16. $\mathbb{Z}/7\mathbb{Z}$:

$$\begin{aligned}
&\text{(i)} && \mathcal{N}(\alpha) = \pm 1,\\
&\text{(ii)} && \mathcal{N}(\alpha - 1) = \pm 1,\\
&\text{(iii)} && \mathcal{N}(\alpha^3 - 2^3\alpha^2 + 5\alpha + 1) = \pm 7^l, \quad 0 \le l \le 2n.
\end{aligned}$$

17. $\mathbb{Z}/11\mathbb{Z}$: *On replacing* β *by* $\beta + 4$, *we get the norm equations*

$$\begin{aligned}
&\text{(i)} && \mathcal{N}(\alpha) = \pm 2^l, && 0 \le l \le 2n,\\
&\text{(ii)} && \mathcal{N}(\alpha - 2^2) = \pm 2^l, && 0 \le l \le 2n,\\
&\text{(iii)} && \mathcal{N}(\beta) = \pm 2^m, && 0 \le m \le 3n,\\
&\text{(iv)} && \mathcal{N}(\beta + 2^3) = \pm 2^m, && 0 \le m \le 3n,\\
&\text{(v)} && \mathcal{N}(2\alpha + \beta) = \pm 2^m, && 0 \le m \le 3n.
\end{aligned}$$

In addition, the (*new*) *parameters satisfy the equation*

$$X_1(11): \ \alpha^3 - 2^2\alpha^2 = \beta^2 + 2^3\beta.$$

18. $\mathbb{Z}/13\mathbb{Z}$:

$$\begin{aligned}
&\text{(i)} && \mathcal{N}(\alpha) = \pm 1,\\
&\text{(ii)} && \mathcal{N}(\alpha - 1) = \pm 1,\\
&\text{(iii)} && \mathcal{N}(\beta) = \pm 1,\\
&\text{(iv)} && \mathcal{N}(\beta - 1) = \pm 1,\\
&\text{(v)} && \mathcal{N}(\alpha - \beta) = \pm 1,\\
&\text{(vi)} && \mathcal{N}((\alpha - 1) + \beta) = \pm 1,\\
&\text{(vii)} && \mathcal{N}(\alpha(\alpha - 1) + \beta) = \pm 1\ ,\\
&\text{(viii)} && \mathcal{N}(\alpha(\alpha^2 - 1) + \beta) = \pm 1,\\
&\text{(ix)} && \mathcal{N}(\alpha^2(\alpha - 1) + \beta) = \pm 1,\\
&\text{(x)} && \mathcal{N}(\alpha^2(\alpha - 1) + (\beta - 1)) = \pm 1.
\end{aligned}$$

In addition, the parameters satisfy the equation

$$X_1(13): \ \beta^2 + (\alpha^3 - \alpha^2 - 1)\beta = \alpha^2 - \alpha\ .$$

These norm equations are not so easy to solve (see the paper of Pethő, Weis, and Zimmer [162]). Let, for instance, $\mathbb{K}$ be a cubic number field so that $n = 3$, and consider the norm equations occurring in Theorem 6.19

$$\begin{aligned}
\mathcal{N}(\alpha) &= n_1, \ (n_1 \in \mathbb{Z}),\\
\mathcal{N}(g(\alpha)) &= n_2, \ (n_2 \in \mathbb{Z}),
\end{aligned} \tag{6.3}$$

where $g(X)$ is a quadratic polynomial over $\mathbb{Q}$:

$$g(X) := X^2 + cX + d, \quad (c, d \in \mathbb{Z})$$

of discriminant

$$D := c^2 - 4d \in \mathbb{Z}.$$

There are two possibilities.

Possibility 1. *$g(X)$ is reducible.* Then we obtain from (6.3) the three norm equations

$$\begin{aligned} \mathcal{N}(\alpha) &= n_1 \quad (n_1 \in \mathbb{Z}), \\ \mathcal{N}(\alpha - a) &= n_2 \quad (n_2 \in \mathbb{Z}), \\ \mathcal{N}(\alpha - b) &= n_3 \quad (n_3 \in \mathbb{Z}), \end{aligned} \tag{6.4}$$

where $g(X) = (X-a)(X-b)$, $(a, b \in \mathbb{Z})$. We may suppose that $a \neq b$ and $ab \neq 0$, since otherwise (6.4) reduces to a system of two norm equations.

We bring in Gröbner basis theory (see Becker, Weispfennig [13]). To this end, the element $\mathcal{N}(\alpha)$ is written as a product of three indeterminates X, Y, Z:

$$\mathcal{N}(\alpha) = XYZ.$$

The system (6.4) then gives rise to an ideal

$$\mathfrak{a} \trianglelefteq \mathbb{Q}[X, Y, Z]$$

with basis

$$\mathcal{B} = \{XYZ - n_1, (X-a)(Y-a)(Z-a) - n_2, (X-b)(Y-b)(Z-b) - n_3\}.$$

This ideal $\mathfrak{a}$ has a Gröbner basis

$$\mathcal{C} = \{ab(a-b)((X+Y+Z) - (a+b+m_1)), G(Y,Z), ab(a-b)H(Z)\}$$

with polynomials

$$G(Y, Z) \in \mathbb{Z}[Y, Z] \text{ of degree } 2 \text{ in } Y,$$
$$H(Z) = Z^3 - (a+b+m_1)Z^2 + (ab+m_2)Z - n_1 \in \mathbb{Z}[Z]$$

and numbers

$$\begin{aligned} m_1 &:= \frac{n_1(a-b) + bn_2 - an_3}{ab(a-b)}, \\ m_2 &:= \frac{n_1(a^2-b^2) + b^2 n_2 - a^2 n_3}{ab(a-b)}. \end{aligned}$$

We restrict to cubic fields $\mathbb{K}$, that is, to the case $n = 3$.

Proposition 6.20. *If the norm equations* (6.4) *have a solution*

$$\alpha \in \mathcal{O}_{\mathbb{K}} \backslash \mathbb{Z}$$

in a cubic number field $\mathbb{K}$, *then* $m_1, m_2 \in \mathbb{Z}$ *and* $H(\alpha) = 0$.

If conversely

$$m_1, m_2 \in \mathbb{Z} \text{ and } H(\alpha) = 0 \text{ for } H(Z) \in \mathbb{Z}[Z] \text{ irreducible},$$

then $\mathbb{K} = \mathbb{Q}(\alpha)$ *is the unique cubic field in which the norm equations* (6.4) *have the solution* α.

Possibility 2. *$g(X)$ is irreducible.* Then we are left with the norm equations (6.3). We add to (6.3) the trace equation

$$\mathrm{Tr}(\alpha) = t \quad (t \in \mathbb{Z}).$$

In this case the ideal

$$\mathfrak{a} \trianglelefteq \mathbb{Q}[X, Y, Z]$$

has the basis

$$\mathcal{B} = \{XYZ - n_1, g(X)g(Y)g(Z) - n_2, X + Y + Z - t\}.$$

Its Gröbner basis is

$$\mathcal{C} = \{H(Z), n_1 dY^2 + G(Y, Z), X + Y + Z - t\}$$

with polynomials

$$G(Y, Z) \in \mathbb{Z}[Y, Z] \quad \text{of degree 1 in } Y,$$

$$\begin{aligned} H(Z) = {} & dZ^6 - 2tdZ^5 + (2d^2 + (t^2 - tc - c^2)d - n_1 c)Z^4 \\ & - (2td^2 - (t^2 c + tc^2 - 2n_1)d - tn_1 c)Z^3 \\ & - (2n_1 d^2 - (tn_1 c + n_1 c^2)d - n_1^2 c)Z + n_1^2 d \in \mathbb{Z}[Z]. \end{aligned}$$

The analogue of Proposition 6.20 is (see Pethő, Weis, Zimmer [162])

Proposition 6.21. *If the norm equations* (6.3) *have a solution*

$$\alpha \in \mathcal{O}_{\mathbb{K}} \backslash \mathbb{Z}$$

in a cubic number field $\mathbb{K}$, *then there are integers* $l, t, t_1 \in \mathbb{Z}$ *such that*

$$4dn_2 = l^2 - D((t + c)d - n_1)^2, \tag{6.5}$$

$$t_1 = \frac{d - c((t + c)d + n_1) + l}{2d}, \tag{6.6}$$

and α has the minimal polynomial

$$H_1(Z) = Z^3 - tZ^2 + t_1 Z - n_1 \in \mathbb{Z}[Z],$$

so that $H_1(\alpha) = 0$.

If conversely there are integers $l, t, t_1 \in \mathbb{Z}$ satisfying (6.5), (6.6) *and if $H_1(Z)$ is the minimal polynomial of an algebraic number α:*

$$H_1(\alpha) = 0,$$

then $\mathbb{K} = \mathbb{Q}(\alpha)$ is the unique cubic field in which the norm equations (6.3) *have the solution α.*

It follows from Proposition 6.21 (see Pethő, Weis, and Zimmer [162]) that in case $D < 0$ or $D > 0$ but $\mathbb{K}$ has negative discriminant, there are only finitely many cubic number fields $\mathbb{K}$ in which the norm equations (6.3) have a solution $\alpha \in \mathbb{K}$. However, in case $D > 0$ and $D \in \mathbb{Z}$ is not a square, the norm equations (6.3) admit an infinity of cubic number fields $\mathbb{K}$ such that

$$\alpha \in \mathbb{K}$$

is a solution of (6.3) provided that there is at least one such solution.

An example of this situation is Case 14 of Theorems 6.18, 6.19, where the torsion group

$$E(\mathbb{K})_{\text{tors}} \cong \mathbb{Z}/5\mathbb{Z}$$

is treated. For cubic fields $\mathbb{K}$, the norm equations (6.3) read

$$\begin{aligned} \mathcal{N}(\alpha) &= \pm 1, \\ \mathcal{N}(\alpha^2 - 11\alpha - 1) &= \pm 5^m, \quad 0 \le m \le 9. \end{aligned} \tag{6.7}$$

This case requires some extra endeavor as the subsequent proposition (see Pethő, Weis, Zimmer [162]) shows.

Proposition 6.22. *An integer*

$$\alpha \in \mathcal{O}_{\mathbb{K}} \backslash \mathbb{Z}$$

with minimal polynomial $H(Z) \in \mathbb{Z}[Z]$ and polynomial discriminant D_α is a solution of the norm equations (6.7) in a cyclic cubic number field

$$\mathbb{K} = \mathbb{Q}(\alpha)$$

of discriminant $D_{\mathbb{K}}$ in exactly the following eight cases:

m	$H(Z)$	D_α	$D_{\mathbb{K}}$
0	$Z^3 - 12Z^2 + 9Z + 1$	$(3^2 \cdot 13)^2$	$(3^2 \cdot 13)^2$
0	$Z^3 - 13Z^2 + 10Z + 1$	139^2	139^2
2	$Z^3 - 14Z^2 + 11Z + 1$	163^2	163^2
3	$Z^3 - 9Z^2 + 6Z + 1$	$(3^2 \cdot 7)^2$	$(3^2 \cdot 7)^2$
4	$Z^3 - 12Z^2 + 35Z + 1$	$(5 \cdot 13)^2$	13^2
4	$Z^3 + 3Z^2 - 160Z + 1$	$(5^2 \cdot 163)^2$	163^2
5	$Z^3 + 3Z^2 - 10Z + 1$	$(5 \cdot 13)^2$	13^2
5	$Z^3 - 17Z^2 - 25Z + 1$	$(2^3 \cdot 5 \cdot 13)^2$	13^2

This way all the norm equations can be solved if the degree of $\mathbb{K}$,

$$n = [\mathbb{K} : \mathbb{Q}],$$

is sufficiently small. Then, there exist only finitely many elliptic curves E (up to isomorphisms) with integral j-invariant over a finite set of number fields $\mathbb{K}$ such that the Mordell–Weil group $E(\mathbb{K})$ contains a certain given torsion group $E(\mathbb{K})_{\text{tors}}$ (aside from groups of very small order) (see Abel-Hollinger and Zimmer [1], [100], Müller, Ströher, Zimmer [149], Pethő, Weis, Zimmer [162], and the theses of Stein [215] and Weis [236]).

6.3 The theorem of Nagell, Lutz, and Cassels

We formulate the theorem of Nagell, Lutz, and Cassels in an extended version according to Folz [64], using results of the second author. Let, in the case of Theorem 6.23, $(\mathfrak{a})$ denote the ideal of $\mathbb{K}$ corresponding to the divisor $\mathfrak{a}$ of $\mathbb{K}$ (see Hasse [92]).

Theorem 6.23. *Let $P = (x, y)$ be a torsion point of order $m > 1$ on the elliptic curve E over an algebraic number field $\mathbb{K}$. Further let φ be the Euler φ-function and*

$$\mathfrak{m} = \prod_{\mathfrak{p} \in M_{\mathbb{K}}} \mathfrak{p}^{\mu_{\mathfrak{p}}}$$

the coefficient divisor of $E|\mathbb{K}$, where the exponents $\lambda_{\mathfrak{p}}$ are defined as in Definition 4.8*:*

$$\mu_{\mathfrak{p}} := \min\left\{\operatorname{ord}_{\mathfrak{p}}(b_2), \frac{1}{2}\operatorname{ord}_{\mathfrak{p}}(b_4), \frac{1}{3}\operatorname{ord}_{\mathfrak{p}}(b_6), \frac{1}{4}\operatorname{ord}_{\mathfrak{p}}(b_8)\right\},$$

where $\operatorname{ord}_{\mathfrak{p}}$ (as always) is the normalized absolute value corresponding to $\mathfrak{p}$.

We use the division polynomial $\psi_2(x, y) = 2y + a_1x + a_3$ with

$$\psi_2^2(x) = 4x^3 + b_2x^2 + 2b_4x + b_6.$$

a) $x \in (\mathfrak{n})$ *and* $\psi_2^2(x) \in (\mathfrak{n}^3)$ *with*

$$\mathfrak{n} = \begin{cases} \mathfrak{m} & \text{if } m \neq p^n, \\ \mathfrak{m} p^{-\frac{2}{\varphi(p^n)}} & \text{if } m = p^n, \end{cases}$$

where $p \in \mathbb{P}, n \in \mathbb{N}$.

b) $\psi_2(x) = 0$ *if* $m = 2$, *else* $\psi_2^{-2}(x) \in (\Delta^{-1}\mathfrak{m}^2\mathfrak{n})$ *with*

$$\mathfrak{n} = \begin{cases} \mathfrak{m} & \text{if } m \neq 2p^n, \\ \mathfrak{m} p^{-\frac{2}{\varphi(p^n)}} & \text{if } m = 2p^n, \end{cases}$$

where $p \in \mathbb{P}, n \in \mathbb{N}$.

Proof. This is the global version of the local Theorem 4.13, hence a combination of Theorem 4.13 for all valuations of $\mathbb{K}$. For $\mathfrak{p} \mid p$, we use $\mathrm{ord}_{\mathfrak{p}}(p) = e_{\mathfrak{p}}$ with $p \in \mathbb{P}$. □

Corollary 6.24. *Let* $E|\mathbb{K}$ *be an elliptic curve in long Weierstraß normal form with integral coefficients over the number field* $\mathbb{K}$. *Let* $P = (x, y) \in E(\mathbb{K})_{\mathrm{tors}}$ *be a point of exact order* $m \geq 2$.

a) *If* m *is not a prime power, then* $x, y \in \mathcal{O}_{\mathbb{K}} = (1)$.

b) *If* $m = p^n$ *with* $p \in \mathbb{P}, n \in \mathbb{N}$, *we define for every prime divisor* $\mathfrak{p}$ *of* $\mathbb{K}$:

$$r_{\mathfrak{p}} = \left\lfloor \frac{\mathrm{ord}_{\mathfrak{p}}(p)}{\varphi(p^n)} \right\rfloor.$$

(*writing this time* $\lfloor\ \rfloor$ *for the greatest integer function*). *Then one has*

$$x \in (\mathfrak{r}^{-2}), \quad y \in (\mathfrak{r}^{-3})$$

for the divisor

$$\mathfrak{r} = \prod_{\mathfrak{p} \in M_{\mathbb{K}}^0} \mathfrak{p}^{r_{\mathfrak{p}}},$$

that is,

$$\mathrm{ord}_{\mathfrak{p}}(x) \geq -2r_{\mathfrak{p}}, \quad \mathrm{ord}_{\mathfrak{p}}(y) \geq -3r_{\mathfrak{p}}$$

for all prime divisors $\mathfrak{p}$ *of* $\mathbb{K}$ *with* $\mathfrak{p} \in M_{\mathbb{K}}^0$.

Proof. This is a corollary of the above theorem, since in the case that the coefficients of the equation are integral we have $\mu_{\mathfrak{p}} \geq 0$ for $\mathfrak{p} \in M_{\mathbb{K}}^0$. If m is not a prime power, the integrality of x follows directly. Using the original equation of the elliptic curve, we get the integrality of y. □

For the short Weierstraß form, Theorem 6.23 reads as follows. For an elliptic curve

$$E : Y^2 = X^3 + a_4 X + a_6 \quad (a_4, a_6 \in \mathbb{K})$$

over an algebraic number field $\mathbb{K}$ and each finite place $\mathfrak{p}$ of $\mathbb{K}$, we put

$$\mu_{\mathfrak{p}} := \min\left\{\frac{1}{2}\mathrm{ord}_{\mathfrak{p}}(a_4), \frac{1}{3}\mathrm{ord}_{\mathfrak{p}}(a_6)\right\}$$

and define the coefficient "divisor" of $E|\mathbb{K}$ by setting

$$\mathfrak{m} = \prod_{\mathfrak{p}} \mathfrak{p}^{\mu_{\mathfrak{p}}},$$

the product being taken over all finite places of $\mathbb{K}$. Of course, the 6-th power of $\mathfrak{m}$ is a proper divisor with respect to $\mathbb{K}$. Then, for a torsion point $P = (x, y)$ of order $m > 1$ of $E|\mathbb{K}$, we obtain the following result (see Zimmer [251]) in which the coordinates x, y of the point P are elements of $\mathbb{K}$, but $\mathfrak{m}$ and $\mathfrak{n}$ are not necessary proper divisors of $\mathbb{K}$. However, $\mathfrak{m}$ and $\mathfrak{n}$ are proper divisors with respect to a finite extension field of $\mathbb{K}$.

Theorem 6.25. *For a torsion point $P = (x, y) \in E(\mathbb{K})$ of order $m > 1$,*

$$x \in (\mathfrak{n}), \quad y^2 \in (\mathfrak{n}^3),$$

where

$$\mathfrak{n} := \begin{cases} \mathfrak{m} & \text{if } m \neq p^n, \\ \mathfrak{m}p^{-\frac{2}{\varphi(p^n)}} & \text{if } m = p^n, \end{cases}$$

with $p \in \mathbb{P}, n \in \mathbb{N}$, and $y = 0$ or

$$y^{-2} \in (\Delta_0^{-1}\mathfrak{m}^2\mathfrak{n}),$$

where

$$\mathfrak{n} = \begin{cases} \mathfrak{m} & \text{if } m \neq 2p^n, \\ \mathfrak{m}p^{-\frac{2}{\varphi(p^n)}} & \text{if } m = 2p^n, \end{cases}$$

with $p \in \mathbb{P}, n \in \mathbb{N}$. Here, $\Delta_0 = 4a_4^3 + 27a_6^2$, where $\Delta = -16\Delta_0$ is the discriminant of the curve $E|\mathbb{K}$.

Proof. The proof is similar to that of Theorem 6.23. For this theorem is simply Theorem 6.23 specialized to the short Weierstraß form. □

Example. We consider the elliptic curve

$$E : Y^2 = X^3 + 4(1 - 2i)X$$

over the Gaussian number field $\mathbb{K} = \mathbb{Q}(i)$ with $i := \sqrt{-1}$. Its discriminant is

$$\Delta = -16\Delta_0 = -16(4^4(-11 + 2i)).$$

The curve has torsion group (see Schmitt [189])

$$E(\mathbb{K})_{\text{tors}} = \langle P\rangle \cong \mathbb{Z}/10\mathbb{Z}$$

with the point

$$P = (x, y) = (2(2 + i), 4(2 + i))$$

of order 10 (so that $5P = (0, 0)$ has order 2). The coefficient divisor of $E|\mathbb{K}$ is

$$\mathfrak{m} \cong 2\sqrt{1 - 2i}.$$

Since $1 - 2i = -i(2 + i)$ and hence

$$x = 2(2 + i) \mid 2^2(2 + i) \cong \mathfrak{m}^2$$

we obtain right away (cf. the remark concerning the coefficient divisor $\mathfrak{m}$ of $E|\mathbb{K}$)

$$x \in (\mathfrak{m}).$$

Furthermore,

$$y^2 = 2^4(2 + i)^2 \mid 2^4(2 + i)^2 \cong \mathfrak{m}^4$$

so that $\mathfrak{m}^3 \cong 2^3(2 + i)\sqrt{2 + i}$ shows (cf. the above remark about $\mathfrak{m}$)

$$y^2 \in (\mathfrak{m}^3).$$

This corroborates the first part of Theorem 6.25.

Of course, once $x = 2(2 + i)$ is known, $y = 2^2(2 + i)$ can be read off from the Weierstraß equation of the curve E.

For $p = 5$, we have

$$p^{\frac{-2}{p-1}} = \frac{1}{\sqrt{p}} = \frac{1}{\sqrt{5}}.$$

Thus,

$$\begin{aligned}\Delta_0\mathfrak{m}^{-3}p^{\frac{2}{\varphi(p)}} &\cong 2^5\frac{-11 + 2i}{(1 - 2i)\sqrt{1 - 2i}}\sqrt{5}\\ &= -2^5(3 + 4i)\sqrt{1 + 2i}\\ &= 2^5(2 + i)^2\sqrt{1 + 2i}\end{aligned}$$

because in $\mathbb{K}$ $(2 + i)^2 = 3 + 4i$.

Hence,

$$y^2 = 2^4(2 + i)^2 \mid 2^5(2 + i)^2\sqrt{1 + 2i} \cong \Delta_0\mathfrak{m}^{-3}\sqrt{5}.$$

This implies

$$y^{-2} \in \left(\Delta_0^{-1}\mathfrak{m}^3\sqrt{5}^{-1}\right)$$

as claimed. Theorem 6.25 is therefore completely verified.

The example shows that Theorem 6.25 is somewhat stronger than the classical Theorem of Nagell, Lutz, Cassels (see Cassels [25], [27]). The same holds true also for Theorem 6.23. On the other hand, for the field $\mathbb{K} = \mathbb{Q}$ of rationals as basic field for the elliptic curve, the classical theorem of Nagell and Lutz is stronger than Theorems 6.23 and 6.25 in that it does not contain any prime numbers p.

Theorem 6.26 (Nagell, Lutz). *Let $E|\mathbb{Q}$ be an elliptic curve in short Weierstraß normal form*

$$E : Y^2 = X^3 + a_4X + a_6$$

with integral coefficients $a_4, a_6 \in \mathbb{Z}$. Let $\mathcal{O} \neq P = (x, y) \in E(\mathbb{Q})_{\text{tors}}$. Then

a) $x, y \in \mathbb{Z}$.

b) *Either $2P = \mathcal{O}$, or y^2 divides $\Delta_0 = \frac{-\Delta}{16} = 4a_4^3 + 27a_6^2$.*

Proof. Let $P = (x, y) \in E(\mathbb{Q})$ be a point of exact order $m > 1$.

a) If $m = 2$, then $y = 0$ and x is a rational root of the polynomial

$$X^3 + a_4X + a_6$$

with coefficients in $\mathbb{Z}$, therefore $x \in \mathbb{Z}$. Suppose that $m > 2$. If m is not a prime power, the assertion follows directly from Part a) of Corollary 6.24. If $m = p^r$ is a prime power, then $p^r > 2$ and

$$\varphi(p^r) = p^r - p^{r-1} \geq 2.$$

Therefore, for every prime number $q \in \mathbb{P}$:

$$0 \leq r_q = \left\lfloor \frac{\operatorname{ord}_q(p)}{p^r - p^{r-1}} \right\rfloor \leq \left\lfloor \frac{1}{2} \right\rfloor = 0.$$

With Corollary 6.24 we get

$$\operatorname{ord}_q(x) \geq -2r_q = 0 \quad \text{and} \quad \operatorname{ord}_q(y) \geq -3r_q = 0,$$

hence also $x \in (\mathfrak{r}^{-2})$ and $y \in (\mathfrak{r}^{-3})$. Because $\mathfrak{r} = (1)$, the numbers x and y are integral.

b) We assume that $2P \neq \mathcal{O}$, therefore $y \neq 0$. Then $2P = (x', y') \in E(\mathbb{K})$ is also a torsion point. From Part a) it follows that $x, y, x', y' \in \mathbb{Z}$.

One has

$$x' = \frac{\phi_2(x)}{\psi_2^2(x)}$$

with the multiplication polynomials

$$\phi_2(X) = X^4 - 2a_4X^2 - 8a_6X + a_4^2, \quad \psi_2^2(X) = 4X^3 + 4a_4X + 4a_6.$$

From Equation (1.7), Chapter 1, we get with

$$g_1(X) = 12X^2 + 16a_4 \quad \text{and} \quad g_2(X) = 3X^3 - 5a_4X - 27a_6,$$

the equation

$$-4g_1(X)\phi_2(X) + 4g_2(X)\psi_2^2(X) = \Delta = -16(4a_4^3 + 27a_6^2),$$

therefore

$$g_1(X)\phi_2(X) - g_2(X)\psi_2^2(X) = 4(4a_4^3 + 27a_6^2),$$

Inserting $X = x$ into this equation and dividing by $\psi_2^2(x) \neq 0$, one gets the equation

$$g_1(x)\frac{\phi_2(x)}{\psi_2^2(x)} - g_2(x) = \frac{4(4a_4^3 + 27a_6^2)}{\psi_2^2(x)}$$

or

$$g_1(x)x' - g_2(x) = \frac{4(4a_4^3 + 27a_6^2)}{\psi_2^2(x)}.$$

Because the left hand side is integral, the right hand side must be integral and the assertion follows because of $\psi_2^2(x) = 4y^2$. □

6.4 Reduction

We recall Definition 4.1 and extend it in the usual manner.

Definition 6.27. Let $\mathfrak{p}$ be a prime ideal of $\mathcal{O}_{\mathbb{K}}$ (or a prime divisor of $\mathbb{K}$) and $E|\mathbb{K}$ an elliptic curve in long Weierstraß normal form over $\mathbb{K}$ with $\mathfrak{p}$-integral coefficients. The Weierstraß normal form is called *minimal at* $\mathfrak{p}$, if $\mathrm{ord}_{\mathfrak{p}}(\Delta)$ is minimal under the condition of $\mathfrak{p}$-integrability of the coefficients. The equation is called *globally minimal*, if it is minimal at all prime ideals of $\mathbb{K}$. The *reduction types at* $\mathfrak{p}$ are given according to the localisation at $\mathfrak{p}$ (see Definition 4.3).

Theorem 6.28. *For every elliptic curve $E|\mathbb{K}$ and every prime ideal $\mathfrak{p}$, there exists a minimal Weierstraß equation at $\mathfrak{p}$. A global minimal equation does not have to exist. It exists, if the class number of the number field $\mathbb{K}$ is $h_{\mathbb{K}} = 1$ (for example for $\mathbb{K} = \mathbb{Q}$).*

Proof. The first part of the theorem follows from Theorem 4.2. For the global minimal equation we refer to Silverman, [204] Chapter VIII, Corollary 8.3. □

Kida outlined the following example (found by Comalada) for an elliptic curve which has no global minimal equation. Consider the quadratic number field $\mathbb{K} = \mathbb{Q}(\sqrt{26})$, which has class number 2, and the elliptic curve

$$\begin{aligned} E : Y^2 + \sqrt{26}XY + (1+\sqrt{26})Y \\ = X^3 + (-1+\sqrt{26})X^2 + (724+144\sqrt{26})X + (-50867 - 9975\sqrt{26}). \end{aligned}$$

The discriminant of this equation is

$$\Delta = -13^6\epsilon^6,$$

where $\epsilon = 5+\sqrt{26}$ is a fundamental unit of $\mathbb{K}$. Let $\mathfrak{p} = (13, \sqrt{26})$ be the prime ideal lying above 13 (which is not a principal ideal). Then, as the ramification index of 13 is 2, we get

$$\operatorname{ord}_{\mathfrak{p}}(\Delta) = 12.$$

Similarly,

$$\operatorname{ord}_{\mathfrak{p}}(c_4) = 4 \quad \text{and} \quad \operatorname{ord}_{\mathfrak{p}}(c_6) = 7.$$

Hence the equation is not minimal at $\mathfrak{p}$. Now we apply a birational transformation with

$$u = \sqrt{26}, \quad r = 9+4\sqrt{26}, \quad s = 0,\ t = 32+8\sqrt{26}.$$

This results in an elliptic curve

$$\begin{aligned} E' : Y'^2 + X'Y' + \left(1+\frac{1}{4}\sqrt{26}\right)Y' \\ = X'^3 + \left(1+\frac{1}{2}\sqrt{26}\right)X'^2 + \left(\frac{13}{4}+\frac{1}{2}\sqrt{26}\right)X' + \left(-\frac{11}{8}-\frac{1}{4}\sqrt{26}\right). \end{aligned}$$

We get for the discriminant of this equation:

$$\operatorname{ord}_{\mathfrak{p}}(\Delta') = 0,$$

hence our elliptic curve has good reduction at $\mathfrak{p}$. That means, if there were a global minimal equation, the discriminant of that equation must generate a trivial ideal (because the curve has everywhere good reduction). But this is impossible, because there is no element $u \in \mathbb{K}$ such that

$$\operatorname{ord}_{\mathfrak{q}}(-13^6\epsilon^6/u^{12}) = 0$$

for every prime ideal $\mathfrak{q}$ of $\mathbb{K}$. Hence there exists no global minimal model.

A (global) minimal model of an elliptic curve can be constructed using the algorithm of Tate [220] (see Chapter 4, section 4.1). Laska [119], [120] also gives a method to construct global minimal models of elliptic curves.

Examples. 1) Consider the elliptic curve

$$E : Y^2 = X^3 + 16$$

over $\mathbb{K} = \mathbb{Q}$. Its discriminant is

$$\Delta = -110592 = -2^{12}3^3.$$

This equation is minimal at all prime numbers except for $p = 2$. For $p = 2$ we take for example the transformation $r = 0, s = 0, t = 4, u = 2$, which we get using the algorithm of Tate. We arrive at the equation

$$E : (Y')^2 + Y' = (X')^3.$$

The discriminant is

$$\Delta' = -3^3.$$

Hence the second equation is a global minimal Weierstraß equation for E over $\mathbb{Q}$.

The discriminant of the minimal equation of the curve E implies that this curve has good reduction for all prime numbers $p \in \mathbb{P}\backslash\{3\}$. Further

$$c_4 = 0.$$

Hence the curve has additive reduction at $p = 3$.

2) Consider the curve

$$E : Y^2 = X^3 + 5^4 \cdot 3X + \frac{5^6}{7}$$

over $\mathbb{Q}$. The equation for E is not integral. With the transformation $r = s = t = 0, u = 1/7$ we get an equation with integral coefficients:

$$E : Y^2 = X^3 + 3 \cdot 5^4 7^4 X + 5^6 7^5$$

with the discriminant

$$\Delta = -2^4 3^3 5^{12} 7^{10} 197.$$

Thus the equation is not minimal at 5. With the transformation

$$r = s = t = 0, \quad u = 5$$

we end up with the equation

$$E : (Y')^2 = (X')^3 + 3 \cdot 7^4 X' + 7^5.$$

The discriminant of this equation is

$$\Delta' = -2^4 3^3 7^{10} 197.$$

This equation is a global minimal equation for E. The curve has good reduction at every prime $p \in \mathbb{P}\backslash\{2, 3, 7, 197\}$. Further

$$c_4 = -345744 = -2^4 3^2 7^4.$$

Therefore the curve has additive reduction at the primes $p = 2$, $p = 3$, and $p = 7$, and multiplicative reduction at $p = 197$.

6.5 Computation of the torsion group

We now explain the different methods for the computation of the torsion group of elliptic curves over number fields. If the curve is given in short Weierstraß normal form over $\mathbb{Q}$ with integral coefficients we can use Theorem 6.26 of Lutz and Nagell. Here we test all $y \in \mathbb{Z}$ with $y^2 \mid \frac{-\Delta}{16}$, if it corresponds to a point $(x, y) \in E(\mathbb{Q})$ and if this point is a torsion point. In the next proposition we show how to test if a given point is a torsion point.

Proposition 6.29. *Let $E|\mathbb{K}$ be an elliptic curve over a number field $\mathbb{K}$ and $P = (x, y) \in E(\mathbb{K})$. The point P is a torsion point of E, if and only if one of the following conditions holds.*

a) *There exists an $m \in \mathbb{N}, m < B(E, \mathbb{K})$ with $mP = \mathcal{O}$.*

b) *There exists an $m \in \mathbb{N}, m < B(E, \mathbb{K})$ with $\psi_m(x, y) = 0$, where ψ_m is the m-th division polynomial (see Chapter 1, Section 1.3).*

c) $\hat{h}(P) = 0$.

Here, $B(E, \mathbb{K})$ is a suitable bound for the number of torsion points of E over $\mathbb{K}$. Such a bound can be computed by using Corollary 6.5, but we shall describe in Theorem 6.30 a method for computing bounds which are generally better.

Proof. a) This is the definition of torsion points.

b) This is Proposition 1.25. But the coefficients of $\psi_m(X, Y)$ are in general very large.

c) This follows from Proposition 5.19 (a). □

A variant of the Nagell–Lutz Theorem 6.26 for elliptic curves over quadratic or cubic number fields is indicated in the paper of Cremona and Whitley [41].

Another method for the computation of the torsion group of elliptic curves over $\mathbb{Q}$ is to use the analytic parametrization of elliptic curves (see Doud [55]). One chooses a lattice Λ corresponding to the elliptic curve. The n-torsion points of $\mathbb{C}/\Lambda$ are simply $\frac{1}{n}\Lambda/\Lambda$. Using the theorem given below, one computes an estimate for the number of torsion points. To test if a torsion point of order n exists over $\mathbb{Q}$, one computes the corresponding point (over $\mathbb{C}$) of an n-torsion point on $\mathbb{C}/\Lambda$ and tests if the coordinates are rational.

A (theoretical) method for computing the torsion group of elliptic curves over general number fields is to estimate the number of torsion points and then to use the division polynomials to compute the torsion points. For the estimate we need reduction theory, more precisely we need the following theorem already stated as Theorem 6.12:

Theorem 6.30. *Let $\mathbb{K}$ be a number field and $\mathrm{ord}_{\mathfrak{p}}$ a non-archimedean normalized (additive) valuation of $\mathbb{K}$ with prime ideal $\mathfrak{p}$. The ideal $\mathfrak{p}$ divides the ideal (p) with* *ramification index $e_{\mathfrak{p}|p}$ and residue class degree $f_{\mathfrak{p}|p}$. Let E be an elliptic curve over $\mathbb{K}$ given by a minimal equation at $\mathfrak{p}$. Then one has for the torsion group $E(\mathbb{K})_{\text{tors}}$ the following estimates:*

$$\sharp(E(\mathbb{K})_{\text{tors}}) \text{ divides}$$

$$\begin{cases} \sharp(\widetilde{E}(k(\mathfrak{p})))p^{2t}, & \text{at good reduction mod } \mathfrak{p}, \\ |\text{ord}_{\mathfrak{p}}(j)|(p^{2f_{\mathfrak{p}|p}}-1)p^{2t}, & \text{at multiplicative reduction mod } \mathfrak{p}, \\ \sharp(E(\mathbb{K})/E_0(\mathbb{K}))p^{2+2t} \le 4p^{2+2t}, & \text{at additive reduction mod } \mathfrak{p}, \end{cases}$$

with

$$t = \begin{cases} 0, & \text{if } \varphi(p) > e_{\mathfrak{p}|p}, \\ \max\{r \in \mathbb{N} : \varphi(p^r) \le e_{\mathfrak{p}|p}\}, & \text{else.} \end{cases}$$

Here $\sharp(\widetilde{E}(k(\mathfrak{p})))$ is the number of points on the reduced curve E modulo $\mathfrak{p}$, j is the j-invariant of the curve E and $E_0(\mathbb{K})$ is the set of points on $E(\mathbb{K})$, which is nonsingular at reduction modulo $\mathfrak{p}$.

As pointed out already in the definition of t, the sign $>$ resp. $\le$ can probably be replaced by the divisibility symbol (see Frey [66]).

Proof. (See Folz [64].) From the implication

$$E(\mathbb{K}) \subset E(\mathbb{K}_{\mathfrak{p}}) \Rightarrow E(\mathbb{K})_{\text{tors}} \subset E(\mathbb{K}_{\mathfrak{p}})_{\text{tors}}$$

we infer the divisibility relation

$$\sharp E(\mathbb{K})_{\text{tors}} \mid \sharp E(\mathbb{K}_{\mathfrak{p}})_{\text{tors}}.$$

It suffices therefore to consider the local torsion group $E(\mathbb{K}_{\mathfrak{p}})_{\text{tors}}$. Clearly we have

$$E(\mathbb{K}_{\mathfrak{p}})_{\text{tors}}/E_1(\mathbb{K}_{\mathfrak{p}})_{\text{tors}} \subset E(\mathbb{K}_{\mathfrak{p}})/E_1(\mathbb{K}_{\mathfrak{p}}).$$

where, according to the filtration,

$$E_1(\mathbb{K}_{\mathfrak{p}})_{\text{tors}} := E_1(\mathbb{K}_{\mathfrak{p}}) \cap E(\mathbb{K}_{\mathfrak{p}})_{\text{tors}}$$

is a finite p-group. We obtain $\sharp\left(E(\mathbb{K}_{\mathfrak{p}})_{\text{tors}}/E_1(\mathbb{K}_{\mathfrak{p}})_{\text{tors}}\right) \mid \sharp\left(E(\mathbb{K}_{\mathfrak{p}})/E_1(\mathbb{K}_{\mathfrak{p}})\right)$ and hence the divisibility relation

$$\sharp E(\mathbb{K})_{\text{tors}} \mid \sharp\left(E(\mathbb{K}_{\mathfrak{p}})/E_1(\mathbb{K}_{\mathfrak{p}})\right) \cdot \sharp E_1(\mathbb{K}_{\mathfrak{p}})_{\text{tors}}.$$

Now if $P = (x, y) \in E_1(\mathbb{K}_{\mathfrak{p}})_{\text{tors}}$, then $P \in E_n(\mathbb{K}_{\mathfrak{p}})$ for a maximal $n \in \mathbb{N}$ such that $\text{ord}_{\mathfrak{p}}(x) = -2n$. Since $P \in E_n(\mathbb{K}_{\mathfrak{p}})$ has order p^{ν} (see Theorem 4.11 c)), by the general Nagell–Lutz–Cassels theorem (Theorem 6.23 resp. Corollary 6.24), we have

$$\text{ord}_{\mathfrak{p}}(x) = -2n \ge \mu_{\mathfrak{p}} - \frac{2e_{\mathfrak{p}|p}}{\varphi(p^{\nu})} \ge -\frac{2e_{\mathfrak{p}|p}}{\varphi(p^{\nu})},$$

so that

$$n \le \frac{e_{\mathfrak{p}|p}}{\varphi(p^\nu)}.$$

If $\varphi(p^\nu) > e_{\mathfrak{p}|p}$ this contradicts $n \in \mathbb{N}$. Therefore we must have

$$\nu \le t$$

for $t \in \mathbb{N}_0$ defined in the theorem above and hence

$$\sharp E_1(\mathbb{K}_\mathfrak{p})_{\text{tors}} \mid p^{2t}.$$

The factor $\sharp\left(E(\mathbb{K}_\mathfrak{p})/E_1(\mathbb{K}_\mathfrak{p})\right)$ is estimated in Theorem 4.12. □

Frey [66] has given for t the better estimates

$$t = \begin{cases} 0, & \text{if } \varphi(p) \nmid e_{\mathfrak{p}|p}, \\ \max\{r \in \mathbb{N} : \varphi(p^r) | e_{\mathfrak{p}|p}\}, & \text{else,} \end{cases}$$

but he has proved it only for $p \neq 2, 3$.

If one wants to test whether an elliptic curve over a number field has a point of order m, one has to compute the roots of ψ_m in the number field.

Because the degrees of the polynomials ψ_m are large, it does not make sense to test all ψ_m. For example let $m = q \cdot r$ with $q, r > 1$, $q \in \mathbb{P}$, and $\gcd(q, r) = 1$. Then one gets the points of order m by adding the points of order q to the points of order r. Hence it is only necessary to consider the division polynomials ψ_{p^n} for prime numbers p and $n \in \mathbb{N}$.

Here for $n > 1$ the torsion points of order p^{n-1} are a subset of the torsion points of order p^n. To use this fact for our algorithm, we do not consider the polynomial ψ_{p^n}, but the polynomial $\psi_{p^n}/\psi_{p^{n-1}}$. After reduction modulo the equation of the curve this is a polynomial in X of degree $p^{2n-2}(\frac{p^2-1}{2})$ for $p > 2$ or $3 \cdot 2^{2n-3}$ for $p = 2, n \ge 2$ (see Proposition 1.20 and Proposition 1.22 and the paper of Cassels [25] and the second author [251]).

6.6 Examples

We first consider the elliptic curve

$$E : Y^2 = X^3 - 10X$$

and the point $P = (-1, 3)$ (see page 16). We show by different methods that this point P is not a torsion point.

- The canonical height of P is $\hat{h}(P) = 1.2815287022\ldots \neq 0$, hence P is not a torsion point.

- Using Theorem 6.26, we see that P has integral coordinates, but $2P \neq \mathcal{O}$ and $3^2 \nmid \frac{-\Delta}{16} = 4 \cdot (-10)^3$. Hence P is not a torsion point.
- Again using Theorem 6.26, we now consider the point $2P$. If P is a torsion point, $2P$ is also a torsion point. But as the coefficients of

$$2P = (121/36, 451/216)$$

are not integral, $2P$ and therefore P is not a torsion point.

Next we consider the elliptic curve

$$E : Y^2 = X(X-2)(X+4)$$

over $\mathbb{Q}$. The discriminant of this curve is $\Delta = 2^{12}3^2$. The curve has good reduction at all prime numbers $p \geq 5$. It is easy to compute $\sharp\widetilde{E}(\mathbb{Z}/5\mathbb{Z}) = 8$ and $\sharp\widetilde{E}(\mathbb{Z}/7\mathbb{Z}) = 12$, hence $\sharp E(\mathbb{Q})_{\text{tors}}$ divides 4. As we can find the torsion points

$$\{T_0 = \mathcal{O}, T_1 = (0,0), T_2 = (2,0), T_3 = (-4,0)\},$$

we see that $E(\mathbb{Q})_{\text{tors}} \cong \mathbb{Z}/4\mathbb{Z}$.

We consider now special short Weierstraß equations.

Proposition 6.31. *Let $k \in \mathbb{Z}$, $k \neq 0$ without 6-th powers,*

$$E_k : Y^2 = X^3 + k$$

a Mordell curve. Then

$$E_k(\mathbb{Q})_{\text{tors}} = \begin{cases} \{\mathcal{O}, (m,0)\} \cong \mathbb{Z}/2\mathbb{Z} & \text{if } -k = m^3 \neq 1, m \in \mathbb{Z}, \\ \{\mathcal{O}, (0, \pm n)\} \cong \mathbb{Z}/3\mathbb{Z} & \text{if } k = n^2 \neq 1, n \in \mathbb{Z}, \\ \{\mathcal{O}, (12, \pm 36)\} \cong \mathbb{Z}/3\mathbb{Z} & \text{if } k = -432, \\ \{\mathcal{O}, (2, \pm 3), (0, \pm 1), (-1, 0)\} \cong \mathbb{Z}/6\mathbb{Z} & \text{if } k = 1, \\ \{\mathcal{O}\} & \text{else.} \end{cases}$$

Proof. Although the result is known (see for example Fueter [72]), we will prove it here as an example for computing torsion groups.

For a Mordell curve E_k we have

$$\Delta = -2^4 3^3 k^2, \quad c_4 = 0.$$

The equation is therefore minimal for all primes $p > 3$. For primes $p \nmid 6k$ the curve has good reduction. For primes $p \mid k$ with $p > 3$ it has additive reduction.

In Theorem 6.30 if $\mathbb{K} = \mathbb{Q}$, the ramification degree is 1 for all cases and therefore $t = 0$. Now we consider the curve $E_k : Y^2 = X^3 + k$ modulo 5, i.e. $\widetilde{E_k}$ over $\mathbb{F}_5$. If E_k

has good reduction at 5, i.e. $5 \nmid k$, we can use Lemma 3.10 and obtain (with Theorem 6.30)

$$\sharp(E_k(\mathbb{Q})_{\text{tors}}) \text{ divides } \sharp\widetilde{E_k}(\mathbb{F}_5) = 6,$$

therefore

$$\sharp(E_k(\mathbb{Q})_{\text{tors}}) \in \{1, 2, 3, 6\}.$$

If E_k has bad reduction at 5, i.e. $5 \mid k$, the reduction is additive and we obtain for the torsion group:

$$\sharp(E_k(\mathbb{Q})_{\text{tors}}) \text{ divides 25 or 50 or 75 or 100},$$

that means

$$\sharp(E_k(\mathbb{Q})_{\text{tors}}) \in \{1, 2, 3, 4, 5, 10, 15, 20, 25, 50, 75, 100\}.$$

Using Theorem 6.10 we see that the cases

$$\sharp(E_k(\mathbb{Q})_{\text{tors}}) = 25, 50, 75, 100$$

are not possible, because in this case, the 5-th roots of unity should lie in $\mathbb{Q}$, which is not the case. Altogether we obtain from the investigation of E_k at $p = 5$:

$$\sharp(E_k(\mathbb{Q})_{\text{tors}}) \in \{1, 2, 3, 4, 5, 6, 10, 15, 20\}.$$

Now we consider the curve at $p = 11$. If the curve has good reduction we get

$$\sharp(E_k(\mathbb{Q})_{\text{tors}}) \text{ divides } \sharp\widetilde{E_k}(\mathbb{F}_{11}) = 12,$$

that means

$$\sharp(E_k(\mathbb{Q})_{\text{tors}}) \in \{1, 2, 3, 4, 6, 12\}.$$

If the curve has bad (additive) reduction, we obtain

$$\sharp(E_k(\mathbb{Q})_{\text{tors}}) \text{ divides 121 or 242 or 363 or 484},$$

that means

$$\sharp(E_k(\mathbb{Q})_{\text{tors}}) \in \{1, 2, 3, 4, 11, 22, 33, 44, 121, 242, 363, 484\}.$$

Again using Theorem 6.10 and the fact that the 11-th roots of unity are not given over $\mathbb{Q}$, we find that the numbers 121, 242, 363, 484 are no possible orders. Altogether we obtain from the investigation of E_k at $p = 11$:

$$\sharp(E_k(\mathbb{Q})_{\text{tors}}) \in \{1, 2, 3, 4, 6, 11, 12, 22, 33, 44\}.$$

Comparing this with the investigation for $p = 5$, we arrive at

$$\sharp(E_k(\mathbb{Q})_{\text{tors}}) \in \{1, 2, 3, 4, 6\}.$$

First we suppose the curve has a point $P = (x_0, 0)$ of order 2, so that

$$X^3 + k = 0$$

must have a solution x_0 in $\mathbb{Q}$. This only exists, if

$$x_0^3 = -k = m^3$$

is a cubic in $\mathbb{Q}$. In this case one has that $(m, 0)$ is the only point of order 2.

For the computation of the points of exact order 4 we consider the polynomial

$$\psi_4(X, Y)/\psi_2(X, Y) = 2(X^6 + 20kX^3 - 8k^2).$$

For a root $x_0 \in \mathbb{C}$ of this polynomial one has

$$x_0^3 = -10k \pm \sqrt{100k^2 + 8k^2} = k(-10 \pm \sqrt{108}).$$

Because this x_0 does not belong to $\mathbb{Q}$, these curves cannot have points of order 4 over $\mathbb{Q}$.

For the determination of points of exact order 3 we look at

$$\psi_3(X, Y) = 3X^4 + 12kX = 3X(X^3 + 4k).$$

A root of this polynomial is $x_0 = 0$. We obtain the equation

$$Y^2 = k.$$

Therefore one only gets a point, if $k = n^2$ is a square in $\mathbb{Q}$. In this case one gets the point

$$(0, n)$$

of order 3. A second root of the polynomial is a root x_1 of

$$X^3 + 4k$$

over $\mathbb{Q}$. This exist if $k = 2a^3$ with $a \in \mathbb{Q}$. In this case $x_1 = -2a$ and one gets the equation

$$Y_1^2 = x_1^3 + k = -8a^3 + 2a^3 = -6a^3.$$

A solution of this equation exists over $\mathbb{Q}$ if and only if $a = -6b^2$ with $b \in \mathbb{Q}$. Then $y_1 = \pm 6b^3$ and $k = -432b^6$. Because we assumed that k is without 6-th powers, we must have $b = 1$. So such a point of order 3 exists if and only if $k = -432$. In this case the two points are

$$(12, \pm 36).$$

We get points of order 6 if points of order 2 and points of order 3 exist. A point of order 2 exists, if $k = m^3$ with $m \in \mathbb{Z}$. Points of order 3 exist, if $k = n^2$ with $n \in \mathbb{Z}$, or if $k = -432 = -2^4 3^3$. The only possibility for points of order 6 is therefore that $k = m^3 = n^2 = c^6$ with non-zero $c \in \mathbb{Z}$. Because k is free of 6-th powers, this can only happen for $k = 1$. □

Proposition 6.32. *Let $k \in \mathbb{Z}$, $k \neq 0$ without 4-th powers, $E_k : Y^2 = X^3 + kX$. Then*

$$E_k(\mathbb{Q})_{\mathrm{tors}} = \begin{cases} \{\mathcal{O}, (2, \pm 4), (0, 0)\} \cong \mathbb{Z}/4\mathbb{Z} & \text{if } k = 4, \\ \{\mathcal{O}, (0, 0), (\pm a, 0)\} \cong \mathbb{Z}/2\mathbb{Z} \times \mathbb{Z}/2\mathbb{Z} & \text{if } k = -a^2, a \in \mathbb{Z}, \\ \{\mathcal{O}, (0, 0)\} \cong \mathbb{Z}/2\mathbb{Z} & \text{else.} \end{cases}$$

Proof. The discriminant of E_k is

$$\Delta_k = -2^6 k^3.$$

Such curves are minimal at all $p > 2$. For $p \nmid 2k$ the reduction is good, for $p | 2k$, $p > 2$ it is additive.

First we consider the curve for $p = 3$. If $3 \nmid k$, we have good reduction. If $k \equiv 1 \bmod 3$, we get for the number of points:

$$\sharp \widetilde{E}_k(\mathbb{F}_3) = 4.$$

For $k \equiv 2 \bmod 3$ one also has

$$\sharp \widetilde{E}_k(\mathbb{F}_3) = 4.$$

So we obtain from the Theorem 6.30

$$\sharp E_k(\mathbb{Q})_{\mathrm{tors}} \in \{1, 2, 4\}.$$

If $3 \mid k$, the reduction is additive. With Theorem 6.30 we get

$$\sharp(E_k(\mathbb{Q})_{\mathrm{tors}}) \text{ divides } 9 \text{ or } 18 \text{ or } 27 \text{ or } 36,$$

therefore

$$\sharp E_k(\mathbb{Q})_{\mathrm{tors}} \in \{1, 2, 3, 4, 6, 9, 12, 18, 27, 36\}.$$

Using Theorem 6.10 and the fact that the third roots of unity are not given over $\mathbb{Q}$, we find that the numbers 9, 18, 27, 36 are not possible.

Altogether we have the same possibilities:

$$\sharp E_k(\mathbb{Q})_{\mathrm{tors}} \in \{1, 2, 3, 4, 6, 12\}.$$

First we consider the torsion points of order 2. These correspond to solutions of

$$X^3 + kX = X(X^2 + k) = 0$$

over $\mathbb{Q}$. The point $(0, 0)$ is, for every k, a torsion point of order 2. If further $k = -a^2$ with $a \in \mathbb{Z}$, there are also the points $(\pm a, 0)$ of order 2 over $\mathbb{Q}$.

For finding points of order 4, we consider the polynomial

$$\psi_4(X, Y)/\psi_2(X, Y) = 2(X^6 + 5kX^4 - 5k^2X^2 - k^3) = 2(X^2 - k)(X^4 + 6kX^2 + k^2)$$

over $\mathbb{Q}$. If $k = b^2$ is a square in $\mathbb{Z}$, there exist the roots $x = \pm b$. For these x we have the equations

$$Y^2 = \pm(b^3 + b^3) = \pm 2b^3.$$

These equations only have a solution for $x = b = 2c^2$. In this case $k = b^2 = 4c^4$. Because k is free of 4-th powers, this case can only occur for $c = 1$, that is, $k = 4$. In this case we have the points

$$(2, \pm 4)$$

of order 4. For the other points of order 4, one has

$$X^4 + 6kX^2 + k^2 = 0,$$

that is tantamount to

$$X^2 = -3k \pm \sqrt{9k^2 - k^2} = k(-3 \pm \sqrt{8}).$$

Hence, over $\mathbb{Q}$, there exist no other points of order 4.

For determining the points of order 3, we consider the equation

$$\psi_3(X, Y) = 3X^4 + 6kX^2 - k^2.$$

For such points one has

$$X^4 + 2kX^2 - \frac{k^2}{3} = 0,$$

that is,

$$X^2 = -k \pm \sqrt{k^2 + \frac{1}{3}k^2} = k\left(-1 \pm \sqrt{\frac{4}{3}}\right).$$

Because this does not exist over $\mathbb{Q}$, there can not exist points of order 3. Therefore, a fortiori, there cannot exist points of order 6 or 12. □

We now give an example which shows that even all local estimates are not enough to determine the torsion group exactly. Consider the elliptic curve

$$E : Y^2 = X^3 + X.$$

For the discriminant of this curve one has

$$\Delta = -2^6.$$

The equation is globally minimal and has good reduction at all primes $p > 2$. With Lemma 3.12 we know for all prime numbers $p > 2$ that

$$4 \mid \sharp\widetilde{E}(\mathbb{F}_p).$$

But this is all we "know" about the torsion group $E(\mathbb{Q})_{\text{tors}}$ using the estimates. Of course, Proposition 6.32 gives us the torsion group $E(\mathbb{Q})_{\text{tors}} \cong \mathbb{Z}/2\mathbb{Z}$.

The following two examples are examples of elliptic curves over quadratic number fields. The first one is taken from the article of Müller, Ströher, and Zimmer [149] (see Section 6.2). In this article, the authors construct elliptic curves which contain given groups in the torsion group. Schmitt [189] computed the exact torsion group for every curve given in the article.

We consider the elliptic curve

$$E: \quad Y^2 = X^3 - (420 + 240\sqrt{3})X + (4576 + 2640\sqrt{3})$$

over $\mathbb{Q}(\sqrt{3})$. In the article [149], it was estimated that

$$\mathbb{Z}/6\mathbb{Z} \subseteq E(\mathbb{Q}(\sqrt{3}))_{\text{tors}}.$$

The discriminant

$$\Delta = 597639168 + 345047040\sqrt{3}$$

of this curves generates an ideal which has the prime ideal factorization

$$(\Delta) = \mathfrak{p}_2^{28}\mathfrak{p}_3^6.$$

Here, $\mathfrak{p}_2$ and $\mathfrak{p}_3$ are the prime ideals with

$$(2) = \mathfrak{p}_2^2, \quad (3) = \mathfrak{p}_3^2.$$

It follows that the curve has good reduction at the prime ideal lying over 5:

$$(5) = \mathfrak{p}_5.$$

From Theorem 6.30 we get

$$\sharp E(\mathbb{Q}(\sqrt{3}))_{\text{tors}} \mid \sharp\widetilde{E}(k(\mathfrak{p}_5)),$$

where a system of representatives of the finite field $k(\mathfrak{p}_5)$ is given by

$$\{a + b\sqrt{3} : 1 \le a, b \le 5\}.$$

The modulo $\mathfrak{p}_5$ reduced elliptic curve is

$$\widetilde{E}: \quad Y^2 = X^3 + 1.$$

Computation of the number of points of $\widetilde{E}(k(\mathfrak{p}_5))$ yields

$$\sharp\widetilde{E}(k(\mathfrak{p}_5)) = 36.$$

Using Theorem 6.9, we get for the possible number of torsion points:

$$\sharp E(\mathbb{Q}(\sqrt{3}))_{\text{tors}} \in \{1, 2, 3, 4, 6, 9, 12, 18\}.$$

The torsion points of order 2 can be found by computing the roots of the polynomial

$$X^3 - (420 + 240\sqrt{3})X + (4576 + 2640\sqrt{3})$$

over $\mathbb{Q}(\sqrt{3})$. We get three torsion points of order 2:

$$P_1 = (8 + 4\sqrt{3}, 0), \quad P_2 = (8 + 6\sqrt{3}, 0), \quad P_3 = (-16 - 10\sqrt{3}, 0).$$

It follows that

$$\sharp E(\mathbb{Q}(\sqrt{3}))_{\text{tors}} \in \{4, 12\}.$$

Now we consider points of order 3. Therefore we compute the roots of the division polynomial

$$\begin{aligned}\psi_3(X) &= 3X^4 - (2520 + 1440\sqrt{3})X^2 + (54912 + 31680\sqrt{3})X \\ &\quad - (349200 + 201600\sqrt{3})\end{aligned}$$

over $\mathbb{Q}(\sqrt{3})$ and get one root:

$$x_0 = 12 + 6\sqrt{3}.$$

This leads to the point of order 3:

$$P_4 = (12 + 6\sqrt{3}, 20 + 12\sqrt{3}).$$

It follows that $\sharp E(\mathbb{Q}(\sqrt{3}))_{\text{tors}} = 12$. The torsion group is then

$$E(\mathbb{Q}(\sqrt{3}))_{\text{tors}} \cong \mathbb{Z}/6\mathbb{Z} \times \mathbb{Z}/2\mathbb{Z}.$$

Generators for this torsion group are for example the points

$$P_1 + P_4 = (-4 - 2\sqrt{3}, 60 + 36\sqrt{3}), \quad P_2 = (8 + 6\sqrt{3}, 0),$$

where the first point is of order 6.

The next example was outlined by Kida. It is taken from Reichert [174]. Consider the quadratic number field $\mathbb{K} = \mathbb{Q}(\sqrt{33})$ and the elliptic curve

$$E : Y^2 = X^3 - (162675 + 28296\sqrt{33})X + 35441118 + 6168312\sqrt{33}.$$

The factorization of the ideal generated by the discriminant

$$\Delta = 157274439344445568 + 27378000938926088\sqrt{33}$$

is

$$(\Delta) = \mathfrak{p}_2^{21}\mathfrak{p}_2'^{30}\mathfrak{p}_3^{24},$$

where the prime ideals are such that

$$(3) = \mathfrak{p}_3^2 \quad \text{and} \quad (2) = \mathfrak{p}_2\mathfrak{p}_2'.$$

It follows that the curve has good reduction above 5. Using Theorem 6.30 for the prime ideal lying over 5:

$$(5) = \mathfrak{p}_5,$$

we get

$$\sharp E(\mathbb{K})_{\text{tors}} \mid \sharp \widetilde{E}(k(\mathfrak{p}_5)).$$

We now compute the number of points of $\widetilde{E}(k(\mathfrak{p}_5))$. A system of representatives of $k(\mathfrak{p}_5)$ is given by

$$\left\{ a + b\left(\frac{1+\sqrt{33}}{2}\right) : 1 \le a, b \le 5 \right\}.$$

We get

$$\sharp \widetilde{E}(k(\mathfrak{p}_5)) = 18.$$

We first consider torsion points of order 2. For these points, the second coordinate is 0 and the first one is a root of the polynomial

$$X^3 - (162675 + 28296\sqrt{33})X + (35441118 + 6168312\sqrt{33}).$$

Over $\mathbb{K}$, this polynomial has exactly one root, hence there is one point of order 2 given by

$$P_1 = \left(\frac{285}{2} + \frac{57}{2}\sqrt{33}, 0\right).$$

Now we consider the 3-th division polynomial. The factorization of this polynomial contains a factor of degree 3 and a factor of degree 1, hence one point of order 3:

$$P_2 = (207 + 36\sqrt{33}, 1296 + 240\sqrt{33}).$$

Using the 9-th division polynomial, we find a torsion point of order 9:

$$P_3 = (219 + 24\sqrt{33}, 108 - 108\sqrt{33}).$$

In sum, we have the torsion group

$$E(\mathbb{K})_{\text{tors}} \cong \mathbb{Z}/18\mathbb{Z}$$

generated by

$$P_3 + P_1 = (-285 - 48\sqrt{33}, -4428 - 756\sqrt{33}).$$

Note that such a torsion group cannot occur over $\mathbb{Q}$ (see Theorem 6.8).

6.7 Exercises

1) Consider the elliptic curve over $\mathbb{Q}$:

$$E : Y^2 = X^3 + aX + b, \quad a, b \in \mathbb{Q}, \quad \Delta = 4a^3 + 27b^2 \neq 0.$$

Show that for a point $P = (x, y) \in E(\mathbb{Q})$ with order 2 one has $y = 0$. Show that there are three cases:

$$E(\mathbb{Q})[2] \cong \mathbb{Z}/2\mathbb{Z} \times \mathbb{Z}/2\mathbb{Z}, \quad E(\mathbb{Q})[2] \cong \mathbb{Z}/2\mathbb{Z}, \quad \text{or} \quad E(\mathbb{Q})[2] = \{\mathcal{O}\}.$$

Verify that in all cases $E(\mathbb{C})[2] \cong \mathbb{Z}/2\mathbb{Z} \times \mathbb{Z}/2\mathbb{Z}$ by explicit computations.

2) Determine the torsion group of the following curves over $\mathbb{Q}$:

   a) $E : Y^2 = X^3 - 2$
   b) $E : Y^2 = X^3 + 8$
   c) $E : Y^2 = X^3 + 4$
   d) $E : Y^2 = X^3 + 4X$
   e) $E : Y^2 - Y = X^3 - X^2$
   f) $E : Y^2 = X^3 + 1$
   g) $E : Y^2 + XY + Y = X^3 - X^2 - 3X + 3$
   h) $E : Y^2 + 7XY = X^3 + 16X$
   i) $E : Y^2 + XY + Y = X^3 - X^2 - 14X + 29$
   j) $E : Y^2 + XY = X^3 - 45X + 81$
   k) $E : Y^2 + 43XY - 210Y = X^3 - 210X^2$
   l) $E : Y^2 = X^3 - 4X$
   m) $E : Y^2 + XY - 5Y = X^3 - 5X^2$
   n) $E : Y^2 + 5XY - 6Y = X^3 - 3X^2$
   o) $E : Y^2 + 17XY - 120Y = X^3 - 60X^2$

3) Show that the curve

$$E : Y^2 + 67Y = X^3 - 21X^2 - 10X + 30$$

has trivial torsion over $\mathbb{Q}$. (Hint: Apply reduction theory.)

4) Show that the curve

$$E : Y^2 = X^3 + iX + (1 - i),$$

where $i = \sqrt{-1}$, has trivial torsion over $\mathbb{K} = \mathbb{Q}(i)$.

5) Show that the Mordell curve

$$E : Y^2 = X^3 + 2(11 + 5\sqrt{5})$$

has trivial torsion over $\mathbb{K} = \mathbb{Q}(\sqrt{5})$.

6) Consider the elliptic curve

$$E : Y^2 = X^3 + 2^4 \cdot 3^3 \cdot 5X + 2^4 \cdot 3^3 \cdot 5 \cdot 79$$

over $\mathbb{Q}$.

a) Show that the non-zero torsion points over $\mathbb{Q}$ of order 5 are

$$P_\pm = (-2^3 \cdot 3, \pm 2^2 \cdot 3^4), \quad Q_\pm = (2^2 \cdot 3 \cdot 7, \pm 2^2 \cdot 3^5).$$

b) Prove that $E|\mathbb{Q}$ has torsion group

$$E(\mathbb{Q})_{\text{tors}} \cong \mathbb{Z}/5\mathbb{Z}.$$

7) Prove by the Theorem of Nagell–Lutz–Cassels that the torsion group $E(\mathbb{K})_{\text{tors}}$ of an elliptic curve E over an imaginary quadratic number field $\mathbb{K}$ is finite.

Chapter 7
The rank

In this chapter we consider the rank of elliptic curves over number fields. This is in general a difficult topic but the rank can determined in special cases (see Frey [68]).

In the first section we introduce the L-series of elliptic curves. Then we define the Tate–Shafarevich group and the Selmer group and discuss some conjectures related to the rank, among them the most famous one, the conjecture of Birch and Swinnerton-Dyer.

There are two general methods to compute the rank of elliptic curves over number fields: descent methods and the application of the conjecture of Birch and Swinnerton-Dyer. These methods are explained in Section 7.4 and Section 7.6.

In the last section we consider the behaviour of the rank in field extensions.

In this chapter, let $\mathbb{K}$ be a number field and $G = G_{\overline{\mathbb{K}}|\mathbb{K}}$ the absolute Galois group of $\mathbb{K}$. The notation is the same as in Chapter 5.

7.1 L-series

The L-series of an elliptic curve is defined as a product over local L-functions. It is conjectured that this combination of local factors reveals global properties of the curve (see Section 7.4, or, for example, the paper of Zagier [244]), a principle which often occurs in number theory.

We first give the definition of the local L-functions.

Definition 7.1. Let $E|\mathbb{K}$ be an elliptic curve over a number field $\mathbb{K}$ and $\mathfrak{p}$ a prime ideal in $\mathcal{O}_{\mathbb{K}}$ with norm $\mathcal{N}(\mathfrak{p})$. If E has good reduction at $\mathfrak{p}$, the *local L-function*

$$L_{\mathfrak{p}}(E;\cdot) : \mathbb{C} \to \mathbb{C}$$

of E at $\mathfrak{p}$ is

$$L_{\mathfrak{p}}(E;T) := 1 - a_{\mathfrak{p}}T + \mathcal{N}(\mathfrak{p})T^2$$

with the number

$$a_{\mathfrak{p}} := \mathcal{N}(\mathfrak{p}) + 1 - \sharp(\widetilde{E}(\mathbb{F}_{\mathfrak{p}})).$$

If E has bad reduction at $\mathfrak{p}$, the *local L-function of E at* $\mathfrak{p}$ is

$$L_{\mathfrak{p}}(E;T) := 1 - a_{\mathfrak{p}}T$$

with

$$a_\mathfrak{p} := \begin{cases} 1, & \text{if } E \text{ has split multiplicative reduction at } \mathfrak{p}, \\ -1, & \text{if } E \text{ has non-split multiplicative reduction at } \mathfrak{p}, \\ 0, & \text{if } E \text{ has additive reduction at } \mathfrak{p}. \end{cases}$$

With help of this local L-function we can now define the global L-series.

Definition 7.2. Let $E|\mathbb{K}$ be an elliptic curve over a number field $\mathbb{K}$. The (global) *L-series* of $E|\mathbb{K}$ is

$$L(E|\mathbb{K}; s) := \prod_{\mathfrak{p}\in M_\mathbb{K}^0} L_\mathfrak{p}(E; \mathcal{N}(\mathfrak{p})^{-s})^{-1},$$

where $s \in \mathbb{C}$ is a complex variable with $\mathrm{Re}(s) > \frac{3}{2}$. Here the product is taken over all prime ideals of $\mathcal{O}_\mathbb{K}$ (or prime divisors of $\mathbb{K}$ lying over primes $p \in \mathbb{P}$).

Proposition 7.3. *The product*

$$\prod_{\mathfrak{p}\in M_\mathbb{K}^0} L_\mathfrak{p}(E; \mathcal{N}(\mathfrak{p})^{-s})^{-1}, \quad s \in \mathbb{C}$$

converges absolutely for $s \in \mathbb{C}$ with $\mathrm{Re}(s) > \frac{3}{2}$.

Proof. Theorem 3.3 of Hasse gives the estimate

$$|a_\mathfrak{p}| \le 2\sqrt{\mathcal{N}(\mathfrak{p})}.$$

Let $\mathfrak{p}$ be a prime ideal of $\mathbb{K}$ where E has good reduction. We estimate

$$\begin{aligned} |L_\mathfrak{p}(E; \mathcal{N}(\mathfrak{p})^{-s})| &\ge 1 - |a_\mathfrak{p}|\mathcal{N}(\mathfrak{p})^{-\mathrm{Re}(s)} + \mathcal{N}(\mathfrak{p})^{1-2\mathrm{Re}(s)} \\ &\ge 1 - 2\mathcal{N}(\mathfrak{p})^{1/2-\mathrm{Re}(s)} + \mathcal{N}(\mathfrak{p})^{1-2\mathrm{Re}(s)} \\ &= (1 - \mathcal{N}(\mathfrak{p})^{1/2-\mathrm{Re}(s)})^2. \end{aligned}$$

Let S be the finite set of prime ideals in $\mathcal{O}_\mathbb{K}$ where the curve E has bad reduction. It follows that

$$\begin{aligned} |L(E|\mathbb{K}; s)| &= \prod_\mathfrak{p} |L_\mathfrak{p}(E; \mathcal{N}(\mathfrak{p})^{-s})^{-1}| \\ &\le \left(\prod_{\mathfrak{p}\in S} \frac{(1 - \mathcal{N}(\mathfrak{p})^{1/2-\mathrm{Re}(s)})^2}{|L_\mathfrak{p}(E; \mathcal{N}(\mathfrak{p})^{-s})|}\right) \prod_\mathfrak{p} \frac{1}{(1 - \mathcal{N}(\mathfrak{p})^{1/2-\mathrm{Re}(s)})^2} \\ &= \left(\prod_{\mathfrak{p}\in S} \frac{(1 - \mathcal{N}(\mathfrak{p})^{1/2-\mathrm{Re}(s)})^2}{|L_\mathfrak{p}(E; \mathcal{N}(\mathfrak{p})^{-s})|}\right) \left(\prod_\mathfrak{p} \frac{1}{1 - \mathcal{N}(\mathfrak{p})^{1/2-\mathrm{Re}(s)}}\right)^2 \\ &= \left(\prod_{\mathfrak{p}\in S} \frac{(1 - N(\mathfrak{p})^{1/2-\mathrm{Re}(s)})^2}{|L_\mathfrak{p}(E; \mathcal{N}(\mathfrak{p})^{-s})|}\right) \left(\zeta_\mathbb{K}\left(\mathrm{Re}(s) - \frac{1}{2}\right)\right)^2. \end{aligned}$$

Here $\zeta_{\mathbb{K}}(s)$ is the zeta function of the number field $\mathbb{K}$ which is known to converge for $\mathrm{Re}(s) > 1$ (see for example Neukirch [157], Chapter VII, Theorem 5.2). The first product is finite. Therefore $L(E|\mathbb{K}; s)$ converges for $\mathrm{Re}(s) > \frac{3}{2}$. □

In the following we write $L(s)$ for $L(E|\mathbb{K}; s)$ and $L_{\mathfrak{p}}(T)$ for $L_{\mathfrak{p}}(E; T)$, assuming that an elliptic curve E over a fixed number field $\mathbb{K}$ is given.

In the domain of convergence, the function $L(s)$ is an analytic function which can be written as a Dirichlet series.

Theorem 7.4. *Let $E|\mathbb{K}$ be an elliptic curve over the number field $\mathbb{K}$. Then the L-series of E at s with $\mathrm{Re}(s) > \frac{3}{2}$ is*

$$L(s) = \sum_{\mathfrak{a}} a(\mathfrak{a}) \mathcal{N}(\mathfrak{a})^{-s},$$

where the sum is over all integral ideals $\mathfrak{a}$ of $\mathcal{O}_{\mathbb{K}}$ (or integral divisors of $\mathbb{K}$). Here $\mathcal{N}(\mathfrak{a})$ is the norm of the ideal (divisor) $\mathfrak{a}$ in $\mathbb{K}|\mathbb{Q}$.

If $\mathfrak{a} = \mathfrak{p}_1^{k_1} \dots \mathfrak{p}_r^{k_r}$ $(k_i \in \mathbb{N})$ is the decomposition of $\mathfrak{a}$ into pairwise different prime ideals (prime divisors) $\mathfrak{p}_1, \dots, \mathfrak{p}_r$, the integer $a(\mathfrak{a})$ is defined as

$$a(\mathfrak{a}) := a(\mathfrak{p}_1^{k_1}) \cdot \ldots \cdot a(\mathfrak{p}_r^{k_r}).$$

Further let $\mathfrak{p}$ be a prime ideal (prime divisor).[1] *Then*

- $a((1)) = 1,$
- *if E has good reduction modulo $\mathfrak{p}$:*

$$\begin{aligned} a(\mathfrak{p}) &= a_{\mathfrak{p}} = \mathcal{N}(\mathfrak{p}) + 1 - \sharp(\widetilde{E}(\mathbb{F}_{\mathfrak{p}})), \\ a(\mathfrak{p}^k) &= a(\mathfrak{p})a(\mathfrak{p}^{k-1}) - \mathcal{N}(\mathfrak{p})a(\mathfrak{p}^{k-2}) \quad \text{for } k \geq 2, \end{aligned}$$

- *if E has split multiplicative reduction modulo $\mathfrak{p}$:*

$$a(\mathfrak{p}^k) = 1 \quad \text{for } k \in \mathbb{N},$$

- *if E has non-split multiplicative reduction modulo $\mathfrak{p}$:*

$$a(\mathfrak{p}^k) = (-1)^k \quad \text{for } k \in \mathbb{N},$$

- *if E has additive reduction modulo $\mathfrak{p}$:*

$$a(\mathfrak{p}^k) = 0 \quad \text{for } k \in \mathbb{N}.$$

[1] We always take the concept of ideal as synonymous to the concept of divisor (see Hasse [92])

Proof. The following equations are only valid in the domain of convergence of the given series.

Let $\mathfrak{p}$ be a prime ideal of $\mathcal{O}_{\mathbb{K}}$, where E has good reduction. If we write

$$L_{\mathfrak{p}}(T)^{-1} = \sum_{k\geq 0} c_k T^k$$

with $c_k \in \mathbb{C}$, we get:

$$\begin{aligned} 1 &= \Big(\sum_{k\geq 0} c_k T^k\Big) L_{\mathfrak{p}}(T) \\ &= \Big(\sum_{k\geq 0} c_k T^k\Big)(1 - a_{\mathfrak{p}} T + \mathcal{N}(\mathfrak{p})T^2) \\ &= c_0 + (c_1 - c_0 a_{\mathfrak{p}})T + \sum_{k\geq 2}(c_k - a_{\mathfrak{p}} c_{k-1} + \mathcal{N}(\mathfrak{p})c_{k-2})T^k. \end{aligned}$$

Comparing the coefficients yields

$$c_0 = 1, \quad c_1 = a_{\mathfrak{p}}$$

and

$$c_k = a_{\mathfrak{p}} c_{k-1} - \mathcal{N}(\mathfrak{p}) c_{k-2} \qquad \text{for } k \geq 2.$$

Defining $a(\mathfrak{p}^k) := c_k$ we proved the assumption for prime ideals $\mathfrak{p}$, where E has good reduction.

The computation of the coefficients $a(\mathfrak{p}^k)$ for prime ideals $\mathfrak{p}$, where E has bad reduction, is carried over in an analogous way (Exercise 7.8.1).

With this notation, in the domain of convergence we can write the L-series as

$$\begin{aligned} L(s) &= \prod_{\mathfrak{p}} L_{\mathfrak{p}}(\mathcal{N}(\mathfrak{p})^{-s})^{-1} \\ &= \prod_{\mathfrak{p}} \Big(\sum_{k\geq 0} a(\mathfrak{p}^k)\mathcal{N}(\mathfrak{p})^{-ks}\Big) \\ &= \sum_{\mathfrak{a}=\mathfrak{p}_1^{k_1}\ldots\mathfrak{p}_r^{k_r}} \Big(\prod_{i=1}^{r}(a(\mathfrak{p}_i^{k_i})\mathcal{N}(\mathfrak{p}_i^{k_i})^{-s})\Big) \\ &= \sum_{\mathfrak{a}} a(\mathfrak{a})\mathcal{N}(\mathfrak{a})^{-s}. \end{aligned}$$

Here, the last sum is to be taken over all integral ideals $\mathfrak{a}$ of $\mathcal{O}_{\mathbb{K}}$, and for $\mathfrak{a} = \mathfrak{p}_1^{k_1} \ldots \mathfrak{p}_r^{k_r}$, the integer $a(\mathfrak{a})$ is defined as $a(\mathfrak{a}) := a(\mathfrak{p}_1^{k_1}) \cdot \ldots \cdot a(\mathfrak{p}_r^{k_r})$. □

7.2 The coefficients of the L-series

For further computations we need the following notation. We write

$$c(m) := \sum_{\mathfrak{a} \text{ integral}, \, \mathcal{N}(\mathfrak{a})=m} a(\mathfrak{a})$$

for $m \in \mathbb{N}$. Then

$$L(s) = \sum_{m=1}^{\infty} c(m) m^{-s}.$$

The computation of the coefficients $c(m)$ is carried out recursively. We have $c(1) = a((1)) = 1$. If $m = m_1 m_2$ with $m_1, m_2 \in \mathbb{N}$ and $\gcd(m_1, m_2) = 1$, then suppose that

$$c(m_1) = \sum_{i=1}^{h_1} a(\mathfrak{a}_i) \quad \text{and} \quad c(m_2) = \sum_{j=1}^{h_2} a(\mathfrak{b}_j).$$

It follows that

$$\begin{aligned} c(m) &= \sum_{\mathfrak{a} \text{ integral}, \, \mathcal{N}(\mathfrak{a})=m} a(\mathfrak{a}) \\ &= \sum_{i=1}^{h_1} \sum_{j=1}^{h_2} a(\mathfrak{a}_i \mathfrak{b}_j) \\ &= \sum_{i=1}^{h_1} \sum_{j=1}^{h_2} a(\mathfrak{a}_i) a(\mathfrak{b}_j) \\ &= c(m_1) c(m_2). \end{aligned}$$

Hence for the computation of the coefficients of the L-series we only have to consider the computation of $c(p^k)$, where p is a prime number and $k \in \mathbb{N}_0$. Therefore we have to determine all ideals in $\mathcal{O}_{\mathbb{K}}$ with norm p^k.

Proposition 7.5. *Let p be a prime number with prime ideal factorization*

$$(p) = \mathfrak{p}_1^{e_1} \dots \mathfrak{p}_\mu^{e_\mu}$$

in $\mathcal{O}_{\mathbb{K}}$. The norm of $\mathfrak{p}_i$ is $\mathcal{N}(\mathfrak{p}_i) = p^{f_i}$, where $f_i = f_{\mathfrak{p}_i|p}$ is the residue class degree of $\mathfrak{p}_i$. Further let $k \in \mathbb{N}$. The ideals $\mathfrak{a}$ of $\mathcal{O}_{\mathbb{K}}$ with norm $\mathcal{N}(\mathfrak{a}) = p^k$ are the ideals given by

$$\mathfrak{a} = \mathfrak{p}_1^{k_1} \dots \mathfrak{p}_\mu^{k_\mu}$$

where the $k_i \in \mathbb{N}_0$ are non-negative integers with

$$k = \sum_{i=1}^{\mu} k_i f_i.$$

Proof. Only the ideals in the prime factorization of (p) can occur in the prime factorization of $\mathfrak{a}$. Therefore,

$$\begin{aligned}\mathcal{N}(\mathfrak{a}) &= \mathcal{N}(\mathfrak{p}_1^{k_1}\dots\mathfrak{p}_\mu^{k_\mu})\\ &= \mathcal{N}(\mathfrak{p}_1)^{k_1}\dots\mathcal{N}(\mathfrak{p}_\mu)^{k_\mu}\\ &= p^{k_1 f_1}\dots p^{k_\mu f_\mu}\\ &= p^k.\end{aligned}$$

□

Algorithm 7.6 (Coefficients of the L-series).

INPUT: An elliptic curve over $\mathbb{K}$, $M \in \mathbb{N}$.
OUTPUT: The coefficients $c(1), \dots, c(M)$ of the L-series $L(E|\mathbb{K}; s)$.
1. $c(1) \leftarrow 1$.
2. For $j = 2$ to M do:
3. Factorize $j = p^k q$ where p is the smallest prime factor of j, $k \in \mathbb{N}$, and $\gcd(p, q) = 1$.
4. If $q = 1$:
5. For all ideals $\mathfrak{a}_i$ of norm $\mathcal{N}(\mathfrak{a}_i) = p^k$ compute $a(\mathfrak{a}_i)$ according to Theorem 7.4.
6. Compute $c(p^k)$.
7. Else $c(j) \leftarrow c(p^k)c(q)$.
8. return $(c(1), \dots, c(M))$.

To illustrate the algorithm we give a short

Example. We consider the number field $\mathbb{K} = \mathbb{Q}(\sqrt{5})$ and the elliptic curve

$$E : Y^2 = X^3 + 2\sqrt{5}X$$

over $\mathbb{K}$. The discriminant of this curve is

$$\Delta = -2^9 \cdot 5 \cdot \sqrt{5}.$$

This curve has bad (additive) reduction at the prime ideal $\mathfrak{p}_2 = (2)$ lying over 2 and at the prime ideal $\mathfrak{p}_5$ with $\mathfrak{p}_5^2 = (5)$ lying over 5.

We compute the first 10 coefficients of the L-series of this curve over $\mathbb{K}$. The first coefficient is $c(1) = 1$. Then we choose $j = 2$. As this is a prime number, we get $p^k = 2, q = 1$. The residue class degree of 2 is $f_2 = 2$, hence there is no ideal with norm 2 in $\mathbb{K}$. We set $c(2) = 0$. For $j = 3$, we also get $f_3 = 0$ and hence $c(3) = 0$.

Now we choose $j = 4 = 2^2$. There is exactly one ideal with norm 4:

$$\mathcal{N}(\mathfrak{p}_2) = 2^{f_2} = 2^2.$$

As the curve has additive reduction at $\mathfrak{p}_2$, we get $a(\mathfrak{p}_2) = 0$ and hence $c(4) = 0$.

For $j = 5$ we get one ideal with norm 5:

$$\mathcal{N}(\mathfrak{p}_5) = 5^{f_5} = 5.$$

Here, the curve has also additive reduction and hence $a(\mathfrak{p}_5) = 0$ and $c(5) = 0$.

Then $c(6) = 0$, because $6 = 2 \cdot 3$ and $c(6) = c(2)c(3)$. The prime number 7 has also residue class degree $f_7 = 2$, therefore $c(7) = 0$. As $f_2 = 2$, there is no ideal with norm 8, because such an ideal would be a power of $\mathfrak{p}_2$ and then

$$\mathcal{N}(\mathfrak{p}_2^k) = 2^{f_2 k} = 2^{2k}$$

is an even power of 2.

Now we have $j = 9 = 3^2$. There is one ideal with norm 9:

$$\mathcal{N}(\mathfrak{p}_3) = 3^{f_3} = 3^2,$$

where $\mathfrak{p}_3 = (3)$ is the prime ideal lying over 3. The curve has good reduction at 3. We have to compute the number of points on the modulo $\mathfrak{p}_3$ reduced curve. Therefore we consider the set of representatives of the finite field $\mathbb{F}_{\mathfrak{p}_3}$:

$$\{a + b\sqrt{5} : a, b \in \{0, 1, 2\}\}.$$

We get $\sharp\widetilde{E}(\mathbb{F}_{\mathfrak{p}_3}) = 8$ and then

$$a(\mathfrak{p}_3) = \mathcal{N}(\mathfrak{p}_3) + 1 - \sharp\widetilde{E}(\mathbb{F}_{\mathfrak{p}_3}) = 9 + 1 - 8 = 2.$$

At last, $c(10) = c(2)c(5) = 0$.

We later need a general estimate for the coefficients of the L-series.

Proposition 7.7. *Let $E|\mathbb{K}$ be an elliptic curve over a number field $\mathbb{K}$ of degree $n = [\mathbb{K} : \mathbb{Q}]$. Let $L(s) = \sum_{m=1}^{\infty} c(m)m^{-s}$ be the L-series of $E|\mathbb{K}$. Then for $m \in \mathbb{N}$:*

$$|c(m)| \leq (4m)^n.$$

Proof. a) For an ideal $\mathfrak{a}$ of $\mathcal{O}_{\mathbb{K}}$, let $\sigma(\mathfrak{a})$ be the number of integral ideals of $\mathbb{K}$ which divide $\mathfrak{a}$. If $(\mathfrak{a}, \mathfrak{b}) = 1$, then it is easy to see $\sigma(\mathfrak{a}\mathfrak{b}) = \sigma(\mathfrak{a})\sigma(\mathfrak{b})$. We shall show that

$$\sigma(\mathfrak{a}) \leq 4^n \sqrt{\mathcal{N}(\mathfrak{a})},$$

where again $n = [\mathbb{K} : \mathbb{Q}]$. If $\mathfrak{p}$ is a prime ideal dividing 2 or 3 and $k \in \mathbb{N}$, then

$$\sigma(\mathfrak{p}^k) = k + 1 \leq 2 \cdot 2^{k/2} \leq 2\sqrt{\mathcal{N}(\mathfrak{p}^k)}.$$

If $\mathfrak{p}$ is a prime ideal dividing a prime number ≥ 5, then

$$\sigma(\mathfrak{p}^k) = k + 1 \leq 5^{k/2} \leq \sqrt{\mathcal{N}(\mathfrak{p}^k)}.$$

(The inequalities are easy to show by induction on k.)

Let $\mathfrak{a}$ be an ideal with prime ideal factorization

$$\mathfrak{a} = \prod_{i=1}^{\nu} \mathfrak{p}_i^{k_i} \prod_{i=\nu+1}^{\mu} \mathfrak{p}_i^{k_i},$$

where the $\mathfrak{p}_1, \ldots, \mathfrak{p}_\nu$ are prime ideals dividing 2 or 3 and the $\mathfrak{p}_{\nu+1}, \ldots, \mathfrak{p}_\mu$ are prime ideals dividing prime numbers ≥ 5. Then

$$\begin{aligned} \sigma(\mathfrak{a}) &= \prod_{i=1}^{\nu} \sigma(\mathfrak{p}_i^{k_i}) \prod_{i=\nu+1}^{\mu} \sigma(\mathfrak{p}_i^{k_i}) \\ &\leq \prod_{i=1}^{\nu} 2\sqrt{\mathcal{N}(\mathfrak{p}_i^{k_i})} \prod_{i=\nu+1}^{\mu} \sqrt{\mathcal{N}(\mathfrak{p}_i^{k_i})} \\ &= 2^{\nu}\sqrt{\mathcal{N}(\mathfrak{a})} \\ &\leq 2^{2n}\sqrt{\mathcal{N}(\mathfrak{a})} \\ &= 4^{n}\sqrt{\mathcal{N}(\mathfrak{a})}. \end{aligned}$$

b) Let $\mathfrak{p}$ be a prime ideal and $k \in \mathbb{N}$. We show that

$$|a(\mathfrak{p}^k)| \leq \sigma(\mathfrak{p}^k)\sqrt{\mathcal{N}(\mathfrak{p}^k)}.$$

If E has bad reduction at $\mathfrak{p}$, then

$$|a(\mathfrak{p}^k)| \leq 1 \leq \sigma(\mathfrak{p}^k)\sqrt{\mathcal{N}(\mathfrak{p}^k)}.$$

If E has good reduction at $\mathfrak{p}$, we write the local L-function as a product

$$L_{\mathfrak{p}}(T) = 1 - a_{\mathfrak{p}}T + \mathcal{N}(\mathfrak{p})T^2 = (1 - \alpha_{\mathfrak{p}}T)(1 - \bar{\alpha}_{\mathfrak{p}}T)$$

with

$$\alpha_{\mathfrak{p}} = \frac{a_{\mathfrak{p}} + \sqrt{a_{\mathfrak{p}}^2 - 4\mathcal{N}(\mathfrak{p})}}{2}, \quad \bar{\alpha}_{\mathfrak{p}} = \frac{a_{\mathfrak{p}} - \sqrt{a_{\mathfrak{p}}^2 - 4\mathcal{N}(\mathfrak{p})}}{2} \in \mathbb{C}.$$

By Theorem 3.3 of Hasse $|a_{\mathfrak{p}}| \leq 2\sqrt{\mathcal{N}(\mathfrak{p})}$, thus $a_{\mathfrak{p}}^2 - 4\mathcal{N}(\mathfrak{p}) \leq 0$, i.e. $\alpha_{\mathfrak{p}}$ and $\bar{\alpha}_{\mathfrak{p}}$ are conjugate complex numbers. This implies

$$|\alpha_{\mathfrak{p}}|^2 = \frac{a_{\mathfrak{p}}^2 - a_{\mathfrak{p}}^2 + 4\mathcal{N}(\mathfrak{p})}{4} = \mathcal{N}(\mathfrak{p}),$$

hence

$$|\alpha_{\mathfrak{p}}| = |\bar{\alpha}_{\mathfrak{p}}| = \sqrt{\mathcal{N}(\mathfrak{p})}.$$

In the domain of convergence of the following sums, we have

$$\begin{aligned}\sum_{k=0}^{\infty} a(\mathfrak{p}^k)T^k &= L_{\mathfrak{p}}(T)^{-1} \\ &= (1-\alpha_{\mathfrak{p}}T)^{-1}(1-\bar{\alpha}_{\mathfrak{p}}T)^{-1} \\ &= \Big(\sum_{i=0}^{\infty} \alpha_{\mathfrak{p}}^i T^i\Big)\Big(\sum_{j=0}^{\infty} \bar{\alpha}_{\mathfrak{p}}^j T^j\Big) \\ &= \sum_{k=0}^{\infty}\Big(\sum_{j=0}^{k} \alpha_{\mathfrak{p}}^j \bar{\alpha}_{\mathfrak{p}}^{k-j}\Big)T^k.\end{aligned}$$

Comparing coefficients we see that

$$|a(\mathfrak{p})^k| = \Big|\sum_{j=0}^{k} \alpha_{\mathfrak{p}}^j \bar{\alpha}_{\mathfrak{p}}^{k-j}\Big| \le (k+1)\sqrt{\mathcal{N}(\mathfrak{p})}^{k} = \sigma(\mathfrak{p}^k)\sqrt{\mathcal{N}(\mathfrak{p}^k)}.$$

c) Applying Part b) for the prime ideal factorization of an ideal $\mathfrak{a}$ of $\mathcal{O}_{\mathbb{K}}$, we get

$$|a(\mathfrak{a})| \le \sigma(\mathfrak{a})\sqrt{\mathcal{N}(\mathfrak{a})}.$$

With Part a) it follows that

$$|a(\mathfrak{a})| \le 4^n \mathcal{N}(\mathfrak{a}).$$

This is also true for $\mathfrak{a} = (1)$.

d) The number of integral ideals $\mathfrak{a}$ of Norm $\mathcal{N}(\mathfrak{a}) = m$ is $\le m^{n-1}$. This can be shown in the following way.

With easy combinatorics we can see that the number of ideals with norm equal to p^k, where p is a prime number, is less than or equal to

$$\binom{n+k-1}{k}.$$

With induction on k, one can prove that

$$\binom{n+k-1}{k} \le 2^{k(n-1)} \le p^{k(n-1)}.$$

Using the prime number factorization of m gives the assertion.

e) Let $m \in \mathbb{N}$. Then with Part c) and Part d) it follows that

$$\begin{aligned} |c(m)| &\leq \sum_{\substack{\mathfrak{a} \text{ intergral} \\ \mathcal{N}(\mathfrak{a})=m}} |a(\mathfrak{a})| \\ &\leq \sum_{\substack{\mathfrak{a} \text{ intergral} \\ \mathcal{N}(\mathfrak{a})=m}} 4^n \mathcal{N}(\mathfrak{a}) \\ &= \sum_{\substack{\mathfrak{a} \text{ intergral} \\ \mathcal{N}(\mathfrak{a})=m}} 4^n m \\ &\leq m^{n-1} 4^n m = (4m)^n. \end{aligned}$$

□

7.3 Continuation of the L-series

The L-series of elliptic curves are a priori defined only for complex numbers $s \in \mathbb{C}$ with $\mathrm{Re}(s) > \frac{3}{2}$. The conjecture of Hasse and Weil gives a functional equation for the L-series which can be used to find a continuation of this function to the whole complex plane. To formulate this conjecture we need the definition of the conductor, which measures how bad the reduction of E at the non-archimedean places v of $\mathbb{K}$ is.

Definition 7.8. Let $E|\mathbb{K}$ be an elliptic curve over a number field. The *conductor* of E is the divisor (see the footnote on page 200)

$$\mathfrak{N}_{E|\mathbb{K}} := \prod_{v \in M_{\mathbb{K}}^0} \mathfrak{p}_v^{f_v}$$

with

$$f_v = \begin{cases} 0, & \text{if } E \text{ has good reduction at } v, \\ 1, & \text{if } E \text{ has multiplicative reduction at } v, \\ 2, & \text{if } E \text{ has additive reduction at } v \text{ and } \mathrm{char}(\mathbb{F}_{\mathfrak{p}_v}) \neq 2, 3, \\ 2 + \delta_v, & \text{if } E \text{ has additive reduction at } v \text{ and } \mathrm{char}(\mathbb{F}_{\mathfrak{p}_v}) = 2 \text{ or } 3. \end{cases}$$

Here δ_v is computed with the formula of Ogg [156]:

$$\delta_v = v(\Delta_{E|\mathbb{K}}) - 1 - m_v,$$

where $v(\Delta_{E|\mathbb{K}})$ is the absolute value of the discriminant of an equation for E which is minimal at v. The integer m_v is the number of connected components of the special fibre of the Néron model of E, which can be determined with the algorithm of Tate [220].

Conjecture 7.9 (Hasse, Weil [235], see Husemöller [103], Chapter 16). *Let E be an elliptic curve over a number field $\mathbb{K}$ with L-series $L(s) = L(E|\mathbb{K}; s)$. Further let $\mathfrak{N} = \mathfrak{N}_{E|\mathbb{K}}$ be the conductor of E, $D_{\mathbb{K}}$ the discriminant of $\mathbb{K}|\mathbb{Q}$, and $n = [\mathbb{K} : \mathbb{Q}]$. We define*

$$A_{E,\mathbb{K}} := \mathcal{N}(\mathfrak{N})|D_{\mathbb{K}}|^2,$$
$$G(s) := (2\pi)^{-s}\Gamma(s)$$

with the usual Γ-function, and the function

$$\Lambda(s) := A_{E,\mathbb{K}}^{s/2} G(s)^n L(s).$$

Then

$$\Lambda(s) = \varepsilon\Lambda(2 - s),$$

with $\varepsilon \in \{\pm 1\}$.

The sign $\varepsilon \in \{\pm 1\}$ is called the *sign of the functional equation* of $E|\mathbb{K}$.

This conjecture has been proved for elliptic curves with complex multiplication by Deuring ([49], [50], [51], [52], [53]).

For elliptic curves over $\mathbb{Q}$ the Conjecture of Hasse and Weil is equivalent to the Conjecture of Shimura, Taniyama and Weil, which states that every elliptic curve over $\mathbb{Q}$ is modular. This conjecture has been proved in 1997 by Conrad, Diamond, and Taylor [38], in connection with the work of Taylor and Wiles [237], [221], for elliptic curves over $\mathbb{Q}$ which have a conductor not divisible by 27. In 1999, it has been proved by Breuil, Conrad, Diamond, and Taylor for all elliptic curves over $\mathbb{Q}$ (see Darmon [44]).

Conjecture 7.9 yields a possibility to construct an extension of the L-series for the whole complex plane, which we will explain now.

Theorem 7.10. *Let $E|\mathbb{K}$ be an elliptic curve over a number field $\mathbb{K}$ and $\Lambda(s)$ the function*

$$\Lambda(s) = A_{E,\mathbb{K}}^{s/2} G(s)^n L(s),$$

defined in the conjecture 7.9 of Hasse and Weil. Then there exists a function $g : \mathbb{R}_0^+ \to \mathbb{C}$ with

$$\Lambda(s) = \int_0^\infty g(t)t^s \frac{dt}{t}$$

for $s \in \mathbb{C}$ with $\mathrm{Re}(s) > 2$.

Hence the function Λ is the *Mellin transform* of the function g.

Proof. (i) We first show that there exists a function $h : \mathbb{R}_0^+ \to \mathbb{C}$ with

$$(2\pi)^{-ns} A_{E,\mathbb{K}}^{s/2} \Gamma(s)^n m^{-s} = \int_0^\infty h(mt)t^s \frac{dt}{t}$$

for all $m \in \mathbb{N}$. For this we use the representation of the Γ-function as Mellin transform

$$\Gamma(s) = \int_0^\infty \exp(-z) z^s \frac{dz}{z}$$

(see for example Hurwitz, Courant [102], I.6, §12). Then

$$\Gamma(s)^n = \int_0^\infty \cdots \int_0^\infty \exp(-z_1 - \cdots - z_n)(z_1 \ldots z_n)^s \frac{dz_1}{z_1} \cdots \frac{dz_n}{z_n}.$$

Substituting $w = z_1 \ldots z_n$ and changing the order of integration (which is allowed for $\mathrm{Re}(s) > 2$) leads to

$$\Gamma(s)^n = \int_0^\infty \cdots \int_0^\infty \exp\left(-\left(\frac{w}{z_2 \ldots z_n} + z_2 + \cdots + z_n\right)\right) \frac{dz_2}{z_2} \cdots \frac{dz_n}{z_n} w^s \frac{dw}{w}.$$

With the substitution $w = (2\pi)^n A_{E|\mathbb{K}}^{-1/2} mt$ we get

$$\Gamma(s)^n = (2\pi)^{ns} A_{E|\mathbb{K}}^{-s/2} m^s \int_0^\infty h(mt) t^s \frac{dt}{t}.$$

Here for $\mathbb{K} = \mathbb{Q}$ we have $n = 1$, and the function is

$$h(u) = \exp(-2\pi A_{E|\mathbb{K}}^{-1/2} u).$$

For $n = [\mathbb{K} : \mathbb{Q}] > 1$ we have

$$h(u) = \int_0^\infty \cdots \int_0^\infty \exp\left(-\left(\frac{(2\pi)^n A_{E|\mathbb{K}}^{-1/2} u}{z_2 \ldots z_n} + z_2 + \cdots + z_n\right)\right) \frac{dz_2}{z_2} \cdots \frac{dz_n}{z_n}$$

and with the substitution $z_i = 2\pi (A_{E|\mathbb{K}}^{-1/2} u)^{1/n} x_i$ (for $i = 2, \ldots, n$) we get

$$h(u) = \int_0^\infty \ldots \int_0^\infty \tilde{h}(x_2, \ldots, x_n) \frac{dx_2}{x_2} \cdots \frac{dx_n}{x_n}$$

with

$$\tilde{h}(x_2, \ldots, x_n) = \exp\left(-2\pi (A_{E|\mathbb{K}}^{-1/2} u)^{1/n} \left(\frac{1}{x_2 \ldots x_n} + x_2 + \cdots + x_n\right)\right).$$

(ii) We have from Part (i)

$$(2\pi)^{-ns} A_{E,\mathbb{K}}^{s/2} \Gamma(s)^n m^{-s} = \int_0^\infty h(mt) t^s \frac{dt}{t}.$$

Multiplying this equation with $c(m)$ and summing over all $m \in \mathbb{N}$ leads to

$$\Lambda(s) = (2\pi)^{-ns} A_{E,\mathbb{K}}^{s/2} \Gamma(s)^n L(s) = \sum_{m=1}^\infty \int_0^\infty c(m) h(mt) t^s \frac{dt}{t}.$$

Interchanging the sum and the integral, which is allowed for $\mathrm{Re}(s) > 2$, leads to

$$\Lambda(s) = \int_0^\infty \Big(\sum_{m=1}^\infty c(m)h(mt)\Big)t^s \frac{dt}{t} = \int_0^\infty g(t)t^s \frac{dt}{t}$$

with

$$g(t) = \sum_{m=1}^\infty c(m)h(mt).$$

□

We shall need the following lemma.

Lemma 7.11. *Let $m \in \mathbb{N}$, $t, \beta, a, b, c \in \mathbb{R}$, $t \geq 1$, $\beta > 0$, $a, c \geq 0$. There exists a constant $D \in \mathbb{R}$, $D > 0$, independent of t and m, such that for all $m \in \mathbb{N}$:*

$$m^a t^b \exp(-\beta(mt)^c) \leq Dt^{b-a}.$$

We may take

$$D = \beta^{-a/c}\left(\frac{a}{c}\right)^{a/c} \exp(-a/c).$$

Proof. a) It is an easy exercise to show that, for $\xi, x \in \mathbb{R}$, $\xi, x > 0$, the function $f(x) := x^\xi \exp(-x)$ takes its maximum at $x = \xi$, hence it can be estimated as

$$x^\xi \exp(-x) \leq \xi^\xi \exp(-\xi).$$

b) We have

$$\begin{aligned} m^a t^b \exp(-\beta(mt)^c) &= (m^c t^c)^{a/c} \exp(-\beta(mt)^c)t^{b-a} \\ &= \beta^{-a/c}(\beta m^c t^c)^{a/c} \exp(-\beta(mt)^c)t^{b-a}. \end{aligned}$$

With $x = \beta m^c t^c$ and $\xi = a/c$ in Part a) we get

$$m^a t^b \exp(-\beta(mt)^c) \leq \beta^{-a/c}\left(\frac{a}{c}\right)^{a/c} \exp(-a/c)t^{b-a} = Dt^{b-a}.$$

□

Proposition 7.12. *Let $E|\mathbb{K}$ be an elliptic curve over a number field $\mathbb{K}$ with L-series $L(s) = L(E|\mathbb{K}; s)$. Further let $g : \mathbb{R}_0^+ \to \mathbb{C}$ be the function obtained in Theorem 7.10. Then for all $z \in \mathbb{C}$ the integral*

$$\int_1^\infty g(t)t^z dt$$

exists.

Proof. a) It is easy to find the following estimate for the function $h(u)$ defined in Part (i) of the proof of Theorem 7.10:

$$h(u) \le \left(\frac{A_{E|\mathbb{K}}^{1/2n}}{\pi}\right)^{n-1} u^{-(n-1)/n} \exp(-\alpha u^{1/n})$$

with the constant

$$\alpha = \begin{cases} 2\pi A_{E|\mathbb{K}}^{-1/2}, & \text{if } n = 1, \\ 2\pi (n-1) A_{E|\mathbb{K}}^{-1/2n}, & \text{if } n > 1. \end{cases}$$

b) Let $z_1 = \mathrm{Re}(z)$ and, as above, $n = [\mathbb{K} : \mathbb{Q}]$.
We have from Proposition 7.7 and Part a)

$$\begin{aligned} |g(t)t^z| &= \Big| \sum_{m=1}^{\infty} c(m)h(mt)t^z \Big| \\ &\le \sum_{m=1}^{\infty} |c(m)||h(mt)|t^{z_1} \\ &\le \sum_{m=1}^{\infty} (4m)^n \left(\frac{A_{E|\mathbb{K}}^{1/2n}}{\pi}\right)^{n-1} (mt)^{-(n-1)/n} \exp(-\alpha (mt)^{1/n}) t^{z_1} \\ &= 4^n \left(\frac{A_{E|\mathbb{K}}^{1/2n}}{\pi}\right)^{n-1} \sum_{m=1}^{\infty} m^{n-(n-1)/n} t^{z_1-(n-1)/n} \exp(-\alpha (mt)^{1/n}) \end{aligned}$$

Now we apply Lemma 7.11 with

$$\begin{aligned} a &= n - \frac{n-1}{n} + 2 = n + 1 + \frac{1}{n}, \\ b &= z_1 - \frac{n-1}{n} + 2 = z_1 + 1 + \frac{1}{n}, \\ c &= \frac{1}{n}, \end{aligned}$$

and

$$\beta = \frac{\alpha}{2}.$$

It follows that there exists a $D_1 \in \mathbb{R}$, $D_1 > 0$, independent of t and m, such that for all $m \in \mathbb{N}$:

$$\begin{aligned} & m^{n-\frac{n-1}{n}+2} t^{z_1-\frac{n-1}{n}+2} \exp\left(-\frac{\alpha}{2}(mt)^{1/n}\right) \le D_1 t^{z_1-n} \\ \Leftrightarrow\ & m^{n-(n-1)/n} t^{z_1-(n-1)/n} \exp(-\alpha (mt)^{1/n}) \\ & \le D_1 t^{z_1-n} \exp\left(-\frac{\alpha}{2}(mt)^{1/n}\right) \frac{1}{m^2 t^2}. \end{aligned}$$

Therefore we get the estimate

$$|g(t)t^z| \leq 4^n \left(\frac{A_{E|\mathbb{K}}^{1/2n}}{\pi}\right)^{n-1} \sum_{m=1}^{\infty} D_1 t^{z_1-n} \exp\left(-\frac{\alpha}{2}(mt)^{1/n}\right) \frac{1}{m^2t^2}.$$

c) As $m \geq 1$, we see that

$$\exp\left(-\frac{\alpha}{2}(mt)^{1/n}\right) \leq \exp\left(-\frac{\alpha}{2}t^{1/n}\right).$$

It is easy to show that there exists a constant $D_2 \in \mathbb{R}$, $D_2 > 0$, D_2 independent of t and m, such that for all $t \geq 1$:

$$t^{z_1-n} \exp\left(-\frac{\alpha}{2}t^{1/n}\right) \leq D_2.$$

d) With Part c) we get for $t \geq 1$:

$$\begin{aligned} |g(t)t^z| &\leq 4^n \left(\frac{A_{E|\mathbb{K}}^{1/2n}}{\pi}\right)^{n-1} \sum_{m=1}^{\infty} D_1 D_2 \frac{1}{m^2t^2} \\ &= 4^n \left(\frac{A_{E|\mathbb{K}}^{1/2n}}{\pi}\right)^{n-1} D_1 D_2 \frac{\pi^2}{6} \frac{1}{t^2} \\ &= C\frac{1}{t^2} \end{aligned}$$

with

$$C = 4^n \left(\frac{A_{E|\mathbb{K}}^{1/2n}}{\pi}\right)^{n-1} D_1 D_2 \frac{\pi^2}{6}.$$

Therefore

$$\int_1^{\infty} |g(t)t^z| dt \leq C \int_1^{\infty} \frac{1}{t^2} dt = C.$$

□

Theorem 7.13. *Let $E|\mathbb{K}$ be an elliptic curve over a number field $\mathbb{K}$ with L-series $L(s) = L(E|\mathbb{K}; s)$. We assume that the conjecture of Hasse and Weil is satisfied. Then the function*

$$g(t) = \frac{1}{2\pi i} \int_{c-i\infty}^{c+i\infty} \Lambda(s) t^{-s} ds$$

has the functional equation

$$g(t) = \varepsilon t^{-2} g\left(\frac{1}{t}\right),$$

where ε is the sign of the functional equation of E.

Proof. The functional equation of Λ is

$$\Lambda(s) = \varepsilon\Lambda(2-s)$$

with $\varepsilon \in \{\pm 1\}$. Using this for the above integral we get

$$\begin{aligned} g(t) &= \frac{1}{2\pi i}\int_{c-i\infty}^{c+i\infty} \Lambda(s)t^{-s}ds \\ &= \varepsilon\frac{1}{2\pi i}\int_{c-i\infty}^{c+i\infty} \Lambda(2-s)t^{-s}ds. \end{aligned}$$

With the substitution $u = 2-s$ it follows that

$$\begin{aligned} g(t) &= \varepsilon\frac{1}{2\pi i}\int_{2-c+i\infty}^{2-c-i\infty} \Lambda(u)t^{u-2}(-1)du \\ &= \varepsilon\frac{1}{2\pi i}t^{-2}\int_{2-c-i\infty}^{2-c+i\infty} \Lambda(u)t^{u}du \\ &= \varepsilon t^{-2}g\left(\frac{1}{t}\right). \end{aligned}$$

□

Theorem 7.14. *Let $E|\mathbb{K}$ be an elliptic curve over a number field $\mathbb{K}$ with L-series $L(s) = L(E|\mathbb{K}; s)$. We assume that the conjecture of Hasse and Weil is satisfied. Then we can define an extension of $L(s)$ over the whole complex plane by*

$$L(s) = A_{E|\mathbb{K}}^{-s/2}G(s)^{-n}\int_1^{\infty} g(t)(\varepsilon t^{1-s} + t^{s-1})dt.$$

In particular, if $s = 1$ is a root of $L(s)$ of order $k \in \mathbb{N}_0$, then

$$L^{(k)}(1) = ((-1)^k\varepsilon + 1)(2\pi)^n A_{E,\mathbb{K}}^{-1/2}\int_1^{\infty} g(t)(\log t)^k dt.$$

Proof. We have with Theorem 7.10 (for $\mathrm{Re}(s) > 2$) and Theorem 7.13

$$\begin{aligned} \Lambda(s) &= \int_0^{\infty} g(t)t^s\frac{dt}{t} \\ &= \int_0^1 g(t)t^s\frac{dt}{t} + \int_1^{\infty} g(t)t^s\frac{dt}{t} \\ &= \int_0^1 \varepsilon t^{-2}g\left(\frac{1}{t}\right)t^s\frac{dt}{t} + \int_1^{\infty} g(t)t^s\frac{dt}{t}. \end{aligned}$$

Substituting in the first integral $u = \frac{1}{t}$ leads to

$$\begin{aligned} \Lambda(s) &= \int_{\infty}^1 \varepsilon u^2 g(u)u^{-s}(-1)\frac{du}{u} + \int_1^{\infty} g(t)t^s\frac{dt}{t} \\ &= \int_1^{\infty} \varepsilon g(u)u^{2-s}\frac{du}{u} + \int_1^{\infty} g(t)t^s\frac{dt}{t}. \end{aligned}$$

Now we write in the first integral t for u and get

$$\Lambda(s) = \int_1^\infty g(t)(\varepsilon t^{2-s} + t^s)\frac{dt}{t}$$
$$= \int_1^\infty g(t)(\varepsilon t^{1-s} + t^{s-1})dt.$$

From Proposition 7.12 we see that the right hand side of the above equation exists for all $s \in \mathbb{C}$. The first part of the theorem follows with the definition

$$\Lambda(s) = H(s)L(s)$$

with $H(s) = A_{E,\mathbb{K}}^{s/2} G(s)^n$ in Conjecture 7.9 of Hasse and Weil.

The formula for $L(1)$ follows directly.

If $s = 1$ is a root of order $k > 0$ of $L(s)$, then the derivatives $L^{(j)}(1)$ are zero for $j = 0, \dots k-1$. We have

$$\Lambda^{(k)}(s) = \sum_{j=0}^{k} \binom{k}{j} H^{(j)}(s) L^{(k-j)}(s)$$

and therefore

$$\Lambda^{(k)}(1) = H(1)L^{(k)}(1).$$

Then the desired formula for $L^{(k)}(1)$ follows directly. □

In this connection we also call attention to the paper [179] of Rubin.

7.4 Conjectures concerning the rank

From the theorem of Mordell and Weil we know, that the rank of an elliptic curve over a number field is finite. It is conjectured that the rank of an elliptic curve (even over $\mathbb{Q}$) can become arbitrarily large.

Conjecture 7.15. *There exist elliptic curves over $\mathbb{Q}$ with arbitrarily large rank.*

The record up to now is an elliptic curve over $\mathbb{Q}$ of rank ≥ 24, which was found by R. Martin and W. McMillen in 2000 (see the announcement from May 2000 in `http://listserv.nodak.edu/archives/nmbrthry.html`). There is a publication of an elliptic curve of rank ≥ 22 over $\mathbb{Q}$ by Fermigier [62].

Brumer and McGuinness [22] studied the behavior of the rank of elliptic curves in order to determine the average rank of elliptic curves over $\mathbb{Q}$.

Birch and Swinnerton-Dyer connected the behaviour of the L-series of an elliptic curve at $s = 1$ with its number-theoretic properties. Similar to L-series of number fields, not only the rank of the curve, but also other quantities are related to the L-series. We first give a weak version of the conjecture. This was first stated by Birch and Swinnerton-Dyer for elliptic curves over $\mathbb{Q}$ in [15] and can be generalized for arbitrary number fields $\mathbb{K}$ and even for abelian varieties (see the articles of Swinnerton-Dyer [218] and of Tate [219]). The generalization to arbitrary number fields $\mathbb{K}$ goes back to Birch and Swinnerton-Dyer and was implemented by Cremona and Serf [43].

Conjecture 7.16 (Birch and Swinnerton-Dyer). *Let E be an elliptic curve over a number field $\mathbb{K}$ with corresponding L-series $L(E|\mathbb{K}; s)$ and rank r. The function $L(E|\mathbb{K}; s)$ has a zero of order r at $s = 1$.*

Here one assumes implicitly the conjecture that the function L has a continuation.

There is also a second, more detailed version of the conjecture of Birch and Swinnerton-Dyer, where the behaviour of the L-series at $s = 1$ is given explicitly (see also Bloch [19]).

Conjecture 7.17 (Birch and Swinnerton-Dyer). *Let E be an elliptic curve of rank r over a number field $\mathbb{K}$ with corresponding L-series $L(E|\mathbb{K}; s)$. Let $R_{E,\mathbb{K}}$ be the regulator of E. Further let $D_{\mathbb{K}}$ be the discriminant of $\mathbb{K}$ and r_2 the number of complex places of $\mathbb{K}$. Then*

$$\lim_{s\to 1} \frac{L(E|\mathbb{K}; s)}{(s-1)^r} = \frac{\sharp Ш [\mathbb{K} : \mathbb{Q}]^r R_{E|\mathbb{K}}}{(\sharp E(\mathbb{K})_{\mathrm{tors}})^2} \frac{2^{r_2}}{|D_{\mathbb{K}}|^{1/2}} \prod_{v \in M_{\mathbb{K}}} c_v .$$

Here for non-archimedean absolute values $v_{\mathfrak{p}}$ corresponding to the prime ideals $\mathfrak{p}$ of $\mathbb{K}$, the values $c_{v_{\mathfrak{p}}}$ are the Tamagawa numbers at $\mathfrak{p}$ (see Theorem 4.11). The Tate–Shafarevich group Ш is explained in Section 7.5.

As pointed out in Theorem 4.11, the Tamagawa number at $\mathfrak{p}$ is equal to 1, if the curve has good reduction at $\mathfrak{p}$. If the curve has bad reduction at $\mathfrak{p}$, it can be computed via the Tate algorithm [220].

Rubin has drawn our attention to the fact that the height used in the original conjecture was not the normalized height. Therefore we have to multiply the constant on the right hand side with the factor $[\mathbb{K} : \mathbb{Q}]^r$ (see [178], [181]).

Until now this conjecture has been numerically checked for a large number of cases. Moreover, the weak version of it was proved for ranks 0 and 1.

In the following, let r' be the order of the zero of the L-series at $s = 1$ and $r = \mathrm{rk}(E(\mathbb{K}))$. The number r' is called the *analytic rank* of $E|\mathbb{K}$.

For elliptic curves with complex multiplication, the weak conjecture is proved in some special cases. If $\mathbb{K}$ is an imaginary quadratic field with class number 1 set $\mathbb{L} = \mathbb{K}$ or $\mathbb{L} = \mathbb{Q}$. Consider an elliptic curve $E|\mathbb{L}$ which has complex multiplication with $\mathrm{O}_{\mathbb{K}}$. Then Coates and Wiles [33] made the first step towards proving Conjecture 7.16. They showed that

$$r > 0 \Rightarrow r' > 0.$$

This proof was extended by Arthaud [4] and Rubin [178] for certain abelian extensions $\mathbb{F}$ of $\mathbb{K}$. Rubin also gives an exposition in [181] of recent results concerning the Birch and Swinnerton-Dyer conjecture for elliptic curves with complex multiplication. Apart from the above result of Coates and Wiles, it is also proved that if $r = 0$, then for every prime $p > 7$ such that E has good reduction above p, the p-part of the Tate–Shafarevich group of E has the order predicted by the Birch and Swinnerton-Dyer conjecture.

Rubin's methods were also used by Gonzalez-Aviles [84] to prove the Birch and Swinnerton-Dyer conjecture for a large class of elliptic curves with complex multiplication by $\mathbb{Q}(\sqrt{-7})$.

For elliptic curves over $\mathbb{Q}$, Kolyvagin [110] has proved

$$r' = 0 \Rightarrow r = 0$$

and

$$r' = 1 \Rightarrow r = 1.$$

Note that for elliptic curves over $\mathbb{Q}$ there have been some results before, which are included in the results of Kolyvagin. For example, Gross and Zagier [86] have shown that in this case $r' = 1$ implies $r \geq 1$. Greenberg [85] has shown that if the elliptic curve has complex multiplication and if the analytic rank r' (which is the order to which $L(s)$ vanishes at $s = 1$) is odd, then $r \geq 1$ or for infinitely many prime numbers p the p-part of the Tate–Shafarevich group is infinite (which is conjectured to be impossible).

We can use the conjecture of Birch and Swinnerton-Dyer to compute the rank of elliptic curves over number fields. The idea is to compute the order of vanishing of the L-series of an elliptic curve at $s = 1$, assuming that $r' = r$.

For elliptic curves over $\mathbb{Q}$ this conditional algorithm first stated by Manin [136], is implemented in SIMATH. The second author [250] proposed to generalize this algorithm for elliptic curves over number fields, which was carried out by the first author [189].

With the formula from Theorem 7.14 we can compute numerical estimates for the values of the L-function and its derivatives at $s = 1$, always assuming the conjecture of Hasse and Weil to be true (see Schmitt [190]). However, it is difficult to decide if a derivative $L^{(k)}(s)$ is 0 at $s = 1$ (see Gebel and Zimmer [78]).

7.5 The Selmer and the Tate–Shafarevich group

To define the Selmer and the Tate–Shafarevich group, we need some results from cohomology. These may be found in the books of Neukirch et al. [155], [158].

Let $E|\mathbb{K}$ and $E'|\mathbb{K}$ be two isogenous elliptic curves over $\mathbb{K}$ with a nonzero isogeny $\phi : E \to E'$ defined over $\mathbb{K}$ and let $G = G_{\overline{\mathbb{K}}|\mathbb{K}}$. Define $E[\phi] = \ker(\phi)$. There is an exact sequence of G-modules and morphisms

$$0 \to E[\phi] \to E \xrightarrow{\phi} E' \to 0.$$

With Galois cohomology we get the long exact sequence:

$$\begin{array}{cccccc} 0 \to & E(\mathbb{K})[\phi] & \to & E(\mathbb{K}) & \xrightarrow{\phi} & E'(\mathbb{K}) \\ \to & H^1(G, E[\phi]) & \to & H^1(G, E) & \to & H^1(G, E') \to \cdots \end{array}$$

From this long exact sequence we can form the fundamental short exact sequence:

$$0 \to E'(\mathbb{K})/\phi(E(\mathbb{K})) \to H^1(G, E[\phi]) \to H^1(G, E)[\phi] \to 0.$$

Now we can carry out the same construction for all completions $\mathbb{K}_v$ of the number field $\mathbb{K}$. In this manner, we get, for every $v \in M_{\mathbb{K}}$, the exact sequence

$$0 \to E'(\mathbb{K}_v)/\phi(E(\mathbb{K}_v)) \to H^1(G_v, E[\phi]) \to H^1(G_v, E)[\phi] \to 0,$$

where $G_v \subset G$ is the decomposition group at v. Hence we obtain the following commutative diagram:

$$\begin{array}{ccccccc} 0 \to & E'(\mathbb{K})/\phi(E(\mathbb{K})) & \to & H^1(G, E[\phi]) & \to & H^1(G, E)[\phi] & \to 0 \\ & \downarrow & & \downarrow & & \downarrow & \\ 0 \to & \prod\limits_{v \in M_{\mathbb{K}}} E'(\mathbb{K}_v)/\phi(E(\mathbb{K}_v)) & \to & \prod\limits_{v \in M_{\mathbb{K}}} H^1(G_v, E[\phi]) & \to & \prod\limits_{v \in M_{\mathbb{K}}} H^1(G_v, E)[\phi] & \to 0. \end{array}$$

Definition 7.18. In the situation of this section the *ϕ-Selmer group of $E|\mathbb{K}$* is

$$S^{\phi}(E|\mathbb{K}) := \ker\left\{H^1(G, E[\phi]) \to \prod_{v \in M_{\mathbb{K}}} H^1(G_v, E)[\phi]\right\}.$$

The *Tate–Shafarevich group* $Ш(E|\mathbb{K})$ is the subgroup of $H^1(G, E)$ defined by setting

$$Ш(E|\mathbb{K}) := \ker\left\{H^1(G, E) \to \prod_{v \in M_{\mathbb{K}}} H^1(G_v, E)\right\}.$$

The *ϕ-torsion* $Ш(E|\mathbb{K})[\phi]$ *of the Tate–Shafarevich group* is the subgroup of $H^1(G, E)[\phi]$ defined by

$$Ш(E|\mathbb{K})[\phi] := \ker\left\{H^1(G, E)[\phi] \to \prod_{v \in M_{\mathbb{K}}} H^1(G_v, E)[\phi]\right\}.$$

There is no known algorithm to compute the Tate–Shafarevich group of an elliptic curve. Lemmermeyer [126] could determine the ϕ-part of the Tate–Shafarevich groups of parametrized elliptic curves. Curves with non-trivial Tate–Shafarevich groups are given for example by Lind, [131], Reichardt [173], Selmer [195], [196], Yoshida [239], and Poonen [171]. We refer also to the article [180] of Rubin.

It is conjectured that the Tate–Shafarevich group $Ш(E|\mathbb{K})$ is finite and that the number of elements in $Ш(E|\mathbb{K})$ is a square (see Cassels [27]).

In the following we give one example for a non-trivial 2-part of the Tate–Shafarevich group of an elliptic curve. We show that the equation

$$2Y^2 = 1 - 17X^4$$

defining an elliptic curve over the number field $\mathbb{K} = \mathbb{Q}(\sqrt{2})$ has no rational solutions over $\mathbb{Q}$. More generally, this can be done for the equation

$$2Y^2 = 1 - pX^4,$$

where $p \in \mathbb{P}$ is an arbitrary prime congruent to 1 modulo 8:

$$p \equiv 1 \pmod 8$$

and 2 is a quartic non-residue modulo p.

We give here the proof of Silverman [204], Chapter X, Proposition 6.5. First we need two elementary lemmata.

Lemma 7.19 (Fermat). *The odd prime $p \in \mathbb{P}$ is the sum of two squares in $\mathbb{Z}$:*

$$p = a^2 + b^2 \quad (a, b \in \mathbb{Z})$$

if and only if

$$p \equiv 1 \pmod 4.$$

Proof (cf. Nathanson [154], proof of Lagrange's Theorem).

We note that Lemma 7.19 holds trivially also for the even prime $p = 2$ since

$$2 = 1^2 + 1^2.$$

If $p = a^2 + b^2$, the condition $p \equiv -1 \pmod 4$ leads to a contradiction since then say

$$a^2 \equiv -1 \pmod 4 \quad \text{and} \quad b^2 \equiv 0 \pmod 4$$

or

$$a^2 \equiv 1 \pmod 4 \quad \text{and} \quad b^2 \equiv 2 \pmod 4.$$

Let therefore

$$p \equiv 1 \pmod 4.$$

Then, -1 is a quadratic residue modulo p because by the first supplementary theorem to the quadratic reciprocity law for the Legendre symbol (see Hasse [92]),

$$\left(\frac{-1}{p}\right) = (-1)^{\frac{p-1}{2}} = 1.$$

We have therefore

$$a^2 \equiv -b^2 \pmod{p} \tag{7.1}$$

for some

$$a, b \in \left\{\pm 1, \pm 2, \dots, \pm \frac{p-1}{2}\right\}$$

since there are as many residues as non-residues modulo p, namely $\frac{p-1}{2}$.

The congruence (7.1) means that

$$a^2 + b^2 = kp$$

for some $k \in \mathbb{Z}$, $k > 0$. It follows that

$$p \le kp = a^2 + b^2 \le 2\left(\frac{p-1}{2}\right)^2 < \frac{p^2}{2} < p^2$$

so that

$$1 \le k < p.$$

Let $k_0 \in \mathbb{Z}$, $k_0 > 0$, be minimal with

$$k_0 p = a^2 + b^2, \quad 1 \le k_0 \le k < p \text{ and } a, b \in \mathbb{Z} \text{ as above.} \tag{7.1$_0$}$$

We shall show that $k_0 = 1$. Assume the contrary. Then

$$1 < k_0 < p. \tag{7.2}$$

We choose $a_0, b_0 \in \mathbb{Z}$ such that

$$a_0 \equiv a \pmod{k_0}, \quad b_0 \equiv b \pmod{k_0}, \tag{7.3}$$

and

$$-\frac{k_0}{2} < a_0, b_0 \le \frac{k_0}{2}$$

(because $\left(-\frac{k_0}{2}\right)^2 = \left(\frac{k_0}{2}\right)^2$). By (7.3) we then have

$$a_0^2 + b_0^2 \equiv a^2 + b^2 = k_0 p \equiv 0 \pmod{k_0}. \tag{7.4}$$

This congruence implies the identity

$$a_0^2 + b_0^2 = x k_0 \quad \text{for some } x \in \mathbb{Z}, x \ge 0. \tag{7.5}$$

If $x = 0$, then $a_0 = b_0 = 0 \Rightarrow k_0 \mid a$ and $k_0 \mid b$ by (7.3) $\Rightarrow k_0^2 \mid a^2$ and $k_0^2 \mid b^2$. This entails by (7.4) that $k_0^2 \mid k_0 p \Rightarrow k_0 \mid p$ contradicting (7.2).

Thus $x \ge 1$. Moreover, by (7.5),

$$x k_0 = a_0^2 + b_0^2 \le 2\left(\frac{k_0}{2}\right)^2 < k_0^2 \Rightarrow x < k_0.$$

In sum,

$$1 \leq x < k_0. \tag{7.6}$$

Now we get by (7.1_0) and (7.5)

$$\begin{aligned} k_0^2 xp &= (k_0 p)(x k_0) \\ &= (a^2 + b^2)(a_0^2 + b_0^2) \\ &= a^2 a_0^2 + a^2 b_0^2 + a_0^2 b^2 + b^2 b_0^2 \\ &= c^2 + d^2 \end{aligned} \tag{7.7}$$

with

$$\begin{aligned} c &:= a a_0 + b b_0, \\ d &:= a b_0 - a_0 b. \end{aligned}$$

The congruences (7.3) yield by means of (7.1_0)

$$c \equiv d \equiv 0 \pmod{k_0}$$

so that

$$e := \frac{c}{k_0}, \; f := \frac{d}{k_0} \in \mathbb{Z}.$$

From (7.7) we obtain then the relation

$$xp = e^2 + f^2$$

contradicting the minimality of k_0 in (7.1_0) by (7.6). Hence $k_0 = 1$. □

Lemma 7.20 (Gauß). *Let $p \in \mathbb{P}$ be a prime satisfying*

$$p \equiv 1 \pmod 8$$

Then, if

$$p = a^2 + b^2 \quad \textit{with } a, b \in \mathbb{Z},$$

$$\left(\frac{2}{4}\right)_4 = (-1)^{\frac{ab}{4}}$$

that is, 2 is a quartic residue modulo p if and only if

$$ab \equiv 0 \pmod 8.$$

Proof. The proof is given by Silverman [204] who in turn follows Mordell:

By Lemma 7.19, there exist $a, b \in \mathbb{Z}$ such that

$$a^2 + b^2 = p \tag{7.8}$$

for a prime $p \in \mathbb{P}$ with $p \equiv 1 \pmod 4$, which is clear from the assumption $p \equiv 1 \pmod 8$. Therefore,

$$\begin{aligned}(a+b)^{\frac{p-1}{2}} &= (a+b)^{2\cdot\frac{p-1}{4}}\\ &\equiv (2ab)^{\frac{p-1}{4}} \pmod p\\ &\equiv 2^{\frac{p-1}{4}} a^{\frac{p-1}{4}} b^{2\cdot\frac{p-1}{8}} \pmod p\\ &\equiv 2^{\frac{p-1}{4}} a^{\frac{p-1}{4}} (-a^2)^{\frac{p-1}{8}} \pmod p\\ &\equiv (-1)^{\frac{p-1}{8}} 2^{\frac{p-1}{4}} a^{\frac{p-1}{2}} \pmod p.\end{aligned}$$

By Euler's criterion and the second supplementary theorem for the Legendre symbol (see Hasse [92]) and for the corresponding quartic symbol (see Lemmermeyer [127]), this means in a mixture of quadratic and quartic symbols, viz.

$$\left(\frac{a+b}{p}\right) = (-1)^{\frac{p-1}{8}} \left(\frac{2}{p}\right)_4 \left(\frac{a}{p}\right). \tag{7.9}$$

In the congruence (7.8), one of the numbers $a, b \in \mathbb{Z}$ must be odd and one even. We suppose that $a \in \mathbb{Z}$ is odd. Then, since $p \equiv 1 \pmod 4$, we conclude from the quadratic reciprocity law for the Jacobi symbol (see Hasse [92]) and from the congruence (7.8) that the Legendre symbol $\left(\frac{a}{p}\right)$ equals 1:

$$\left(\frac{a}{p}\right) = \left(\frac{p}{a}\right) = \left(\frac{b^2}{a}\right) = 1.$$

(Here, $\left(\frac{p}{a}\right)$ and $\left(\frac{b^2}{a}\right)$ are, of course, Jacobi symbols.) By the quadratic reciprocity law for the Jacobi symbol and because of the relation $\left(\frac{2^2}{a+b}\right) = 1$, we have furthermore

$$\left(\frac{a+b}{p}\right) = \left(\frac{p}{a+b}\right) = \left(\frac{2}{a+b}\right)\left(\frac{2p}{a+b}\right) = \left(\frac{2}{a+b}\right), \tag{7.10}$$

the latter identity following from (7.8):

$$2p = (a+b)^2 + (a-b)^2 = 2a^2 + 2b^2.$$

The relation (7.9) thus yields

$$\left(\frac{2}{a+b}\right) = (-1)^{\frac{p-1}{8}} \left(\frac{2}{p}\right)_4. \tag{7.11}$$

But another application of the second supplementary theorem shows that

$$\left(\frac{2}{a+b}\right) = (-1)^{\frac{(a+b)^2-1}{8}}. \tag{7.12}$$

Together, (7.8) and (7.12) render

$$\left(\frac{2}{p}\right)_4 = (-1)^{\frac{(a+b)^2-1}{8} - \frac{p-1}{8}} = (-1)^{\frac{ab}{4}} \tag{7.13}$$

as asserted. □

Since $17 = 1^2 + 4^2$ and $1 \cdot 4 \not\equiv 0 \pmod 8$, Lemma 7.20 shows that 2 is a quartic non-residue modulo 17. Now we are able to prove (see Silverman [204], Chapter X, Proposition 6.5)

Theorem 7.21. *The equation*

$$E:\ 2Y^2 = 1 - pX^4 \tag{7.14}$$

in which $p \in \mathbb{P}$ is a prime satisfying

$$p \equiv 1 \pmod 8$$

and such that

2 is a quartic non-residue modulo p

has no rational solutions over $\mathbb{Q}$.

Proof. We remark first that the prime $p = 17$ fulfills the hypothesis of the theorem.

It is easy to show that a rational solution (x, y) of E can be written in the form (see Chahal [31]

$$x = \frac{r}{t}, \quad y = \frac{s}{t^2}$$

with $r, s, t \in \mathbb{Z}$, $t > 0$ and $\gcd(r, s, t) = 1$. Hence, we obtain from (7.14) the new identity

$$2s^2 = t^4 - pr^4. \tag{7.15}$$

Let $q \in \mathbb{P}$ be an odd prime which divides s:

$$q \mid s, \quad q \neq 2,\ q \in \mathbb{P}.$$

Then, by Equation (7.15) and by quadratic reciprocity,

$$\left(\frac{q}{p}\right) = \left(\frac{p}{q}\right) = 1.$$

This is valid for each $q \mid s$, $q \neq 2$, $q \in \mathbb{P}$.

Since, by the second supplementary law, also

$$\left(\frac{2}{p}\right) = (-1)^{\frac{p^2-1}{8}} = 1,$$

we obtain

$$\left(\frac{s}{p}\right) = 1. \tag{7.16}$$

Hence

$$\left(\frac{s^2}{p}\right)_4 = 1 \tag{7.17}$$

that is, s^2 is a quartic residue modulo p which is equivalent to (7.16). But the Equation (7.15) shows that

$$\left(\frac{2s^2}{p}\right)_4 = 1$$

so that

$$\left(\frac{2}{p}\right)_4 = 1.$$

This is a contradiction to the choice of $p \in \mathbb{P}$ by which

$$\left(\frac{2}{p}\right)_4 \neq 1.$$

Thus,

$$E(\mathbb{Q}) = \emptyset.$$ □

On the other hand, the curve

$$2Y^2 = 1 - pX^4 \tag{7.18}$$

with a prime $p \in \mathbb{P}$ satisfying the congruence $p \equiv 1 \pmod{16}$ has a rational point everywhere locally, that is, over all local fields $\mathbb{Q}_q$ including $\mathbb{Q}_\infty = \mathbb{R}$. We shall show this following Reichardt [173]. Then this assertion holds in particular for the curve

$$2Y^2 = 1 - 17X^4.$$

We know in addition that, by Lemma 7.20, the number 2 is not a quartic residue modulo 17, since $17 = 1^2 + 4^4$ and $1 \cdot 4 \not\equiv 0 \pmod 8$.

For $q = \infty$, the curve (7.18) has, e.g., the two rational points

$$\pm P = \pm(x, y) := \pm\left(0, \frac{\sqrt{2}}{2}\right)$$

over $\mathbb{Q}_\infty = \mathbb{R}$.

For $q = 2$, one verifies first that p is a 2-adic fourth power by proving that the congruence

$$z^4 \equiv p \pmod{2^n} \tag{7.19}$$

is solvable for each $n \in \mathbb{N}$. By Hasse [92], for $n \geq 3$,

$$z \equiv (-1)^\xi (1 + 2^2)^\eta \pmod{2^n}$$

and

$$p \equiv (-1)^{\alpha}(1+2^2)^{\beta} \pmod{2^n}$$

with exponents in $\mathbb{Z}$

$$\alpha, \xi \pmod 2, \quad \beta, \eta \pmod{2^{n-2}}.$$

Hence, the congruence (7.19) is solvable if and only if

$$4\xi \equiv \alpha \pmod 2 \quad \text{and} \quad 4\eta \equiv \beta \pmod{2^{n-2}}. \tag{7.20}$$

Since $p \equiv 1 \pmod{16}$,

$$\alpha \equiv 0 \pmod 2 \quad \text{and} \quad \beta \equiv 4 \pmod{2^{n-2}}$$

as can be shown. Therefore the necessary and sufficient solvability condition (see Hasse [92]) for the congruence system (7.20) is that

$$\gcd(4,2) = 2 \mid \alpha \quad \text{and} \quad \gcd(4, 2^{n-2}) \mid \beta,$$

where $\gcd(4, 2^{n-2}) = 2$ for $n = 3$ and $\gcd(4, 2^{n-2}) = 4$ for $n \geq 4$. The latter conditions are trivially fulfilled for $n \geq 3$, so that (7.19) admits a solution for each $n \in \mathbb{N}$ (the cases of $n = 1, 2$ being trivial since $p \equiv 1 \pmod 2$ and $p \equiv 1 \pmod{2^2}$).

The curve (7.18) admits then a solution $P = (x, y)$ over $\mathbb{Q}_2$ with $y = 0$ and x from the equation

$$pX^4 = 1$$

over $\mathbb{Q}_2$.

For $q = p$, the number 2 is a p-adic square. This is true because for $p \equiv 1 \pmod{16}$, the Legendre symbol, by the second supplementary theorem (see Hasse [92]), is

$$\left(\frac{2}{p}\right) = (-1)^{\frac{p^2-1}{8}} = 1$$

which implies that there is a p-adic number $a \in \mathbb{Z}_p$ such that the generalized Legendre symbol is also 1 (see Hasse [92]):

$$\left(\frac{a}{p}\right) = \left(\frac{2}{p}\right) = 1.$$

Therefore, the p-adic number $a \in \mathbb{Z}_p$ is a square in $\mathbb{Z}_p$:

$$a = b^2$$

and $b \in \mathbb{Z}_p$ is a solution of the congruence

$$z^2 \equiv 2 \pmod p,$$

so that in particular

$$a \equiv 2 \pmod p.$$

A rational point over $\mathbb{Q}_p$ on the curve (7.18) is then given for instance as

$$P = (x, y) = (0, b^{-1}).$$

For $q \neq 2, p, \infty$, we argue as follows.

Remark 7.22. If $x = 0$, we must show that

$$2Y^2 = 1$$

has a solution $Y = c \in \mathbb{Q}_p$, so that $P = (0, c)$ is a rational point of the curve (7.18) over $\mathbb{Q}_q$. To this end, it is necessary that

$$\left(\frac{c}{q}\right) = \left(\frac{2}{q}\right) = (-1)^{\frac{q^2-1}{8}} = 1$$

again by the second supplementary theorem.

This latter condition is satisfied if and only if $q \equiv \pm 1 \pmod 8$.

However, we have also to deal with primes $q \equiv \pm 5 \pmod 8$. For arbitrary primes $q \neq 2, p$, the curve (7.18) defines a function field

$$\mathbb{F} := \mathbb{F}_q(X, Y)$$

over the finite field $\mathbb{F}_q$. By a theorem of F.K.Schmidt [188] such a function field $\mathbb{F}|\mathbb{F}_q$ always has a prime divisor of degree 1, hence there exists a solution $P = (x, y)$ of (7.18) in $\mathbb{F}_q$, i.e.

$$2y^2 \equiv 1 - px^4 \pmod q. \tag{7.21}$$

Now let

$$x = \sum_{i=0}^{\infty} \xi_i q^i, \quad y = \sum_{j=0}^{\infty} \eta_j q^j \quad \text{with } \xi_i, \eta_j \in \{0, 1, \dots, q-1\}.$$

Then,

$$2\eta_0^2 \equiv 1 - p\xi_0^4 \pmod q \tag{7.21$_0$}$$

and

$$\gcd(\xi_0, \eta_0) = 1.$$

If we assume by induction that, for $n \geq 1$,

$$\xi_0, \xi_1, \dots, \xi_{n-1}, \eta_0, \eta_1 \dots, \eta_{n-1} \in \{0, 1, \dots, q-1\}$$

have been determined such that

$$2(\eta_0 + \eta_1 q + \cdots + \eta_{n-1}q^{n-1})^2 \\ \equiv 1 - p(\xi_0 + \xi_1 q + \cdots + \xi_{n-1}q^{n-1})^4 \pmod{q^n} \qquad (7.21_{n-1})$$

and

$$\gcd(\xi_0 + \xi_1 q + \cdots + \xi_{n-1}q^{n-1}, \eta_0 + \eta_1 q + \cdots + \eta_{n-1}q^{n-1}) = 1,$$

then

$$\xi_n, \eta_n \in \{0, 1, \ldots, q-1\}$$

can be found such that

$$2(\eta_0 + \eta_1 q + \cdots + \eta_{n-1}q^{n-1} + \eta_n q^n)^2 \\ \equiv 1 - p(\xi_0 + \xi_1 q + \cdots + \xi_{n-1}q^{n-1} + \xi_n q^n)^4 \pmod{q^{n+1}} \qquad (7.21_n)$$

and

$$\gcd(\xi_0 + \xi_1 q + \cdots + \xi_{n-1}q^{n-1} + \xi_n q^n, \eta_0 + \eta_1 q + \cdots + \eta_{n-1}q^{n-1} + \eta_n q^n) = 1.$$

Indeed the congruence (7.21_n) means that

$$\frac{2(\eta_0 + \eta_1 q + \cdots + \eta_{n-1}q^{n-1})^2}{q^n} + 4(\eta_0 + \eta_1 q + \cdots + \eta_{n-1}q^{n-1})\eta_n \\ \equiv \frac{1 - p(\xi_0 + \xi_1 q + \cdots + \xi_{n-1}q^{n-1})^4}{q^n} \\ -4p(\xi_0 + \xi_1 q + \cdots + \xi_{n-1}q^{n-1})^3\xi_n \pmod{q}$$

in which the fraction

$$\frac{2(\eta_0 + \cdots + \eta_{n-1}q^{n-1})^2}{q^n} - \frac{1 - p(\xi_0 + \cdots + \xi_{n-1}q^{n-1})^4}{q^n} =: -k_n \in \mathbb{Z}$$

is an integer by (7.21_{n-1}), so that the linear congruence in ξ_n, η_n

$$4p(\xi_0 + \xi_1 q + \cdots + \xi_{n-1}q^{n-1})^3\xi_n + 4(\eta_0 + \eta_1 q + \cdots + \eta_{n-1}q^{n-1})\eta_n \equiv k_n \pmod{q}$$

can be solved for ξ_n, η_n because $q \neq 2, p$ and $p \nmid (\eta_0 + \eta_1 q + \cdots + \eta_{n-1}q^{n-1})$ by (7.21_{n-1}). Then trivially

$$\gcd(\xi_0 + \xi_1 q + \cdots + \xi_{n-1}q^{n-1}, \eta_0 + \eta_1 q + \cdots + \eta_{n-1}q^{n-1}) = 1 \mid k_n.$$

One then chooses $\xi_n, \eta_n \in \{0, 1, \ldots, q-1\}$. From (7.21_n) it follows that also

$$\gcd(\xi_0 + \xi_1 q + \cdots + \xi_{n-1}q^{n-1} + \xi_n q^n, \eta_0 + \eta_1 q + \cdots + \eta_{n-1}q^{n-1} + \eta_n q^n) = 1.$$

Remark 7.23. $E|\mathbb{K}$, where $\mathbb{K} = \mathbb{Q}(\sqrt{2})$, is an elliptic curve in the usual sense (see e.g. Roquette [177]). The Jacobian of the elliptic curve

$$Y^2 = \frac{1}{2} - \frac{p}{2}X^4$$

is (see Connell [37]) the elliptic curve over $\mathbb{Q}$

$$E_p : Y^2 = X^3 + pX.$$

It has the 2-part of the Tate–Shafarevich group $Ш(E_p|\mathbb{Q})[2] \cong \mathbb{Z}/2\mathbb{Z} \times \mathbb{Z}/2\mathbb{Z}$. The rank of E_p over $\mathbb{Q}$ is $r_p = 0$.

Proof. Silverman [204] Chapter X, Proposition 6.5. □

The Tate–Shafarevich group is also important for certain applications in elementary number theory, see for example Villegas and Zagier [228] and Cox [40].

We now consider the Selmer group.

Theorem 7.24. *Let $\phi : E|\mathbb{K} \to E'|\mathbb{K}$ be an isogeny of elliptic curves defined over a number field $\mathbb{K}$.*

a) *There is an exact sequence*

$$0 \to E'(\mathbb{K})/\phi(E(\mathbb{K})) \to S^{\phi}(E|\mathbb{K}) \to Ш(E|\mathbb{K})[\phi] \to 0.$$

b) *The Selmer group $S^{\phi}(E|\mathbb{K})$ is finite.*

Proof. a) This follows immediately from the definition.

b) This is essentially analogous to the proof of the Weak Mordell–Weil Theorem 5.1 (see, e.g., Silverman [204] Chapter X, Theorem 4.2). □

In the application, one often uses $E = E'$ and $\phi =$ multiplication by 2, or isogenies ϕ of degree 2, 3, or 4 (2-descent, 3-descent (Quer [172]), or 4-descent (Merriman, Siksek, Smart [145])). We explain 2-descent further in the next section.

The existing methods for determining the rank r and basis points of $E(\mathbb{Q})$ all depend more or less on the Birch and Swinnerton-Dyer conjecture 7.16 and rely on the fundamental exact sequence

$$0 \to E(\mathbb{Q})/mE(\mathbb{Q}) \to S^m(E|\mathbb{Q}) \to Ш(E|\mathbb{Q})[m] \to 0.$$

Since the Selmer group $S^m(E|\mathbb{Q})$ can be computed and – depending on $m \in \mathbb{N}, m \geq 2$ – essentially

$$r = \text{rank}(E(\mathbb{Q})/mE(\mathbb{Q})),$$

it all is a matter of computing the Tate–Shafarevich group $Ш(E|\mathbb{Q})[m]$. However, this is a difficult task as the example of

$$Ш(E|\mathbb{Q})[2] \cong \mathbb{Z}/2\mathbb{Z} \times \mathbb{Z}/2\mathbb{Z}$$

shows.

7.6 2-descent

There are different descent methods to compute the Selmer group and to estimate the rank of elliptic curves. The most common methods are 2*-descent methods.* For elliptic curves over $\mathbb{Q}$, SIMATH offers computation of the rank using those 2-descent methods.

The 2-descent methods can be divided into three categories, depending on the 2-part of the torsion group.

Let $E|\mathbb{K}$ be an elliptic curve over a number field $\mathbb{K}$. If

$$E(\mathbb{K})[2] \cong \mathbb{Z}/2\mathbb{Z} \times \mathbb{Z}/2\mathbb{Z},$$

then one can apply complete 2-descent. If there exists at least one nontrivial torsion point of order 2 in $E(\mathbb{K})$, then one can apply 2-descent via 2-isogeny. These two methods are explained in full detail in the book of Silverman [204], Chapter X.

We explain here the third method, *general* 2*-descent*, which can be applied for an arbitrary elliptic curve over a number field. This method was developed by Birch and Swinnerton-Dyer [14] for elliptic curves over $\mathbb{Q}$. Cremona [42] implemented this method to compute the rank of elliptic curves over $\mathbb{Q}$. It was extended by P. Serf (see [197] and [43]) to elliptic curves over real quadratic number fields with class number 1. Simon [210] used the description of Cassels [29] to develop the general 2-descent method for elliptic curves over arbitrary number fields.

We first need the following definition.

Definition 7.25. a) A *quartic* is a curve given by an equation of the form

$$C : Y^2 = g(X) = aX^4 + bX^3 + cX^2 + dX + e$$

with $a, b, c, d, e \in \mathbb{K}$. We assume $b \neq 0$ if $a = 0$.

b) A quartic $Y^2 = g(X)$ is called *trivial*, if $g(X)$ has a root in $\mathbb{K}$.

c) For a quartic $Y^2 = g(X)$ with coefficients $a, b, c, d, e \in \mathbb{K}$, we define the invariants

$$\begin{aligned} I(g) &:= 12ae - 3bd + c^2 \\ \text{and} \quad J(g) &:= 72ace - 27ad^2 - 27b^2e + 9bcd - 2c^3. \end{aligned}$$

d) Two quartics $Y^2 = g_1(X)$ and $Y^2 = g_2(X)$ over $\mathbb{K}$ are called *equivalent* over $\mathbb{K}$, if there are $\alpha, \beta, \gamma, \delta \in \mathbb{K}$ with $\alpha\delta - \beta\gamma \neq 0$ and $\mu \in \mathbb{K}^*$ such that

$$g_2(X) = \mu^2(\gamma X + \delta)^4 g_1\left(\frac{\alpha X + \beta}{\gamma X + \delta}\right).$$

Proposition 7.26. *Let*

$$\begin{aligned} C_1 : Y^2 &= g_1(X) \\ C_2 : Y^2 &= g_2(X) \end{aligned}$$

be two equivalent quartics. Then

$$\begin{aligned} I(g_2) &= \mu^4(\alpha\delta - \beta\gamma)^4 I(g_1) \\ J(g_2) &= \mu^6(\alpha\delta - \beta\gamma)^6 J(g_1). \end{aligned}$$

Proof. This is an easy but lengthy computation. □

Definition 7.27. Let $E|\mathbb{K}$ be an elliptic curve over the number field $\mathbb{K}$. A *quartic associated to E* is a quartic

$$C : Y^2 = g(X)$$

with invariants

$$\begin{aligned} I(g) &= \lambda^4 c_4, \\ J(g) &= 2\lambda^6 c_6, \end{aligned}$$

where $\lambda \in \mathbb{K}^*$ and c_4, c_6 are the Tate values of E.

Proposition 7.28. *Let $E|\mathbb{K}$ be an elliptic curve.*

a) *The trivial quartics associated to E form an equivalence class. They are elliptic curves over $\mathbb{K}$ which are birationally isomorphic to E over $\mathbb{K}$.*

b) *The non-trivial quartics associated to E are curves of genus 1 over $\mathbb{K}$. If such a quartic has a $\mathbb{K}$-rational point, it is birationally isomorphic to E over the algebraic closure $\overline{\mathbb{K}}$ of $\mathbb{K}$, but in general not over $\mathbb{K}$.*

Proof. a) Let

$$C : Y^2 = g(X) = aX^4 + bX^3 + cX^2 + dX + e$$

be a trivial quartic associated to E. Suppose g has a root $x_0 \in \mathbb{K}$. Then C is (stepwise) equivalent to the following quartics:

$$\begin{aligned} C^{(1)} : Y^2 &= g^{(1)}(X) = g(X + x_0) && \text{with } e^{(1)} = 0 \\ C^{(2)} : Y^2 &= g^{(2)}(X) = X^4 g^{(1)}\left(\frac{1}{X}\right) && \text{with } a^{(2)} = 0 \\ C^{(3)} : Y^2 &= g^{(3)}(X) = \frac{1}{b^{(2)4}} g^{(2)}(b^{(2)}X) && \text{with } a^{(3)} = 0, b^{(3)} = 1 \\ C^{(4)} : Y^2 &= g^{(4)}(X) = g^{(3)}\left(X - \frac{c^{(3)}}{3}\right) && \text{with } a^{(4)} = 0, b^{(4)} = 1, c^{(4)} = 0 \end{aligned}$$

Hence C is equivalent to the "quartic"

$$C^{(4)} : Y^2 = X^3 + d^{(4)}X + e^{(4)},$$

which defines an elliptic curve over $\mathbb{K}$. This quartic has invariants

$$I(g^{(4)}) = -3d^{(4)} \quad \text{and} \quad J(g^{(4)}) = -27e^{(4)}.$$

The values of the elliptic curve are

$$\begin{aligned} c_4(C^{(4)}) &= 2^4 I(g^{(4)}) = 2^4\xi^4 I(g), \\ c_6(C^{(4)}) &= 2^5 J(g^{(4)}) = 2^5\xi^6 J(g) \end{aligned}$$

with some non-zero $\xi \in \mathbb{K}$. As the quartic C is associated to the elliptic curve E with values c_4, c_6, we can assume that $E|\mathbb{K}$ is given by the equation (see the proof of Theorem 1.7)

$$E : Y^2 = X^3 - 27c_4X - 54c_6.$$

We find that

$$c_4(C^{(4)}) = u^4c_4, \quad c_6(C^{(4)}) = u^6c_6$$

with non-zero $u \in \mathbb{K}$. Hence the elliptic curve $C^{(4)}$ is birationally isomorphic over $\mathbb{K}$ to the elliptic curve E.

Now assume that C' is another trivial quartic of the form

$$C' : Y^2 = g'(X) = X^3 + d'X + e'$$

with

$$I(g') = -3d' \quad \text{and} \quad J(g') = -27e'.$$

As both quartics $C^{(4)}$ and C' are associated to E, their invariants differ by a 4th or a 6th power. This implies that there exists a non-zero $\xi \in \mathbb{K}$ with

$$d' = \xi^4 d^{(4)}, \quad e' = \xi^6 e^{(4)}.$$

Therefore

$$g'(X) = \xi^6 g^{(4)}\left(\frac{X}{\xi^2}\right).$$

b) As the polynomial g is of degree 3 or 4, the quartic

$$C : Y^2 = g(X)$$

is a curve of genus 1. If this curve has a $\mathbb{K}$-rational point, it is an elliptic curve over $\mathbb{K}$.

Over the algebraic closure $\overline{\mathbb{K}}$ of $\mathbb{K}$ this quartic becomes trivial, so it is isomorphic to E over $\overline{\mathbb{K}}$. □

There is a bijection between the set of all equivalence classes of quartics associated to E and $H^1(G, E[2])$ (see, for example, Birch and Swinnerton-Dyer [14] or Cassels

[27]). With this bijection, the 2-Selmer group and the 2-torsion of the Tate Shafarevich group correspond to the following groups.

$$\begin{aligned} S^2(E|\mathbb{K}) &= \text{ the group of the classes of quartics associated to } E \\ &\quad\ \text{which have a point over } \mathbb{K}, \\ Ш(E|\mathbb{K})[2] &= \text{ the group of the classes of quartics associated to } E \text{ which} \\ &\quad\ \text{have a point over all completions of } \mathbb{K} \text{ but not over } \mathbb{K} \text{ itself.} \end{aligned}$$

Let

$$\begin{aligned} G = &\text{ the group of the classes of quartics associated to } E \\ &\text{ which have a point over all completions of } \mathbb{K} \end{aligned}$$

For the general 2-descent we consider the exact sequence

$$0 \to E(\mathbb{K})/2E(\mathbb{K}) \to S^2(E|\mathbb{K}) \to Ш(E|\mathbb{K})[2] \to 0.$$

It follows that every element in $S^2(E|\mathbb{K})$ has order 2. As $S^2(E|\mathbb{K})$ is finite, we have

$$\sharp S^2(E|\mathbb{K}) = 2^k, \quad k \in \mathbb{N}.$$

Then we have

$$\sharp E(\mathbb{K})/2E(\mathbb{K}) = 2^{r+t},$$

where $r = \mathrm{rk}(E(\mathbb{K}))$ and $2^t = \sharp E(\mathbb{K})[2]$. Therefore, $k \geq r + t$ and

$$2^{k-t-r} = \sharp Ш(E|\mathbb{K})[2] = \sharp\big(S^2(E|\mathbb{K})/(E(\mathbb{K})/2E(\mathbb{K}))\big).$$

The general 2-descent now proceeds as follows. First one determines all quartics associated to E which are everywhere locally soluble, discarding equivalent quartics (i.e., one determines the elements of G). After that one tries to find a global solution on those quartics.

In many cases, there are everywhere soluble quartics on which one cannot find global points. Then one cannot decide whether the quartic has no global point (i.e. is an element of $Ш(E|\mathbb{K})[2]$) or whether one has not searched long enough (i.e., to find out that the quartic is an element of $S^2(E|\mathbb{K})$). Thus the 2-descent usually gives an upper bound of the rank. To obtain the actual rank, an other method (such as the infinite descent, see page 253) is needed.

For the determination of all quartics (modulo equivalence) which are associated to E, there are results for elliptic curves over $\mathbb{Q}$ (see the article of Birch and Swinnerton-Dyer [14] and the implementation of Cremona [42]) and over real quadratic number fields of class number 1 (see the thesis of Serf [197] and the article of Cremona and Serf [43]).

Note again that Simon [210] developed the 2-descent for elliptic curves over arbitrary number fields.

At the end of this section we give an example of the elliptic curve

$$\begin{aligned} E : Y^2 + (1+\sqrt{13})XY + (2+\sqrt{13})X \\ = X^3 + \left(-\frac{1}{2} - \frac{1}{2}\sqrt{13}\right) X^2 + \left(\frac{1}{2} + \frac{1}{2}\sqrt{13}\right) X + \left(\frac{1}{2} + \frac{1}{2}\sqrt{13}\right) \end{aligned}$$

over $\mathbb{K} = \mathbb{Q}(\sqrt{13})$. The Tate values are

$$\begin{aligned} c_4 &= -240 - 96\sqrt{13} = \left(\frac{1}{2} + \frac{1}{2}\sqrt{13}\right)^4 (8 - 8\sqrt{13}), \\ c_6 &= 1080 + 432\sqrt{13} = \frac{1}{2}\left(\frac{1}{2} + \frac{1}{2}\sqrt{13}\right)^6 (-80 + 32\sqrt{13}). \end{aligned}$$

Hence we can take the invariants

$$I = 8 - 8\sqrt{13}, \quad J = -80 + 32\sqrt{13}.$$

One non-trivial quartic with these invariants is

$$Y^2 = -\sqrt{13}X^4 + (11+\sqrt{13})X^3 - (8+4\sqrt{13})X^2 + (8+2\sqrt{13})X - \frac{3}{2} - \frac{1}{2}\sqrt{13}.$$

The solution of this quartic leads to the point

$$P_1 = \left(-\frac{3}{4}, -\frac{1}{2} - \frac{1}{8}\sqrt{13}\right),$$

which is a point of infinite order.

As there is no non-equivalent non-trivial quartic, we get

$$\sharp S^2(E|\mathbb{K}) = 2.$$

Further $\sharp E(\mathbb{K})[2] = 1$, because there is no non-trivial point of order 2 on this curve over $\mathbb{K}$. We get

$$2^{1-0-r} = \sharp Ш(E|\mathbb{K})[2],$$

where r is the rank of $E(\mathbb{K})$. As the given quartic has a global solution, the 2-part of the Tate–Shafarevich group is trivial, hence

$$2^{1-r} = 1.$$

This implies $r = \mathrm{rk}(E(\mathbb{K})) = 1$.

7.7 The rank in field extensions

In this section we consider the following problem. Given an elliptic curve over a number field $\mathbb{K}$ and a finite extension $\mathbb{L}|\mathbb{K}$ of $\mathbb{K}$, can we express the rank of $E(\mathbb{L})$ in terms of objects which are defined over $\mathbb{K}$ and which are related to the extension $\mathbb{L}|\mathbb{K}$?

The theory given in this section was elaborated on by Sato in [183]. He proved his results in a more general context of abelian varieties. See also the thesis of Schneiders [191], [192] for quadratic field extensions, or the paper of Kida [107] for more general results. Kida [108] gave also an algorithm to explicitly construct the group $E(\mathbb{K})$ from the groups $E(\mathbb{Q})$ and $E^m(\mathbb{Q})$, where E is an elliptic curve over $\mathbb{Q}$, $\mathbb{K}$ a quadratic number field and E^m the twist corresponding to $\mathbb{K}$.

Throughout this section, we assume that $\mathbb{K}$ is an algebraic number field (of finite degree) and $E|\mathbb{K}$ an elliptic curve defined over $\mathbb{K}$ satisfying the following condition

(C) There exists a ring homomorphism $\iota : \mathbb{Z}[\mathbb{U}_m] \to \mathrm{End}_{\mathbb{K}}(E)$, where $\mathbb{U}_m$ is the group of the m-th roots of unity in $\overline{\mathbb{K}}$.

Note that the condition (C) is always satisfied if $m = 2$. If $m > 2$, it implies that the elliptic curve has complex multiplication. As $\mathrm{End}_{\mathbb{K}}(E) \otimes_{\mathbb{Z}} \mathbb{Q}$ is isomorphic to $\mathbb{Q}$ or an imaginary quadratic field, it follows that m must be 2, 3, 4, or 6. Hence condition (C) implies the following:

a) If $m = 2$: $\mathbb{K}$ and E are arbitrary;

b) if $m = 3$ or $m = 6$: $\sqrt{-3} \in \mathbb{K}$ and $j(E) = 0$;

c) if $m = 4$: $\sqrt{-1} \in \mathbb{K}$ and $j(E) = 1728 = 12^3$.

We now define the twists of elliptic curves.

Definition 7.29. Let $E|\mathbb{K}$ be an elliptic curve in short Weierstraß normal form

$$E : Y^2 = X^3 + aX + b$$

over the number field $\mathbb{K}$. Let $0 \neq d \in \mathbb{K}$.

a) The *d-2-twists* of E are

$$E^{2,j} : Y^2 = X^3 + ad^{2j}X + bd^{3j} \quad \text{for } j = 0, 1.$$

b) If $a = 0$, the *d-3-twists* of E are

$$E^{3,j} : Y^2 = X^3 + bd^{2j} \quad \text{for } j = 0, 1, 2.$$

c) If $b = 0$, the *d-4-twists* of E are

$$E^{4,j} : Y^2 = X^3 + ad^{j}X \quad \text{for } j = 0, 1, 2, 3.$$

d) If $a = 0$, the *d-6-twists* of E are

$$E^{6,j} : Y^2 = X^3 + bd^j \quad \text{for } j = 0, \dots, 5.$$

e) Define the function

$$f_m(n) := \begin{cases} 1 & \text{if } m \mid n \\ 0 & \text{if } m \nmid n. \end{cases}$$

With this function, we can write the *d-m-twists* of

$$E : Y^2 = X^3 + f_m(4)aX + f_m(6)b$$

for $m = 2, 3, 4, 6$, as

$$E^{m,j} : Y^2 = X^3 + f_m(4)d^{4j/m}aX + f_m(6)d^{6j/m}b$$

for $j = 0, \dots, m-1$.

In this section, let $E|\mathbb{K}$ be an elliptic curve in short Weierstraß normal form

$$E : Y^2 = X^3 + f_m(4)aX + f_m(6)b$$

over the number field $\mathbb{K}$, satisfying the condition (C) with $m = 2, 3, 4$, or 6. We fix a non-zero $d \in \mathbb{K}$ such that the field extension $\mathbb{K}(\sqrt[m]{d})|\mathbb{K}$ is of degree m.

Recall (see Chapter 2, Section 2.4) that the m-th root of unity ζ_m induces an endomorphism on the curve $E|\mathbb{K}$ (for $P = (x, y) \in E(\mathbb{K})$) by

$$\zeta_m \mathcal{O} = \mathcal{O}, \quad \zeta_m(x, y) = (\zeta_m^{m-2}x, \zeta_m^{m-3}y).$$

For $0 \le j \le m - 1$ we get therefore

$$\zeta_m^j(x, y) = (\zeta_m^{mj-2j}x, \zeta_m^{mj-3j}y) = (\zeta_m^{m-2j}x, \zeta_m^{m-3j}y).$$

Definition 7.30. a) Let

$$\sigma : \mathbb{K}(\sqrt[m]{d}) \to \mathbb{K}(\sqrt[m]{d})$$

be the Galois automorphism with

$$\sigma(x) = x \quad \text{for } x \in \mathbb{K},$$
$$\sigma(\sqrt[m]{d}) = \zeta_m \sqrt[m]{d},$$

where ζ_m is a primitive m-th root of unity. This Galois automorphism defines an automorphism on $E(\mathbb{K}(\sqrt[m]{d}))$ by

$$\sigma(\mathcal{O}) = \mathcal{O}, \quad \sigma((x, y)) = (\sigma(x), \sigma(y)),$$

where this time $P = (x, y) \in E(\mathbb{K}(\sqrt[m]{d})$.

We define for $j = 0, 1, \ldots, m-1$ the homomorphisms

$$S_{m,j}\colon E(\mathbb{K}(\sqrt[m]{d})) \to E(\mathbb{K}(\sqrt[m]{d}))$$
$$P \mapsto \sigma(P) - \zeta_m^j P.$$

b) Furthermore, for $j = 1, \ldots, m-1$, we define the map

$$\alpha_{m,j}\colon E^{m,j} \to E^{m,0} = E$$
$$\mathcal{O} \mapsto \mathcal{O}$$
$$P = (x, y) \mapsto \left(xd^{-2j/m}, yd^{-3j/m}\right).$$

Proposition 7.31. a) *The map $\alpha_{m,j}$ is an injective group homomorphism.*

b) *We have* $\ker(S_{m,j}) = \alpha_{m,j} E^{m,j}(\mathbb{K})$. *In particular,*

$$\sigma(\alpha_{m,j}(P)) = \zeta_m^j(\alpha_{m,j}(P))$$

for $P \in E^{m,j}(\mathbb{K})$.

Proof. a) It is an easy exercise to show that $\alpha_{m,j}$ is well-defined and injective. We show that it is a homomorphism. Let

$$P_1, P_2 \in E^{m,j}.$$

We must show that

$$\alpha_{m,j}(P_1 + P_2) = \alpha_{m,j}(P_1) + \alpha_{m,j}(P_2).$$

We only consider the case that none of these points is the point at infinity. Let $P_i = (x_i, y_i)$, $i = 1, 2, 3$. Then, on $E^{m,j}$, $P_1 + P_2 = (x_3, y_3)$ with

$$x_3 = \lambda^2 - x_1 - x_2,$$
$$y_3 = \lambda(x_1 - x_3) - y_1,$$

and

$$\lambda = \begin{cases} \dfrac{y_2 - y_1}{x_2 - x_1} & \text{if } x_1 \neq x_2, \\[2ex] \dfrac{3x_1^2 + f_m(4)d^{4j/m}a}{2y_1} & \text{if } x_1 = x_2. \end{cases}$$

On the other hand, we have

$$\alpha_{m,j}(P_i) = \left(\frac{x_i}{d^{2j/m}}, \frac{y_i}{d^{3j/m}}\right) \in E^{m,0} = E$$

and therefore $\alpha_{m,j}(P_1) + \alpha_{m,j}(P_2) = (x_3', y_3')$ with

$$x_3' = \lambda'^2 - \frac{1}{d^{2j/m}}(x_1 + x_2),$$
$$y_3' = \lambda' \frac{1}{d^{2j/m}}(x_1 - x_3) - \frac{y_1}{d^{3j/m}},$$

and

$$\lambda' = \begin{cases} \dfrac{(y_2 - y_1)/d^{3j/m}}{(x_2 - x_1)/d^{2j/m}} & \text{if } x_1 \neq x_2, \\ \dfrac{3x_1^2/d^{4j/m} + f_m(4)a}{2y_1/d^{3j/m}} & \text{if } x_1 = x_2. \end{cases}$$

One can see that $\lambda' = \frac{1}{d^{j/m}}\lambda$ and hence

$$\begin{aligned} x_3' &= \frac{1}{d^{2j/m}} x_3 \\ y_3' &= \frac{1}{d^{3j/m}} y_3. \end{aligned}$$

The assertion is proved.

b) First we show that for $P \in \alpha_{m,j} E^{m,j}(\mathbb{K}) \subset E^{m,0}(\mathbb{K}) = E(\mathbb{K})$,

$$\sigma(P) = \zeta_m^j(P).$$

Let

$$P = \left(\frac{x}{\sqrt[m]{d}^{2j}}, \frac{y}{\sqrt[m]{d}^{3j}} \right) \in \alpha_{m,j} E^{m,j}(\mathbb{K})$$

with $(x, y) \in E^{m,j}(\mathbb{K})$. Then

$$\sigma(P) = \left(\zeta_m^{m-2j} \frac{x}{\sqrt[m]{d}^{2j}}, \zeta_m^{m-3j} \frac{y}{\sqrt[m]{d}^{3j}} \right) = \zeta_m^j P.$$

Second we show that for every $Q \in E(\mathbb{K}(\sqrt[m]{d}))$ such that $S_{m,j}(Q) = \mathcal{O}$, there exists a point $R \in E^{m,j}(\mathbb{K})$ such that $\alpha_{m,j}(R) = Q$.

The case $Q = \mathcal{O}$ is trivial. Let $Q = (x, y)$. Then we can write

$$x = \sum_{k=0}^{m-1} x_k \sqrt[m]{d}^k, \quad y = \sum_{k=0}^{m-1} y_k \sqrt[m]{d}^k \quad \text{with } x_k, y_k \in \mathbb{K}.$$

Then

$$\sigma(Q) = (\sigma(x), \sigma(y)) = \Big(\sum_{k=0}^{m-1} \zeta_m^k x_k \sqrt[m]{d}^k, \sum_{k=0}^{m-1} \zeta_m^k y_k \sqrt[m]{d}^k \Big).$$

Further

$$\zeta_m^j Q = (\zeta_m^{m-2j} x, \zeta_m^{m-3j} y) = \Big(\sum_{k=0}^{m-1} \zeta_m^{m-2j} x_k \sqrt[m]{d}^k, \sum_{k=0}^{m-1} \zeta_m^{m-3j} y_k \sqrt[m]{d}^k \Big).$$

As Q is in the kernel of $S_{m,j}$, these two points must be equal. Comparing the coefficients, we see that

$$x = x_{k_x} \sqrt[m]{d}^{k_x} = \frac{\tilde{x}}{d^{2j/m}}, \quad y = y_{k_y} \sqrt[m]{d}^{k_y} = \frac{\tilde{y}}{d^{3j/m}},$$

with $k_x \equiv m-2j \bmod m$ and $k_y \equiv m-3j \bmod m$, $\tilde{x}, \tilde{y} \in \mathbb{K}$. Since $Q \in E\big(\mathbb{K}(\sqrt[m]{d})\big)$, it is easy to show that $R = (\tilde{x}, \tilde{y}) \in E^{m,j}(\mathbb{K})$. □

Definition 7.32. We define the ("componentwise") homomorphism

$$\begin{aligned} \tau : \prod_{j=0}^{m-1} E^{m,j}(\mathbb{K}) &\to E(\mathbb{K}(\sqrt[m]{d})) \\ (P_0, \dots, P_{m-1}) &\mapsto \sum_{j=0}^{m-1} \alpha_{m,j}(P_j). \end{aligned}$$

Proposition 7.33. a) *The kernel of τ is finite.*

b) $mE(\mathbb{K}(\sqrt[m]{d})) \subseteq \tau\big(\prod_{j=0}^{m-1} E^{m,j}(\mathbb{K})\big)$.

Proof. a) We prove that for $(P_0, \dots, P_{m-1})$ in the kernel of τ the point P_j is in the torsion group of $E^{m,j}(\mathbb{K})$. Then it follows that the kernel is finite.

Let $(P_0, \dots, P_{m-1})$ be in the kernel of τ. Using Proposition 7.31, Part b), we get for $j = 0, \dots, m-1$:

$$\begin{aligned} \sum_{k=0}^{m-1} \sigma^k(\alpha_{m,j}(P_j)) &= \Big(\sum_{k=0}^{m-1} \zeta_m^{kj}\Big)\alpha_{m,j}(P_j) \\ &= \begin{cases} m\alpha_{m,0}(P_0) = mP_0 & \text{if } j = 0, \\ \mathcal{O} & \text{else.} \end{cases} \end{aligned}$$

Therefore, as $(P_0, \dots, P_{m-1})$ is in the kernel of τ,

$$\mathcal{O} = \sum_{k=0}^{m-1} \sigma^k(\tau(P_0, \dots, P_{m-1})) = \sum_{j=0}^{m-1}\sum_{k=0}^{m-1} \sigma^k(\alpha_{m,j}(P_j)) = mP_0.$$

Hence $P_0 \in E(\mathbb{K})_{\text{tors}}$.

We now consider the points $P_j^1 := mP_j$ for $j = 0, \dots, m-1$. We have $P_0^1 = \mathcal{O}$. Furthermore, since $(P_0, \dots, P_{m-1}) \in \ker(\tau)$,

$$\sum_{j=1}^{m-1} \alpha_{m,j}(P_j^1) = \mathcal{O}.$$

If $m = 2$ it follows that

$$\alpha_{2,1}(P_1^1) = \mathcal{O}.$$

Since by Part a) of Proposition 7.31 the $\alpha_{m,j}$ are injective, we get

$$P_1^1 = mP_1 = \mathcal{O}.$$

Therefore P_1 is in $E^{2,1}(\mathbb{K})_{\mathrm{tors}}$.

If $m = 3$, then

$$\alpha_{3,1}(P_1^1) = -\alpha_{3,2}(P_2^1).$$

The case $P_1^1 = \mathcal{O}$ or $P_2^1 = \mathcal{O}$ is trivial. Let $P_1^1 = (x, y) \in E^{3,1}(\mathbb{K})$ and $P_2^1 = (u, v) \in E^{3,2}(\mathbb{K})$. Then it follows from the equation above that

$$\frac{x}{\sqrt[3]{d}^2} = \frac{u}{\sqrt[3]{d}^4}.$$

This is only possible, if $x = u = 0$. The points with first coordinate 0 on the curve

$$\begin{aligned} E^{3,j} : Y^2 &= X^3 + f_3(4)d^{4j/3}aX + f_3(6)d^{6j/3}b \\ &= X^3 + d^{2j}b \end{aligned}$$

are points of order 3. Therefore P_j^1 and hence P_j are in $E^{3,j}(\mathbb{K})_{\mathrm{tors}}$ for $j = 1, 2$.

If $m = 4$, consider the relation

$$\begin{aligned} \mathcal{O} &= \tau(\mathcal{O}, P_1^1, P_2^1, P_3^1) + \sigma^2(\tau(\mathcal{O}, P_1^1, P_2^1, P_3^1)) \\ &= \sum_{j=1}^{3} \alpha_{4,j}(P_j^1) + \sum_{j=1}^{3} \sigma^2(\alpha_{4,j}(P_j^1)) \\ &= \sum_{j=1}^{3} \alpha_{4,j}(P_j^1) + \sum_{j=1}^{3} \zeta_4^{2j}(\alpha_{4,j}(P_j^1)) \\ &= \sum_{j=1}^{3} (1 + (-1)^j)\alpha_{4,j}(P_j^1) \\ &= 2\alpha_{4,2}(P_2^1). \end{aligned}$$

Hence P_2^1 and therefore P_2 is a torsion point on $E^{4,2}(\mathbb{K})$. Write for $j = 1, 3$: $P_j^2 := 2P_j^1$. Then we clearly obtain

$$\tau(\mathcal{O}, P_1^2, \mathcal{O}, P_3^2) = \mathcal{O}.$$

It follows that $\alpha_{4,1}(P_1^2) = -\alpha_{4,3}(P_3^2)$. In an analogous way as for $m = 3$ comparing the second coordinates of the two points P_1^2, P_2^2 shows that they are torsion points.

The last case is $m = 6$. Here we first take the relation

$$\begin{aligned}
\mathcal{O} &= \tau(\mathcal{O}, P_1^1, P_2^1, P_3^1, P_4^1, P_5^1) + \sigma^3(\tau(\mathcal{O}, P_1^1, P_2^1, P_3^1, P_4^1, P_5^1)) \\
&= \sum_{j=1}^{5} \alpha_{6,j}(P_j^1) + \sum_{j=1}^{5} \sigma^3(\alpha_{6,j}(P_j^1)) \\
&= \sum_{j=1}^{5} \alpha_{6,j}(P_j^1) + \sum_{j=1}^{5} \zeta_6^{3j}(\alpha_{6,j}(P_j^1)) \\
&= \sum_{j=1}^{6} (1 + (-1)^j)\alpha_{6,j}(P_j^1) \\
&= 2\alpha_{6,2}(P_2^1) + 2\alpha_{6,2}(P_4^1).
\end{aligned}$$

In the same way as for $m = 3$ one can see that P_2^1 and P_4^1 and thus P_2 and P_4 are torsion points of order 3. Now write for $j = 1, 3, 5$: $P_j^2 := 3P_j^1$. Then we get

$$\begin{aligned}
\mathcal{O} &= \tau(\mathcal{O}, P_1^2, \mathcal{O}, P_3^2, \mathcal{O}, P_5^2) + \sigma^2(\tau(\mathcal{O}, P_1^2, \mathcal{O}, P_3^2, \mathcal{O}, P_5^2)) \\
&\quad + \sigma^4(\tau(\mathcal{O}, P_1^2, \mathcal{O}, P_3^2, \mathcal{O}, P_5^2)) \\
&= (1 + \zeta_6^2 + \zeta_6^4)\alpha_{6,1}(P_1^2) + (1 + 1 + 1)\alpha_{6,3}(P_3^2) \\
&\quad + (1 + \zeta_6^4 + \zeta_6^2)\alpha_{6,5}(P_5^2) \\
&= 3\alpha_{6,3}(P_3^2).
\end{aligned}$$

Therefore P_3^2 is a torsion point of order 3 and hence P_3 itself is a torsion point of order 9. With $P_j^3 := 3P_j^2$ for $j = 1, 5$ we obtain

$$\tau(\mathcal{O}, P_1^3, \mathcal{O}, \mathcal{O}, \mathcal{O}, P_5^3) = \mathcal{O}.$$

Then one can see as above that P_1^3 and P_5^3 and hence P_1 and P_5 are torsion points.

b) Let $P \in E(\mathbb{K}(\sqrt[m]{d}))$ and consider

$$\begin{aligned}
mP &= \sum_{k=0}^{m-1} \sum_{j=0}^{m-1} \zeta_m^{(m-k)j} \sigma^k(P) \\
&= \sum_{j=0}^{m-1} \Big(\sum_{k=0}^{m-1} \zeta_m^{(m-k)j} \sigma^k(P) \Big) \\
&= \sum_{j=0}^{m-1} Q_j
\end{aligned}$$

with

$$Q_j = \sum_{k=0}^{m-1} \zeta_m^{(m-k)j} \sigma^k(P).$$

Then

$$\begin{aligned}\sigma(Q_j) &= \sum_{k=0}^{m-1} \zeta_m^{(m-k)j} \sigma^{k+1}(P) \\ &= \zeta_m^j \sum_{k=0}^{m-1} \zeta_m^{(m-(k+1))j} \sigma^{k+1}(P) \\ &= \zeta_m^j \sum_{k=1}^{m} \zeta_m^{(m-k)j} \sigma^k(P) = \zeta_m^j Q_j.\end{aligned}$$

Hence Q_j is in the kernel of the function $S_{m,j}$. From Proposition 7.31 a) it follows that

$$Q_j \in \alpha_{m,j} E^{m,j}(\mathbb{K}).$$

That means there exist an $R_j \in E^{m,j}(\mathbb{K})$ with $\alpha_{m,j}(R_j) = Q_j$. Therefore,

$$mP = \sum_{j=0}^{m-1} \alpha_{m,j}(R_j) = \tau(R_0, \ldots, R_{m-1}). \qquad \square$$

Theorem 7.34. *Let $E|\mathbb{K}$ be an elliptic curve in short Weierstraß normal form*

$$E : Y^2 = X^3 + f_m(4)aX + f_m(6)b$$

over the number field $\mathbb{K}$, satisfying the condition (C) with $m = 2, 3, 4,$ or 6. Let $0 \neq d \in \mathbb{K}$ such that the field extension $\mathbb{K}(\sqrt[m]{d})|\mathbb{K}$ is abelian of degree m. Then

$$\mathrm{rk}(E(\mathbb{K}(\sqrt[m]{d})) = \sum_{j=0}^{m-1} \mathrm{rk}(E^{m,j}(\mathbb{K})).$$

Proof. This follows from Proposition 7.33 which implies that the kernel and the cokernel of the homomorphism τ are finite. $\square$

7.8 Exercises

1) Complete the proof of Theorem 7.4.
2) Consider the coefficients $a(\mathfrak{p}^k)$ of the L-series in the proof of Theorem 7.4. In the case of good reduction, these coefficients are defined by the equations

$$\begin{aligned} a((1)) &= 1, \\ a(\mathfrak{p}) &= \mathcal{N}(\mathfrak{p}) + 1 - \sharp(\widetilde{E}(\mathbb{F}_{\mathfrak{p}})), \\ a(\mathfrak{p}^k) &= a(\mathfrak{p})a(\mathfrak{p}^{k-1}) - \mathcal{N}(\mathfrak{p})a(\mathfrak{p}^{k-2}) \quad \text{for } k \geq 2. \end{aligned}$$

a) Find the solutions x_1, x_2 of the quadratic equation

$$X^2 - a(\mathfrak{p})X + \mathcal{N}(\mathfrak{p}) = 0.$$

b) Compute the numbers b_1, b_2, such that

$$a((1)) = b_1 + b_2, \quad a(\mathfrak{p}) = b_1 x_1 + b_2 x_2.$$

c) Show that for all $k \in \mathbb{N}_0$:

$$a(\mathfrak{p}^k) = b_1 x_1^k + b_2 x_2^k.$$

3) Compute the next 10 coefficients of the L-series in the example after Algorithm 7.6.

4) Show that the rank of the following curves is

$$\text{a) } r = 5, \qquad \text{b) } r = 6, \qquad \text{c) } r = 7.$$

a) $E: \ Y^2 + 67Y = X^3 - 21X^2 - 10X + 30$

b) $E: \ Y^2 + 351Y = X^3 - 63X^2 + 56X + 22$

c) $E: \ Y^2 + 1641Y = X^3 - 168X^2 + 161X - 8$

5) Use the Birch and Swinnerton-Dyer conjecture to prove that the curves

$$E_1 : Y^2 = X^3 + (22 + 10\sqrt{5})$$

over $\mathbb{K}_1 = \mathbb{Q}(\sqrt{5})$ and

$$E_2 : Y^2 = X^3 + iX + (1 - i)$$

over the Gaussian field $\mathbb{K}_2 = \mathbb{Q}(i)$, where $i = \sqrt{-1}$, both have rank $r = 2$.

6) Compute a Weierstraß normal form for the elliptic curve given by the equation

$$17 - X^4 = 2Y^2.$$

Show that this curve leads to a non-trivial element of the Tate-Shafarevich group Ш. (The Hasse principle, therefore, does not hold for elliptic curves.)

7) a) Solve the "easy exercise" mentioned at the beginning of the proof of Lemma 7.11 a).
b) Verify the estimate for the function h at the top of page 211 (in the proof of Part a) of Proposition 7.12).

8) Complete the proof of Proposition 7.31.

Chapter 8
Basis

In this chapter we focus on the computation of a basis of an elliptic curve over a number field. To this end we first consider linearly independent points. Then we use an estimate of Siksek to compute a basis. In the last section we explain the Heegner point method to compute a generator of an elliptic curve of rank one.

The notation is again the same as in Chapter 5.

We mention in this connection a paper of Gebel and the second author [78], where $E(\mathbb{K})$ is computed in some cases for $\mathbb{K} = \mathbb{Q}$ and the dissertation of the first author [189], where it is shown how to compute $E(\mathbb{K})$ both by assuming the truth of the conjecture of Birch and Swinnerton-Dyer. In this connection, we should also mention work of Cremona and Serf [43] as well as the dissertation of Simon [210] (see Chapter 7).

8.1 Linearly independent points

How do we find linearly independent points on $E|\mathbb{K}$? A straightforward method is to search for arbitrary points $P_1, \ldots, P_n$ and to test if these points are linearly independent. Such a test can be done according to the next theorem.

Theorem 8.1. *Let $E|\mathbb{K}$ be an elliptic curve over the number field $\mathbb{K}$.*

a) *Let $\{P_1, \ldots, P_n\}$ be a set of points on $E(\mathbb{K})$, $n > 0$. Then the regulator of these points satisfies $R_{P_1,\ldots,P_n} = 0$ if and only if the points $P_1, \ldots, P_n$ are linearly independent.*

b) *Let $\{P_1, \ldots, P_r\}$ be a maximal set of linearly independent points on $E(\mathbb{K})$, that means $r = \mathrm{rk}(E(\mathbb{K})) > 0$. Then the points $P_1, \ldots, P_r$ are a basis of $E(\mathbb{K})$ or there exists an integer $m \in \mathbb{N}$, $m \geq 2$, with*

$$R_{P_1,\ldots,P_r} = m^2 R_{E|\mathbb{K}},$$

where $R_{E|\mathbb{K}}$ is the regulator of $E|\mathbb{K}$.

Proof. a) Let $P_1, \ldots, P_n$ be linearly dependent over $\mathbb{Z}$. Then there exist integers $z_i \in \mathbb{Z}$, not all equal to 0, such that

$$z_1 P_1 + \cdots + z_n P_n = \mathcal{O}.$$

Assume $z_n \neq 0$. If $n = 1$, the above equation means that $P_n = P_1$ is a torsion point. Then

$$R_{P_1} = \langle P_1, P_1 \rangle = 0.$$

Now, let $n > 1$. Using

$$z_n P_n = -\sum_{j=1}^{n-1} z_j P_j$$

we get

$$\begin{aligned} z_n R_{P_1,\dots,P_n} &= z_n \det \begin{pmatrix} \langle P_1, P_1 \rangle, & \dots, & \langle P_1, P_n \rangle \\ \vdots & & \vdots \\ \langle P_n, P_1 \rangle, & \dots, & \langle P_n, P_n \rangle \end{pmatrix} \\ &= \det \begin{pmatrix} \langle P_1, P_1 \rangle, & \dots, & z_n \langle P_1, P_n \rangle \\ \vdots & & \vdots \\ \langle P_n, P_1 \rangle, & \dots, & z_n \langle P_n, P_n \rangle \end{pmatrix}. \end{aligned}$$

In the last column we write now for $i = 1, \dots, n$

$$z_n \langle P_i, P_n \rangle = \langle P_i, z_n P_n \rangle = \Big\langle P_i, -\sum_{j=1}^{n-1} z_j P_j \Big\rangle = -\sum_{j=1}^{n-1} z_j \langle P_i, P_j \rangle.$$

In this way, we obtain

$$z_n R_{P_1,\dots,P_n} = \det \begin{pmatrix} \langle P_1, P_1 \rangle, & \dots, & -\sum\limits_{j=1}^{n-1} z_j \langle P_1, P_j \rangle \\ \vdots & & \vdots \\ \langle P_n, P_1 \rangle, & \dots, & -\sum\limits_{j=1}^{n-1} z_j \langle P_n, P_j \rangle \end{pmatrix}.$$

Adding to the last column the sum of z_j-times the j-th column for $j = 1, \dots, n-1$ yields (since the determinant is fixed)

$$z_n R_{P_1,\dots,P_n} = \det \begin{pmatrix} \langle P_1, P_1 \rangle, & \dots, & \langle P_1, P_{n-1} \rangle, & 0 \\ \vdots & & \vdots & \vdots \\ \langle P_n, P_1 \rangle, & \dots, & \langle P_n, P_{n-1} \rangle, & 0 \end{pmatrix} = 0.$$

On the other hand let $R_{P_1,\dots,P_n} = 0$. This means that the columns of the matrix $(\langle P_i, P_j \rangle)_{1 \le i,j \le n}$ are linearly dependent. Hence there are $z_j \in \mathbb{Z}$, not all equal to 0, such that

$$\begin{pmatrix} 0 \\ \vdots \\ 0 \end{pmatrix} = \sum_{j=1}^{n} z_j \begin{pmatrix} \langle P_1, P_j \rangle \\ \vdots \\ \langle P_n, P_j \rangle \end{pmatrix} = \begin{pmatrix} \sum_{j=1}^{n} z_j \langle P_1, P_j \rangle \\ \vdots \\ \sum_{j=1}^{n} z_j \langle P_n, P_j \rangle \end{pmatrix} = \begin{pmatrix} \left\langle P_1, \sum_{j=1}^{n} z_j P_j \right\rangle \\ \vdots \\ \left\langle P_n, \sum_{j=1}^{n} z_j P_j \right\rangle \end{pmatrix}.$$

Hence we have for all $i = 1, \ldots, n$

$$\left\langle P_i, \sum_{j=1}^{n} z_j P_j \right\rangle = 0.$$

Multiplying these equations by z_i and adding leads to

$$0 = \sum_{i=1}^{n} z_i \left\langle P_i, \sum_{j=1}^{n} z_j P_j \right\rangle = \left\langle \sum_{i=1}^{n} z_i P_i, \sum_{j=1}^{n} z_j P_j \right\rangle = 2\hat{h}\left(\sum_{i=1}^{n} z_i P_i \right).$$

According to Proposition 5.19 $\sum_{i=1}^{n} z_i P_i$ is a torsion point. Hence there is an integer $m \in \mathbb{N}$ such that

$$m \sum_{i=1}^{n} z_i P_i = \sum_{i=1}^{n} m z_i P_i = \mathcal{O}.$$

Because not all z_i are equal to 0 it follows that the points $P_1, \ldots, P_n$ are linearly dependent.

b) Let $\{Q_1, \ldots, Q_r\}$ be a basis of $E(\mathbb{K})$, and $\{P_1, \ldots, P_r\}$ the set of linearly independent points of $E(\mathbb{K})$. The points P_i can be represented as linear combinations:

$$P_i = \sum_{j=1}^{r} m_{ij} Q_j \quad (i = 1, \ldots, r)$$

with $m_{ij} \in \mathbb{Z}$. Let M be the matrix $M = (m_{ij})_{1 \le i,j \le r}$. Then the determinant is integral: $\det(M) \in \mathbb{Z}$.

Because the points P_j are linearly independent, $\det(M) \neq 0$. If the points $P_1, \ldots, P_r$ are not a basis of $E(\mathbb{K})$, then M is not a matrix of basis change, and it follows $\det(M) \neq \pm 1$. One has

$$\begin{pmatrix} \langle P_1, P_1 \rangle, & \ldots, & \langle P_1, P_r \rangle \\ \vdots & & \vdots \\ \langle P_r, P_1 \rangle, & \ldots, & \langle P_r, P_r \rangle \end{pmatrix} =$$

$$= \begin{pmatrix} \left\langle \sum_{k=1}^r m_{1k} Q_k, \sum_{l=1}^r m_{1l} Q_l \right\rangle, & \dots, & \left\langle \sum_{k=1}^r m_{1k} Q_k, \sum_{l=1}^r m_{rl} Q_l \right\rangle \\ \vdots & & \vdots \\ \left\langle \sum_{k=1}^r m_{rk} Q_k, \sum_{l=1}^r m_{1l} Q_l \right\rangle, & \dots, & \left\langle \sum_{k=1}^r m_{rk} Q_k, \sum_{l=1}^r m_{rl} Q_l \right\rangle \end{pmatrix}$$

$$= \begin{pmatrix} \sum_{k,l=1}^r m_{1k} m_{1l} \langle Q_k, Q_l \rangle, & \dots, & \sum_{k,l=1}^r m_{1k} m_{rl} \langle Q_k, Q_l \rangle \\ \vdots & & \vdots \\ \sum_{k,l=1}^r m_{rk} m_{1l} \langle Q_k, Q_l \rangle, & \dots, & \sum_{k,l=1}^r m_{rk} m_{rl} \langle Q_k, Q_l \rangle \end{pmatrix}$$

$$= M \begin{pmatrix} \langle Q_1, Q_1 \rangle, & \dots, & \langle Q_1, Q_r \rangle \\ \vdots & & \vdots \\ \langle Q_r, Q_1 \rangle, & \dots, & \langle Q_r, Q_r \rangle \end{pmatrix} M^T.$$

It follows that

$$R_{P_1,\dots,P_r} = \det(M) R_{E|\mathbb{K}} \det(M^T) = m^2 R_{E|\mathbb{K}}$$

with $m = \det(M)$. □

With this theorem we have a tool to test the linearly dependence of points on elliptic curves. Therefore we compute, for given points $P_1, \dots, P_n \in E(\mathbb{K})$, the regulator $R = R_{P_1,\dots,P_n}$. If this is not equal to 0, the points are linearly independent.

With this method one can also search for r linearly independent points, if r is the rank. Here the elements $x \in \mathbb{K}$ are ordered in a certain way. For every x, one tests if there is a corresponding point $(x, y) \in E(\mathbb{K})$. Then one tests if this new point is linearly independent to the previously found points.

If $E|\mathbb{K}$ is an elliptic curve of rank r over $\mathbb{K}$, one can find r linearly independent points on $E(\mathbb{K})$. Those points are not necessary a basis. They generate a subgroup of finite index in $E(\mathbb{K})$. Siksek [203] gives an estimate for this index. For the proof of this estimate we need a lemma from geometry of numbers which goes back already to Hermite and Minkowski.

Lemma 8.2. *Let $\mathcal{F} = (f_{i,j})_{1 \le i,j \le r}$ be a symmetric positive definit matrix with determinant $D = \det(\mathcal{F}) > 0$. For a vector $\boldsymbol{x} = (x_1, \dots, x_r) \in \mathbb{R}^r$ let*

$$f(\boldsymbol{x}) := \boldsymbol{x} \mathcal{F} \boldsymbol{x}^T = \sum_{i,j=1}^r f_{i,j} x_i x_j.$$

Then there exists a positive constant γ_r (Hermite constant), such that

$$\inf_{\boldsymbol{m} \neq \boldsymbol{0} \text{ integral}} \{f(\boldsymbol{m})\} \le \gamma_r D^{1/r}.$$

Here $\boldsymbol{m}$ *integral means that the coordinates of the vectors* $\boldsymbol{m}$ *are integers. Further*

$$\gamma_1^1 = 1, \quad \gamma_2^2 = \frac{4}{3}, \quad \gamma_3^3 = 2, \quad \gamma_4^4 = 4,$$
$$\gamma_5^5 = 8, \quad \gamma_6^6 = \frac{64}{3}, \quad \gamma_7^7 = 64, \quad \gamma_8^8 = 2^8$$

and for $r \geq 9$ *one has*

$$\gamma_r = \frac{4}{\pi}\Gamma\left(\frac{r}{2}+1\right)^{2/r}.$$

Proof. See Siksek [203]. A first version of this lemma was due to Hermite. Minkowski gave the formula with $\gamma_r = \frac{4}{\pi}\Gamma\left(\frac{r}{2}+1\right)^{2/r}$ for all r (see Cassels [28] and Siegel [202]). The constants $\gamma_1, \ldots, \gamma_r$ given in the lemma are, for $1 \leq r \leq 8$, the smallest constants which make the lemma valid. This can be seen from Cassels [26]. □

This lemma is used together with the fact that the Néron–Tate height is a positive definit quadratic form on $\mathbb{R} \otimes E(\mathbb{K})$.

Theorem 8.3. *Let* E *be an elliptic curve of rank* $r > 0$ *over the number field* $\mathbb{K}$. *Let*

$$0 < \lambda \leq \inf\{\hat{h}(P) : P \in E(\mathbb{K})\backslash E(\mathbb{K})_{\mathrm{tors}}\}$$

be an estimate for the minimal Néron–Tate height of non torsion points. Furthermore let $P_1, \ldots, P_r$ *be linearly independent points on* $E(\mathbb{K})$. *Then one has for the index* n *of the subgroup generated by* $P_1, \ldots, P_r$ *in the free part of* $E(\mathbb{K})$

$$n \leq R_{P_1,\ldots,P_r}^{1/2}\left(\frac{1}{2}\frac{\gamma_r}{\lambda}\right)^{r/2}.$$

Here γ_r *is defined as in the lemma above and* $R_{P_1,\ldots,P_r}$ *is the regulator of the points* $P_1, \ldots, P_r$.

Proof. Let $Q_1, \ldots, Q_r$ be a basis of $E(\mathbb{K})$. The regulator matrix

$$\mathcal{R}_{E|\mathbb{K}} = (\langle Q_i, Q_j\rangle)_{1\leq i,j\leq r}$$

is symmetric and positive definite with the determinant $R_{E|\mathbb{K}} = \det(\mathcal{R}_{E|\mathbb{K}}) > 0$ from Theorem 5.23.

For a point $Q = \sum_{i=1}^r m_i Q_i + T$ with $m_i \in \mathbb{Z}$, $T \in E(\mathbb{K})_{\mathrm{tors}}$, one has according to Proposition 5.22:

$$2\hat{h}(Q) = \sum_{i,j=1}^{r} m_i m_j \langle Q_i, Q_j\rangle.$$

We consider $\mathbb{R} \otimes E(\mathbb{K})$ as an r-dimensional $\mathbb{R}$-vector space and define for vectors $\boldsymbol{x} = (x_1, \ldots, x_r) \in \mathbb{R}^r$

$$\begin{aligned} f(\boldsymbol{x}) &:= \boldsymbol{x} \mathcal{R}_{E|\mathbb{K}} \boldsymbol{x}^T \\ &= \sum_{i,j=1}^{r} x_i x_j \langle Q_i, Q_j \rangle. \end{aligned}$$

For $\boldsymbol{m} = (m_1, \ldots, m_r)$ and Q as above one has with this definition

$$2\hat{h}(Q) = f(\boldsymbol{m}).$$

Hence we can use the lemma and obtain

$$\begin{aligned} \lambda &\leq \inf\{\hat{h}(Q) : Q \in E(\mathbb{K}) \backslash E(\mathbb{K})_{\text{tors}}\} \\ &= \frac{1}{2} \inf_{\boldsymbol{m} \neq \boldsymbol{0} \text{ integral}} \{f(\boldsymbol{m})\} \\ &\leq \frac{1}{2} \gamma_r R_{E|\mathbb{K}}^{1/r} \end{aligned}$$

Let n be the index of the subgroup generated by $P_1, \ldots, P_r$ in the free part of $E(\mathbb{K})$. One has $R_{P_1,\ldots,P_r} = n^2 R_{E|\mathbb{K}}$. It follows that

$$\lambda \leq \frac{1}{2} \gamma_r \frac{R_{P_1,\ldots,P_r}^{1/r}}{n^{2/r}} \Leftrightarrow n \leq R_{P_1,\ldots,P_r}^{1/2} \left(\frac{1}{2} \frac{\gamma_r}{\lambda} \right)^{r/2}. \qquad \square$$

8.2 Computation of a basis

We give here a simple method to compute a basis using the index estimate of Siksek. A more sophisticated method can be found in the article of Siksek [203].

For estimating the index of the subgroup generated by those points in the free part of $E(\mathbb{K})$, we need an estimate

$$0 < \lambda \leq \inf\{\hat{h}(P) : P \in E(\mathbb{K}) \backslash E(\mathbb{K})_{\text{tors}}\}.$$

To achieve this, we compute the difference δ between the canonical height and the ordinary height for points in $E(\mathbb{K})$ as described in Section 5.5. Then we choose an $\varepsilon > 0$ small and compute the set

$$M(\varepsilon) = \{P \in E(\mathbb{K}) : h(P) \leq \delta + \varepsilon\}.$$

When we search a point $E(\mathbb{K})$ with bounded height, we usually have to use Lemma 5.12 (see Section 5.4). This makes, if the number field is large, the computation to take a long time. Therefore it is better to choose the number ε not too large ($\varepsilon < 0.1$). On the other hand one should choose ε not too small, so that the estimate of the index does not grow too large, because then the test if the index is divisible by a prime number must be used for large primes. In the implementations of the first author [189] the test took several hours for prime numbers ≥ 17.

With the definition of the set $M(\varepsilon)$ one has for all points $P \in M(\varepsilon)$

$$\hat{h}(P) \leq \varepsilon.$$

Thus we set

$$\lambda := \min\{\varepsilon, \hat{h}(P) : P \in M(\varepsilon) \text{ and } P \notin E(\mathbb{K})_{\text{tors}}\}.$$

Algorithm 8.4 (Computation of λ).

```
INPUT: An elliptic curve E|K.
OUTPUT: An estimate 0 < λ ≤ inf{ĥ(P) : P ∈ E(K)\E(K)_tors}.
1.  Compute the difference δ of the canonical height and the
    ordinary height of points in E(K) as described in
    Section 5.5 .
2.  Choose ε > 0.
3.  Find all points in M(ε) = {P ∈ E(K) : h(P) ≤ δ + ε} as
    described in Section 5.4 .
4.  Return λ := min{ε, ĥ(P) : P ∈ M(ε) and P ∉ E(K)_tors}.
```

Suppose that we are given r linearly independent points $P_1, \ldots, P_r \in E(\mathbb{K})$ of infinite order, where $r = \operatorname{rk}(E|\mathbb{K})$. Theorem 8.3 gives for the index n the estimate

$$n \leq \alpha := R_{P_1,\ldots,P_r}^{1/2} \left(\frac{1}{2}\frac{\gamma_r}{\lambda}\right)^{r/2}.$$

If $\alpha < 2$, the index is equal to 1 and the points $P_1, \ldots, P_r$ are a basis. Otherwise we have to test for all prime numbers less than or equal to α if these numbers divide the index.

Therefore we have to test for a prime number $p \leq \alpha$ if there are $a_i \in \mathbb{Z}$, not all divisible by p, and a point $Q \in E(\mathbb{K})$ such that

$$\sum_{i=1}^{r} a_i P_i = pQ.$$

Here we can restrict the a_i to a representative system modulo p. For $p = 2$ it is sufficient to choose $a_i \in \{0, 1\}$. For $p > 2$ we choose the absolutely smallest system (see Hasse [92])

$$a_i \in \left\{\frac{-(p-1)}{2}, \ldots, -1, 0, 1, \ldots, \frac{p-1}{2}\right\}.$$

We apply the multiplication formulas to test if a point $R = (x_0, y_0)$ in $E(\mathbb{K})$ has the representation $R = pQ$ with $Q \in E(\mathbb{K})$ or not. Therefore we take $Q = (x, y)$ and get

$$R = (x_0, y_0) = pQ = \left(\frac{\phi_p(x, y)}{\psi_p(x, y)^2}, \frac{\Omega_p(x, y)}{\psi_p(x, y)^3} \right).$$

This leads to a solution $Q = (x, y)$ of the two equations

$$x_0 \psi_p(X, Y)^2 - \phi_p(X, Y) = 0,$$
$$y_0 \psi_p(X, Y)^3 - \omega_p(X, Y) = 0.$$

Hence there exists a point $Q \in E(\mathbb{K})$ with $R = pQ$ if and only if there exists a solution $(x, y) \in \mathbb{K}^2$ of the above equations.

From Proposition 1.20 we see that $\phi_p(X, Y)$ is a polynomial in X of degree p^2 and that, for $p > 2$, $\psi_p^2(X, Y)$ is a polynomial in X of degree $p^2 - 1$. Hence, we have to find a root $x \in \mathbb{K}$ of the polynomial

$$x_0 \psi_p(X)^2 - \phi_p(X)$$

of degree p^2. Plugging this root in the second equation

$$y_0 \psi_p(x)^3 - \Omega_p(x, Y) = 0$$

and reducing Y^2 modulo the equation of the elliptic curve, we get an equation in degree 1 for Y. Another possibility to find y (which has to be used if $\Omega_p(x, Y) = 0$) is to check from the equation of the curve if there exists a point on E in $\mathbb{K}^2$ with x as first coordinate.

Algorithm 8.5 (Test if a point R is divisible by a prime number).

INPUT: An elliptic curve $E|\mathbb{K}$, a point $R = (x_0, y_0) \in E(\mathbb{K})$, and a prime number p.
OUTPUT: A point $Q \in E(\mathbb{K})$ with $R = pQ$ (if possible).
1. If $p = 2$:
2. Compute the polynomials $\phi_2(X), \psi_2(X, Y), \Omega_2(X, Y)$
3. If there exist solutions (x, y) of the equations $x_0 \psi_p(X, Y)^2 - \phi_p(X, Y) = 0$ and $y_0 \psi_p(X, Y)^3 - \Omega_p(X, Y) = 0$, return (x, y).
4. Return ''no such representation''.
5. Compute the polynomials $\phi_p(X), \psi_p(X), \Omega_p(X, Y)$ (see Section 1.3). Reduce Y^2 modulo the curve equation such that $\Omega_p(X, Y)$ is of degree 1 in Y.
6. Compute all roots of $x_0 \psi_p(X)^2 - \phi_p(X)$ over $\mathbb{K}$.
7. For all roots x computed in Step 6:
8. If $y_0 = 0$ (then $\Omega_p(x, Y) = 0$), use the curve equation to

test if there exists a point $Q = (x, y) \in E(\mathbb{K})$.
If such a point exists, return Q.
9. If $y_0 \neq 0$, compute $y \in \mathbb{K}$ as the root of the polynomial $y_0 \psi_p(x)^3 - \Omega_p(x, Y)$ of degree 1 in Y. Return (x, y).
10. Return ''no such representation''.

The test if, for given $R \in E(\mathbb{K})$, there exists a point $Q \in E(\mathbb{K})$ with $R = pQ$ can be done in polynomial time in p by using the algorithm of Ronyai [176]. This algorithm is designed for elliptic curves over finite fields. As we consider here elliptic curves over number fields, we use specialisation. Take a prime q and test, whether the modulo q reduced point has such a representation on the modulo q reduced curve. If not, there can not be such a representation over the number field. If one finds enough such representations modulo primes, one can use the Chinese Remainder Theorem to find a global solution.

If the index of the group generated by $\{P_1, \ldots, P_r\}$ in $E(\mathbb{K})$ is not divisible by a prime number $p \leq \alpha$, the points $P_1, \ldots, P_r$ are a basis. If the index is divisible by a prime number p, we find a point Q with the above representation. We choose from the set of points $\{P_1, \ldots, P_r, Q\}$ the r linearly independent points with minimal regulator. By construction this set must contain the point Q. This new set of points leads to a new estimate for the index, which is smaller than the previous one.

Algorithm 8.6 (Computation of a basis).

INPUT: An elliptic curve $E|\mathbb{K}$
OUTPUT: A basis of the free part of $E(\mathbb{K})$
1. Compute the rank r of $E(\mathbb{K})$ as described in Chapter 7.
2. If $r = 0$, return $\{\ \}$.
3. Compute λ with Algorithm 8.4.
4. Search for r linearly independent points $\{P_1, \ldots, P_r\}$ of $E(\mathbb{K})$ using the test of Section 8.1 .
5. Compute $\alpha := R_{P_1,\ldots,P_r}^{1/2} \left(\frac{1}{2}\frac{\gamma_r}{\lambda}\right)^{r/2}$.
6. If $\alpha < 2$, return $\{P_1, \ldots, P_r\}$.
7. For $a_i \in \{0, 1\}$:
8. If there exists a $Q \in E(\mathbb{K})$ with $\sum_{i=1}^{r} a_i P_i = 2Q$ let $\{P_1, \ldots, P_r\}$ be the set of r linearly independent points with minimal regulator from the set $\{P_1, \ldots, P_r, Q\}$ and go to Step 5.
9. For all prime numbers $p \geq 3$ with $p \leq \alpha$:
10. For $a_i \in \left\{\frac{-(p-1)}{2}, \ldots, -1, 0, 1, \ldots, \frac{p-1}{2}\right\}$:
11. If there exists $Q \in E(\mathbb{K})$ with $\sum_{i=1}^{r} a_i P_i = pQ$ let $\{P_1, \ldots, P_r\}$ be the set of r linearly independent points with minimal regulator from the set $\{P_1, \ldots, P_r, Q\}$ and go to Step 5.
12. Return $\{P_1, \ldots, P_r\}$.

A similar method to compute a basis of an elliptic curve is the algorithm of Manin [136] (see also the generalization of this algorithm for number fields by the second author [250]). For this method, one uses the explicit version of the conjecture of Birch and Swinnerton-Dyer, Conjecture 7.17, to estimate the regulator of the elliptic curve. With this estimate one gets an estimate for the height of the basis points. As in the algorithm of Siksek, the critical point is here to find all points with given height. Then it is relatively easy to determine a basis from this set of points. In general, the height estimate of Manin is larger than the height estimate of Siksek, hence the method of Manin would take much longer than the method of Siksek.

8.3 Examples

1) We consider first an example of an elliptic curve over $\mathbb{Q}$.

Let

$$E : Y^2 = X(X-2)(X+4)$$

be given over $\mathbb{Q}$. One can compute

$$E(\mathbb{Q})_{\mathrm{tors}} = \{T_0 = \mathcal{O}, T_1 = (0,0), T_2 = (2,0), T_3 = (-4,0)\}$$

(see page 188). Then one can apply complete 2-descent (see Section 7.6) to get $\mathrm{rk}(E(\mathbb{Q})) = 1$ (using for example SIMATH). The point $P = (-2, 4)$ is a point of infinite order. Employing that one can compute, using the estimate of Silverman (Theorem 5.35),

$$|\hat{h}(Q) - h(Q)| \leq 7.9$$

for all $Q \in E(\mathbb{Q})$, we choose $\varepsilon = 0.1$ and search for all points from $E(\mathbb{Q})$ with $\hat{h}(Q) \leq 0.1$. Hence we search for all points $Q \in E(\mathbb{Q})$ with $h(Q) \leq 8.0$. We consider $x = \frac{a}{b} \in \mathbb{Q}$ with

$$H(x) = \max\{|a|, |b|\} \leq e^8 = 2980.96.$$

One has $b = (b')^2$ with $b' \in \mathbb{N}$ (see for example Lemma 5.4). We consider

$$-2980 \leq a \leq 2980, \quad 1 \leq b' \leq 54$$

with $\gcd(a, b') = 1$. For those a, b' we set $x = \frac{a}{b'^2}$ and test if

$$x(x-2)(x+4)$$

is a square in $\mathbb{Q}$. We find the following points:

$$\begin{aligned}
T_1,\ T_2,\ T_3,\ P&,\\
-P + T_2 &= (-1, 3),\\
P + T_1 &= (4, 8),\\
2P + T_3 &= (-4/25, 144/125),\\
-P + T_3 &= (8, 24),\\
-2P &= (9/4, 15/8),\\
-2P + T_1 &= (-32/9, 80/27),\\
-2P + T_2 &= (50, 360),\\
-3P &= (-98/289, 8372/4913),\\
3P + T_2 &= (-529/169, 8211/2197),\\
3P + T_1 &= (1156/49, -40664/343),\\
3P + T_3 &= (1352/529, 37128/12167).
\end{aligned}$$

Further

$$\begin{aligned}
\hat{h}(T_i) &= 0,\\
\hat{h}(\pm P + T_i) &= 0.676 = \hat{h}(\pm P),\\
\hat{h}(\pm 2P + T_i) &= 2.7 = \hat{h}(\pm 2P),\\
\hat{h}(\pm 3P + T_i) &= 6.1 = \hat{h}(\pm 3P)
\end{aligned}$$

for $i = 0, 1, 2, 3$. Then we choose $\lambda = 0.1$. We obtain the estimate

$$n \le \alpha := \sqrt{1.352}/\sqrt{0.2} = 2.6.$$

We only have to test if the index is divisible by 2, that means, if there is a point $Q = (x, y) \in E(\mathbb{Q})$ such that $2Q = P = (-2, 4)$. Using the multiplication formulas (Chapter 1, Section 1.3) we test if there exists an $x \in \mathbb{Q}$ with

$$-2 = \frac{\phi_2(x)}{\psi_2(x)^2} = \frac{(x^2+8)^2}{4x(x-2)(x+4)},$$

that is, if the polynomial

$$X^4 + 8X^3 + 32X^2 - 64X + 64$$

has a root in $\mathbb{Q}$. This root should lie in $\mathbb{Z}$. Because the polynomial has no root in $\mathbb{F}_5$, it can not have a root in $\mathbb{Z}$. Therefore the index is not divisible by 2. Hence the point P is a basis point.

Another method to compute a basis for curves of rank 1 is the following. Let P be a point of infinite order. If P is not a basis point, there exists a basis point Q with

$P = mQ + T$, where $m \in \mathbb{N}$, $m > 1$, and T is a torsion point. This relation implies that

$$\hat{h}(P) = \hat{h}(mQ) = m^2\hat{h}(Q),$$

hence

$$\hat{h}(Q) = \frac{\hat{h}(P)}{m^2} \leq \frac{\hat{h}(P)}{4}.$$

Hence we search for all points Q from $E(\mathbb{Q})$ with

$$\hat{h}(Q) \leq \frac{\hat{h}(P)}{4}.$$

If we do not find a point, then P is a basis point. If we find points, we choose as a new starting point P from this set with minimal Néron–Tate height. After finitely many steps we have found a basis point. This process is called *infinite descent.*

2) The second example is an elliptic curve over a quadratic number field. It is taken from the article of Pethő and Schmitt [166]. Consider $\mathbb{K} = \mathbb{Q}(\sqrt{13})$ and the elliptic curve

$$\begin{aligned} E : Y^2 + (1+\sqrt{13})XY + (2+\sqrt{13})X \\ = X^3 + \left(-\frac{1}{2} - \frac{1}{2}\sqrt{13}\right)X^2 + \left(\frac{1}{2} + \frac{1}{2}\sqrt{13}\right)X + \left(\frac{1}{2} + \frac{1}{2}\sqrt{13}\right). \end{aligned}$$

This curve is an example in the article of Cremona and Serf [43] (see also the example in Chapter 7 Section 7.6). They computed the rank by general 2-descent:

$$\mathrm{rk}(E(\mathbb{K})) = 1.$$

They also found a point of infinite order:

$$P_1 = \left(-\frac{3}{4}, -\frac{1}{2} - \frac{1}{8}\sqrt{13}\right).$$

Using the estimate of Siksek, we find that

$$|\hat{h}(Q) - h(Q)| \leq 0.2090249966$$

for all $Q \in E(\mathbb{K})$. We choose $\varepsilon = 0.3$ and search for all points $Q \in E(\mathbb{K})$ with $h(Q) \leq 0.5090249966$. We come up with the points $\pm Q_1$, where

$$Q_1 = \left(-\frac{1}{2} + \frac{1}{2}\sqrt{13}, -\frac{3}{2} + \frac{1}{2}\sqrt{13}\right)$$

and

$$\hat{h}(\pm Q_1) = 0.0876826508.$$

The index estimate gives

$$n \leq \alpha := 5.0000000006.$$

It is easy to see that the index is divisible by 5 and that

$$P_1 = 5Q_1.$$

Hence $\{Q_1\}$ is a basis of $Q(\mathbb{K})$. If $E/\mathbb{K}$ has rank $r = 1$, one can also (at least theoretically) apply the Heegner point method.

8.4 Heegner point method

If an elliptic curve E over the field $\mathbb{Q}$ of rational numbers has rank one there are at least *two* other possibilities of determining a generator. One possibility is via the L-function of $E|\mathbb{Q}$. Then one needs to improve the standard algorithm for finding a generator (see Silverman [209]). The search length is improved in this way from $O(D^{3/2})$ to $O(D^{1/2})$, where D is a prechosen search bound.

The second possibility is by constructing a Heegner point on $E|\mathbb{Q}$ (see Birch [16], Birch–Stephens [17]). It is restricted to rank one elliptic curves because the Heegner point is trivial, i.e. it is a torsion point, if the rank of $E|\mathbb{Q}$ is greater than one. Since there are Mordell curves of rank one (see, e.g., Gebel, Pethő, Zimmer [80]), we confine ourselves to considering Mordell equations for which the rank is one. Then we follow Birch and Stephens [17] in our sketch of the Heegner procedure. Unfortunately, the Heegner procedure could not be applied to all Mordell curves

$$Y^2 = X^3 + k \quad (0 \neq k \in \mathbb{Z})$$

of rank one, since – in some cases – the coefficients involved were too large in the range

$$|k| \leq 100.000 .$$

We mention that, for $k = 7823$, the missing generator was found by employing 4-descent (see [145]) rather than the Heegner procedure (see Stoll in `http://listserv.nodak.edu/archives/nmbrthry.html`, January 2002).

The procedure is outlined here for curves with complex multiplication. The Mordell curves have, of course, complex multiplication (see the example in Chapter 1, Section 1.5).

Let $\alpha \in \mathbb{C}, \alpha \neq 0$, be a complex quadratic number. Then α is a root of an irreducible quadratic polynomial

$$aX^2 + bX + c \in \mathbb{Z}[X] \tag{8.1}$$

whose rational integral coefficients have no common factor:

$$\gcd(a, b, c) = 1 .$$

The discriminant of this polynomial must be negative:

$$d(\alpha) = d := b^2 - 4ac < 0\,.$$

We introduce the ring

$$R(\alpha) = R := \mathbb{Z}\left[\frac{1}{2}(d + \sqrt{d})\right]$$

and the primitive ideal of this ring

$$\mathfrak{a}(\alpha) = \mathfrak{a} := \{m + n\alpha : m, n \in \mathbb{Z}\}.$$

The element α determines a class of binary quadratic forms equivalent to (8.1) as the ideal $\mathfrak{a}$ defines an ideal class of the ring R (see Zagier's book [242]).

The modular invariant $j(\alpha) = j(\mathfrak{a})$ depending only on the ideal class of $\mathfrak{a}$ is an algebraic integer (see Deuring [54] or Lang [116]). The field

$$\mathbb{K}(\alpha) = \mathbb{K} := \mathbb{Q}(\alpha, j(\alpha))$$

is the ring class field belonging to $R(\alpha)$. We have the following Hasse diagram

$$\begin{array}{l}
\mathbb{K}(\alpha) \\
\big|\, h \qquad \Big]\ G \\
\mathbb{Q}(\alpha) \qquad\qquad \Big]\ H \\
\big|\, 2 \\
\mathbb{Q}
\end{array}$$

with the Galois groups

$$\begin{aligned} G(\alpha) = G &:= \mathrm{Gal}(\mathbb{K}(\alpha)|\mathbb{Q}(\alpha)),\\ H(\alpha) = H &:= \mathrm{Gal}(\mathbb{K}(\alpha)|\mathbb{Q}). \end{aligned}$$

The class number of the ring $R(\alpha)$ or the field $\mathbb{K}(\alpha)$ is $h(\alpha) = h$. We remark that the field $\mathbb{K}(\alpha)$ depends only on the discriminant $d(\alpha)$ of the polynomial (8.1).

To the complex number α, there corresponds the lattice $\mathbb{Z} \oplus \alpha\mathbb{Z}$ and thus the ideal $\mathfrak{a}$. Let the complex numbers $\alpha_1 = \alpha, \ldots, \alpha_h \in \mathbb{C}$ be chosen in such a way that the lattices

$$\mathbb{Z} \oplus \alpha_1\mathbb{Z}, \ldots, \mathbb{Z} \oplus \alpha_h\mathbb{Z}$$

represent the ideal classes of the primitive ideals

$$\mathfrak{a} = \mathfrak{a}_1, \ldots, \mathfrak{a}_h$$

of the ring $R(\alpha)$, then

$$j(\mathfrak{a}_1), \ldots, j(\mathfrak{a}_h)$$

(depending only on the classes of $\mathfrak{a}_1, \dots, \mathfrak{a}_h$) is a complete set of $G(\alpha)$ conjugates of $j(\mathfrak{a})$.

It has been shown recently that every elliptic curve $E|\mathbb{Q}$ is a Weil curve (Wiles and others [38], [221]), that is, for every elliptic curve $E|\mathbb{Q}$ there exist a natural number $N \in \mathbb{N}$ and a surjective morphism

$$\varphi : X_0(N) \to E.$$

This proves the conjecture of Shimura, Taniyama and Weil (see Chapter 7, Section 7.3). Here, when $\Gamma_0(N)$ is the group of linear fractional transformations

$$z \mapsto \frac{\alpha z + \beta}{\gamma z + \delta}, \quad \alpha, \beta, \gamma, \delta \in \mathbb{Z}, \quad \alpha\delta - \beta\gamma = 1, \text{ with } \gamma \equiv 0 \pmod{N}.$$

We apply this group to the upper half plane $\mathbb{H} \subseteq \mathbb{C}$, then

$$Y_0(N) := \mathbb{H}/\Gamma_0(N)$$

is a representative of the equivalence classes of $\mathbb{H}$ with respect to $\Gamma_0(N)$ and $X_0(N)$ is the completion of $Y_0(N)$ resulting from filling in the cusps of $\Gamma_0(N)$.

We observe that the modular invariants $j = j(z)$ and $j_N = j(Nz)$ are connected by the class equation so that (cf. Birch [16], Lay [123], [124], Weber [231])

$$F_N(j, j_N) = 0.$$

Now $\alpha \in \mathbb{H}$ is a *Heegner point* of $X_0(N)$ and hence $\varphi(\alpha) = P_\alpha \in E(\mathbb{K}(\alpha))$ a *Heegner point* of E, when $d(N\alpha) = d(\alpha)$. One takes then the *trace*

$$P(\alpha) = \sum_{\sigma \in G} P_\alpha^\sigma$$

to obtain a *Heegner point*

$$P(\alpha) \in E(\mathbb{Q}(\alpha)).$$

Applied to the Mordell curve

$$E : Y^2 = X^3 - 12^3$$

and its d-twist

$$E^{(d)} : Y^2 = X^3 - 12^3 d^3$$

for $d \in \mathbb{Z}\backslash\{0\}$, one obtains a *Heegner point* on the twist $E^{(\pm d)}(\mathbb{Q})$ in the following way.

With

$$q := e^{\pi i z} \quad \text{for } z \in \mathbb{H}$$

we introduce the functions (see Definition 2.5)

$$\mathfrak{f}(z) = q^{-\frac{1}{24}} \prod_{\nu=1}^{\infty} (1 + q^{2\nu-1}),$$

$$\mathfrak{f}_1(z) = q^{-\frac{1}{24}} \prod_{\nu=1}^{\infty} (1 - q^{2\nu-1})$$

and

$$\mathfrak{f}_2(z) = \sqrt{2} q^{\frac{1}{12}} \prod_{\nu=1}^{\infty} (1 + q^{2\nu}).$$

These functions are intimately related to the η-function

$$\eta(z) = q^{\frac{1}{12}} \prod_{\nu=1}^{\infty} (1 - q^{2\nu}) :$$

$$\mathfrak{f}(z) = \frac{e^{-\frac{\pi i}{24}} \eta\left(\frac{z+1}{2}\right)}{\eta(z)},$$

$$\mathfrak{f}_1(z) = \frac{\eta\left(\frac{z}{2}\right)}{\eta(z)}$$

and

$$\mathfrak{f}_2(z) = \sqrt{2} \frac{\eta(2z)}{\eta(z)} .$$

We observe the trivial relation

$$\prod_{\nu=1}^{\infty} (1 + q^{\nu})(1 - q^{2\nu-1}) = \prod_{\nu=1}^{\infty} (1 + q^{\nu}) \frac{(1 - q^{\nu})}{(1 - q^{2\nu})} = 1 .$$

We use $\mathfrak{f}(z)$, $\mathfrak{f}_1(z)$ and $\mathfrak{f}_2(z)$ to define Weber's functions (see Weber [231], Lay [123])

$$\gamma_2(z) := \frac{\mathfrak{f}(z)^{24} - 16}{\mathfrak{f}(z)^8},$$

$$\gamma_3(z) := \frac{(\mathfrak{f}(z)^{24} + 8)(\mathfrak{f}_1(z)^8 - \mathfrak{f}_2(z)^8)}{\mathfrak{f}(z)^8} .$$

Their relations to the modular invariant are (see Proposition 2.6)

$$\gamma_2(\alpha)^3 = j(\alpha) \quad \text{and} \quad \gamma_3(\alpha)^2 = j(\alpha) - 12^3$$

so that

$$\gamma_3(z)^2 = \gamma_2(z)^3 - 12^3.$$

Then

$$(\gamma_2(\alpha), \gamma_3(\alpha)) \in E$$

is a point of the elliptic curve E with coordinates

$$\gamma_2(\alpha), \gamma_3(\alpha) \in \mathbb{K}(\alpha) \quad \text{for } d < 0,\ 3 \nmid d,\ 3 \mid k,\ d \equiv 1 \pmod 4 \tag{8.2}$$

or

$$\gamma_2(\alpha), i\gamma_3(\alpha) \in \mathbb{K}(\alpha) \quad \text{for } d < 0,\ 3 \nmid d,\ 3 \mid k,\ d \equiv 4, 8 \pmod{16}. \tag{8.3}$$

(The integer k is the coefficient of the elliptic curve $E : Y^2 = X^3 + k$. In our case, $k = -12^3$.)

We remark that the discriminant d is not fundamental in the case of $d \equiv 4 \pmod{16}$, but this fact does not really matter (see Birch–Stephens [17]).

On choosing

$$\alpha := \frac{1}{2}\big(3 + \sqrt{d}\big) \quad \text{in the first case, that is, in (8.2)}$$

and

$$\alpha := \sqrt{d} \quad \text{in the second case, that is, in (8.3),}$$

we finally arrive at a rational point in $\mathbb{Q}$ on the twist $E^{(\pm d)}$.

For let $\mathbb{K}_0(\alpha)$ denote the maximal real subfield of $\mathbb{K}(\alpha)$ so that

$$[\mathbb{K}_0(\alpha) : \mathbb{Q}] = [\mathbb{K}(\alpha) : \mathbb{Q}(\alpha)] = h.$$

The field extension $\mathbb{K}_0(\alpha)$ need not be Galois. This is a disadvantage circumvented in the following way.

When $\alpha = \frac{1}{2}(3 + \sqrt{d})$, $d < 0$, $d \equiv 1 \pmod 4$, that is, in the case (8.2),

$$Q(\alpha) := \big(d\gamma_2(\alpha), \sqrt{d^3}\gamma_3(\alpha)\big) \in E^{(d)}(\mathbb{K}_0(\alpha))$$

is a point on the twisted elliptic curve $E^{(d)}$ with coordinates in the field $\mathbb{K}_0(\alpha)$ and when $\alpha = \sqrt{d}$, $d < 0$, $d \equiv 4, 8 \pmod{16}$, that is in the case (8.3),

$$Q(\alpha) := \big(-d\gamma_2(\alpha), \sqrt{-d^3}\gamma_3(\alpha)\big) \in E^{(-d)}(\mathbb{K}_0(\alpha))$$

is a point on the twisted elliptic curve $E^{(-d)}$ with coordinates in the field $\mathbb{K}_0(\alpha)$.

Since $\mathbb{K}_0(\alpha)|\mathbb{Q}$ is not always Galois, instead of the $\mathbb{K}_0(\alpha)|\mathbb{Q}$-trace, one takes the $\mathbb{K}(\alpha)|\mathbb{Q}(\alpha)$-trace of the point $Q(\alpha)$ and shows that it is in fact a rational point in

$$E^{(\pm d)}(\mathbb{Q}).$$

In this connection, Birch and Stephens [17] come up with the following conjecture:

Conjecture 8.7. *If $E|\mathbb{Q}$ is a modular elliptic curve and $\mathbb{K}|\mathbb{Q}$ is a complex quadratic field extension so that $E|\mathbb{K}$ has odd rank, then the Heegner point on $E|\mathbb{K}$ has Néron–Tate height measured by the first derivative of the L-function $L(E|\mathbb{K}; s)$ of $E|\mathbb{K}$ at $s = 1$.*

We refer here to the paper of Gross and Zagier [86]. In fact they prove the following. Let $L(E|\mathbb{Q}, s)$ be the L-series of $E|\mathbb{Q}$ (see Chapter 7 Section 7.1) and $\mathfrak{N}_{E|\mathbb{Q}}$ be the conductor of $E|\mathbb{Q}$ (see Definition 7.8). Define

$$\Lambda(E|\mathbb{Q}, s) := \mathfrak{N}_{E|\mathbb{Q}}^{\frac{s}{2}}(2\pi)^{-s}\Gamma(x)L(E|\mathbb{Q}, s)$$

with the usual Γ-function. The conjecture of Hasse and Weil (Conjecture 7.9) implies the functional equation

$$\Lambda(E|\mathbb{Q}, 2-s) = \pm\Lambda(E|\mathbb{Q}, s).$$

Assume this relation to be true for the elliptic curves E over $\mathbb{Q}$ considered here. If $L(E|\mathbb{Q}, 1) = 0$, there exists a point $P \in E(\mathbb{Q})$ such that the derivative with respect to s is

$$L'(E|\mathbb{Q}, 1) = \alpha\omega\hat{h}(P)$$

with a certain number $\alpha \in \mathbb{Q}^*$, where ω is the real period of E over $\mathbb{Q}$ and $\hat{h}$ as usual denotes the Néron–Tate height on $E(\mathbb{Q})$ (see the conjecture of Birch and Swinnerton-Dyer 7.17). Since $\hat{h}(P) \neq 0$ for non-torsion points $P \in E(\mathbb{Q})$ it follows in the rank 1 case that $L'(E|\mathbb{Q}, 1) \neq 0 \;\Rightarrow\; \exists P \in E(\mathbb{Q})$ of infinite order so that the rank of $E|\mathbb{Q}$ is at least one. In the rank one case, this should be the Heegner point $P \in E(\mathbb{Q})$. It satisfies then the above relation $L'(E|\mathbb{Q}, 1) = \alpha\omega\hat{h}(P)$.

In this regard the conjecture of Birch and Stephens is corroborated. At least one knows by Gross and Zagier that, if $L'(E|\mathbb{Q}, 1) \neq 0$ and $E|\mathbb{Q}$ has rank 1, then

$$L^{(r')}(E|\mathbb{Q}, 1) = \alpha\omega R_{E|\mathbb{Q}},$$

where r' is the analytic rank and again

$$R_{E|\mathbb{Q}} = \det(\langle P_i, P_j\rangle_{i,j=1,\dots,r})$$

is the regulator of $E|\mathbb{Q}$. The set $(r = r')$

$$\{P_1, \dots, P_r\},$$

as usual, denotes a basis of $E(\mathbb{Q})$.

We note that the Birch and Swinnerton-Dyer conjecture predicts equality of the analytic and the algebraic rank of the elliptic curve E over $\mathbb{Q}$:

$$r' = \mathrm{ord}_{s=1}L(E|\mathbb{Q}, s) = \mathrm{rk}(E|\mathbb{Q}) = r.$$

We mention in this connection that Gross and Zagier [86] prove that the special elliptic curve

$$-139Y^2 = X^3 + 10X^2 - 20X + 8$$

has rank 3:

$$\mathrm{ord}_{s=1}(L(E|\mathbb{Q}, s) = \mathrm{rk}(E|\mathbb{Q}) = 3.$$

This turns out to be a rather important fact.

8.5 Exercises

1) By assuming the Birch and Swinnerton-Dyer conjecture and using, e.g. SIMATH, prove the following assertions.

   a) The curve

   $$E : Y^2 + 67Y = X^3 - 21X^2 - 10X + 30$$

   has rank 5 and basis points

   $$\begin{array}{lll} P_1 = (9, -24), & P_2 = (8, -18), & P_3 = (6, -10), \\ P_4 = (5, -7), & P_5 = (18, -33) & \end{array}$$

   over $\mathbb{Q}$.

   b) The curve

   $$E : Y^2 + 351Y = X^3 - 63X^2 + 56X + 22$$

   has rank 6 and basis points

   $$\begin{array}{lll} P_1 = (33, -175), & P_2 = (32, -146), & P_3 = (31, -133), \\ P_4 = (50, -142), & P_5 = (24, -77), & P_6 = (22, -65) \end{array}$$

   over $\mathbb{Q}$.

   c) The curve

   $$E : Y^2 + 1641Y = X^3 - 168X^2 + 161X - 8$$

   has rank 7 and basis points

   $$\begin{array}{lll} P_1 = (103, -806), & P_2 = (102, -766), & P_3 = (101, -743), \\ P_4 = (120, -784), & P_5 = (100, -724), & P_6 = (99, -707), \\ P_7 = (122, -730) & & \end{array}$$

   over $\mathbb{Q}$.

2) The Mordell curve

   $$E : Y^2 = X^3 + (22 + 10\sqrt{5})$$

   over $\mathbb{K} = \mathbb{Q}(\sqrt{5})$ has trivial torsion and rank 2 (see Exercise 5 in Section 6.7 and Exercise 5 in Section 7.8). Check that the following points are in $E(\mathbb{K})$. (Here, $(a, b) = a + b\sqrt{5}$.)

point	canonical height
$P_1 = (2, (5, 1))$	$\hat{h}(P_1) = 0.6542$
$P_2 = ((-1, -1), (1, 1))$	$\hat{h}(P_2) = 0.4827$
$P_3 = ((3, -2), (-7, 6))$	$\hat{h}(P_3) = 1.9084$
$P_4 = ((3, 1), (7, 3))$	$\hat{h}(P_4) = 0.4771$
$P_5 = ((4, -2), (-9, 7))$	$\hat{h}(P_5) = 1.2654$
$P_6 = ((13, -5), (-63, 25))$	$\hat{h}(P_6) = 1.7969$
$P_7 = ((31/2, 13/2), (82, 37))$	$\hat{h}(P_7) = 1.9309$
$P_8 = ((19, 7), (99, 47))$	$\hat{h}(P_8) = 1.7800$
$P_9 = ((-13/10, -9/10), (35/25, 16/25))$	$\hat{h}(P_9) = 2.6170$
$P_{10} = ((49/5, 15/5), (775/25, 407/25))$	$\hat{h}(P_{10}) = 3.0192$
$P_{11} = ((527/121, 195/121), (15755/1331, 6895/1331))$	$\hat{h}(P_{11}) = 3.0023$

Show that P_2 and P_4 constitute a basis of $E(\mathbb{K})$. Find the representation of the other points with respect to this basis.

3) The elliptic curve

$$E : Y^2 = X^3 + iX + (1 - i)$$

over $\mathbb{K} = \mathbb{Q}(i)$ with $i = \sqrt{-1}$ has trivial torsion and rank 2 (see Exercise 4 in Section 6.7 and Exercise 5 in Section 7.8). Check that the following points are in $E(\mathbb{K})$. (Here $(a, b) = a + ib$.)

point	canonical height
$P_1 = (-1, (-1, 1))$	$\hat{h}(P_1) = 0.5700$
$P_2 = ((0, 1), (-1, 1))$	$\hat{h}(P_2) = 0.5329$
$P_3 = ((1, -1), (-1, 1))$	$\hat{h}(P_3) = 0.8136$
$P_4 = ((-3, -1), (-3, 5))$	$\hat{h}(P_4) = 1.3923$
$P_5 = ((5, -89), (543, 643))$	$\hat{h}(P_5) = 4.4984$
$P_6 = ((3/4, -1), (-3/4, 5/8))$	$\hat{h}(P_6) = 2.1317$
$P_7 = ((5/4, 1), (9/8, 7/4))$	$\hat{h}(P_7) = 2.2802$
$P_8 = ((7/25, 24/25), (-81/125, 133/125))$	$\hat{h}(P_8) = 2.1231$
$P_9 = ((32/25, 1/25), (219/125, 17/125))$	$\hat{h}(P_9) = 2.2344$
$P_{10} = ((-152/25, -39/25), (-753/125, 1829/125))$	$\hat{h}(P_{10}) = 3.3917$
$P_{11} = ((-2, -9/8), (-93/32, 83/32))$	$\hat{h}(P_{11}) = 3.2545$
$P_{12} = ((255/169, 64/169), (4387/2197, 1673/2197))$	$\hat{h}(P_{12}) = 3.2804$
$P_{13} = ((-3784/5329, 105/5329), (-427493/389017, 297413/389017))$	$\hat{h}(P_{13}) = 4.7963$
$P_{14} = ((-1723/4225, 1239/4225), (-290021/274625, 167303/274625))$	$\hat{h}(P_{14}) = 4.7953$

Show that P_1 and P_2 constitute a basis of $E(\mathbb{K})$. Find the representation of the other points with respect to this basis.

Chapter 9

S-integral points

In this chapter we first give an overview over the work on S-integral points on elliptic curves. In the second section we explain elliptic logarithms which we need later when we consider S-integral points on elliptic curves over $\mathbb{Q}$.

For a number field $\mathbb{K}$ let $M_{\mathbb{K}}$ as always be the set of all places of $\mathbb{K}$.

9.1 Overview

Definition 9.1. Let S be a finite set of primes (places) $\mathfrak{p}$ of a number field $\mathbb{K}$ containing the archimedean primes. A point $P = (x, y)$ in the Mordell–Weil group $E(\mathbb{K})$ of an elliptic curve E with integral coefficients over $\mathbb{K}$ is said to be *S-integral* if its first coordinate x (and hence also its second coordinate y) has this property

$$v_{\mathfrak{p}}(x) \geq 0 \quad \text{for each } \mathfrak{p} \in M_{\mathbb{K}} \setminus S.$$

We denote the subset of $E(\mathbb{K})$ consisting of all S-integral points by $E(\mathcal{O}_{\mathbb{K},S})$.

It follows from a theorem of Siegel [201] that the set of all integral points of E over $\mathbb{K}$ is finite. In this case, of course, the finite set S consists only of the archimedean places of $\mathbb{K}$. Mahler [133] showed that $E(\mathcal{O}_{\mathbb{K},S})$ is finite for an arbitrary finite set S of places of $\mathbb{K}$ (including the places over ∞). The proofs of these theorems are not constructive: No hints for the determination of the set of all S-integral points of E over $\mathbb{K}$ are given.

Baker [6] was the first who found explicit (but high) bounds for the height of ordinary integral points in the group $E(\mathbb{K})$. Later on Coates [32] proved for Mordell's elliptic curves E over $\mathbb{K} = \mathbb{Q}$ that the number of S-integral points is finite and can be effectively determined. He used number-theoretic methods reducing the problem to that of solving finitely many S-unit equations.

We follow here a completely different approach suggested independently by Lang [112], [115] and Zagier [243]. The field $\mathbb{K}$ will be restricted here to $\mathbb{Q}$, the field of rational numbers. The approach suggested by Lang and Zagier requires the knowledge of a basis of the Mordell–Weil group $E(\mathbb{K})$. This is a drawback of the approach, but the group $E(\mathbb{K})$ can be computed in certain special cases (see Chapter 8).

For determining ordinary integral points in $E(\mathbb{K})$ complex elliptic logarithms are needed. More precisely, one requires an explicit lower bound for linear forms in complex elliptic logarithms. Such an explicit lower bound, which is difficult to obtain, was given recently by David [45]. This bound opened the door for computing integral points on elliptic curves E over the number field $\mathbb{K} = \mathbb{Q}$. The problem was then solved independently by Stroeker and Tzanakis [217] on the one hand, and by Gebel, Pethő and the second author [77] on the other. The solution can be applied e.g. to the question of sums of consecutive squares and cubes (see the article of Bremner, Stroeker, and Tzanakis [20]) and to two simultaneous Pell equations (see Tzanakis [226] and the thesis [96] of Herrmann).

For computing all S-integral points in the group $E(\mathbb{K})$ however, p-adic elliptic logarithms are needed in addition to ordinary complex elliptic logarithms. In this general case, explicit lower bounds for linear forms in p-adic elliptic logarithms are additionally required. Unfortunately, such lower bounds, which are likewise difficult to obtain, exist only when the rank r of the elliptic curve E over $\mathbb{K} = \mathbb{Q}$ is at most 2. The explicit bound for $r \leq 2$ was obtained by Rémond and Urfels [175]. The application to computing all S-integral points on the curve E over $\mathbb{K} = \mathbb{Q}$ was carried out by Gebel, Pethő and the second author [81]. An application to Mordell's equations [80] and to the curve $y^2 = x^3 - 228x + 848$ studied earlier by de Weger [233] is also made (Herrmann–Pethő, [95]).

However, in order to circumvent the restriction to ranks $r \leq 2$ and also in general, it is interesting that if $\mathbb{K} = \mathbb{Q}$, a sufficient upper bound for S-integral points in $E(\mathbb{K})$ is now available through a paper of Hajdu and Herendi [90] - generalized to arbitrary number fields $\mathbb{K}$ by Hajdu, Herendi, Herrmann, and Pethő (see Herrmann [96]). This bound is used by Pethő, Gebel, Herrmann and the second author [164] to determine, in the special case of $\mathbb{K} = \mathbb{Q}$, all S-integral points on E over $\mathbb{Q}$. Herrmann [94], starting from an idea of Tzanakis, was able to generalize the foregoing procedure to quartic elliptic curves and to certain curves given by general cubics which can be transformed into a Weierstraß form over an arbitrary number field $\mathbb{K}$. For ordinary integer points, a generalisation to arbitrary number fields $\mathbb{K}$, but sticking to elliptic curves E over $\mathbb{K}$ in short Weierstraß form, was carried through by Smart and Stephens [212]. However, the approach which Smart [211] suggested for $\mathbb{K} = \mathbb{Q}$ depends on an assumed lower bound for linear forms in p-adic elliptic logarithms. As pointed out already, this dependence is avoided here. In fact, we take a combination of the methods of Siegel–Baker–Coates and Lang–Zagier and use the explicit upper bound for S-integral points on curves E over $\mathbb{K} = \mathbb{Q}$ of Hajdu and Herendi [90].

In this way, we circumvent the problem created by the restriction mentioned. Of course, the explicit bound, found by Hajdu and Herendi, is large. But it can be reduced by a suitable LLL-procedure given by de Weger [232].

As is immediately clear the explicit bound by Hajdu and Herendi can be generalized to arbitrary number fields $\mathbb{K}$. This generalisation was carried out by Hajdu, Herendi, Herrmann, and Pethő (see Herrmann [96]) in connection with his determination of S-integral points on cubics and quartics of genus one over $\mathbb{K}$.

9.2 Elliptic logarithms

Let $E|\mathbb{Q}$ be an elliptic curve in long Weierstraß normal form. We consider the following two situations.

$\boldsymbol{p = \infty}$**.** Let E' be defined as the elliptic curve in short Weierstraß form into which E can be transformed if the characteristic of the base field is different from two or three (see Theorem 1.7):

$$E' : \quad Y^2 = X^3 - 27c_4X - 54c_6.$$

Definition 9.2. Consider $\varphi : E \to E'$, a birational transformation to the elliptic curve E' in short Weierstraß normal form over $\mathbb{Q}$. Let $\wp(u) = \wp(u|\Lambda)$ denote the Weierstraß function supplying the parametrization of the elliptic curve E' over $\mathbb{Q}$, where $\Lambda \subset \mathbb{C}$ is the corresponding homogeneous lattice generated by a real and a complex fundamental period.

For a point $P = (x, y) \in E(\mathbb{Q})$ we have, according to Theorem 2.15 (with two different but obvious meanings of the symbol prime)

$$\varphi(P) = (x', y') = \left(\wp(u), \frac{1}{2}\wp'(u)\right)$$

for $u \in \mathbb{C}$. The residue class u mod Λ is the *complex elliptic logarithm of* P. We write $u = u(P)$. Hence the number $u(P)$ is determined only up to the period lattice Λ.

By Zagier [243],

$$u \equiv \int_{x'}^{\infty} \frac{d\xi}{\sqrt{\xi^3 - 27c_4\xi - 54c_6}} \mod \Lambda,$$

where $x' = 36x + 3b_2$.

Over the complex number field $\mathbb{C}$ the cubic polynomial in the denominator of the integral decomposes into a product of linear factors according to

$$\xi^3 - 27c_4\xi - 54c_6 = (\xi - e_1)(\xi - e_2)(\xi - e_3)$$

with $e_1, e_2, e_3 \in \mathbb{C}$ and

$$e_1 + e_2 + e_3 = 0.$$

If all the roots are real, we assume

$$e_1 < e_2 < e_3.$$

If there is only one real root, then we assume that $e_3 \in \mathbb{R}$ and e_2 is the complex conjugate of e_1:

$$e_2 = \overline{e}_1.$$

We now introduce the number (occasionally writing $|\dots|$ instead of $|\dots|_\infty$)

$$M = \left\{ \begin{array}{cc} 0 & \text{in case } e_3 \geq 0 \\ \dfrac{3}{\sqrt[3]{2}-1} \max\{\sqrt{|c_4|}, \sqrt[3]{2|c_6|}\} & \text{in case } e_3 < 0 \end{array} \right\}$$

(cf. the definition of μ_∞) and put

$$e_0 = \left\{ \begin{array}{l} 2e_3 \\ 2\max\left\{\dfrac{e_1+e_2}{2}, e_3\right\} \end{array} \right\} + M \quad \text{in case} \left\{ \begin{array}{l} e_1, e_2 \in \mathbb{R} \\ e_1, e_2 \in \mathbb{C} \setminus \mathbb{R} \end{array} \right\}.$$

Of course,

$$\frac{e_1+e_2}{2} = -\frac{e_3}{2}.$$

With the above notation we first state and prove the following

Lemma 9.3. *Suppose that $x' \in \mathbb{R}$ fulfills the condition*

$$x' > \max\{0, e_0\}.$$

Then

$$\int_{x'}^{\infty} \frac{d\xi}{\sqrt{\xi^3 - 27c_4\xi - 54c_6}} < \frac{2\sqrt{2}}{\sqrt{x'}}.$$

Proof. In the proof of the lemma, we shall write x instead of x'.

We may assume that the largest real root of

$$f(X) := X^3 - 27c_4X - 54c_6$$

is non-negative:

$$e_3 \geq 0.$$

For if

$$e_3 < 0,$$

we have

$$e_3 + M \geq 0$$

because by a result of Zassenhaus [246] (see also Mignotte's book [147]),

$$|e_3| \leq M.$$

If we set

$$Y = X + M,$$

the polynomial

$$g(Y) := f(X) = f(Y - M)$$

has the greatest real root $e_3 + M \geq 0$. The integral then becomes, for

$$x > \max\{0, e_0\} \Leftrightarrow y > \max\{0, e_0\} + M > 0,$$

$$\int_x^\infty \frac{d\xi}{\sqrt{f(\xi)}} = \int_y^\infty \frac{d\eta}{\sqrt{g(\eta)}}.$$

Next we introduce the variable

$$Z := Y - (e_3 + M)$$

in order to move the greatest real root $e_3 + M$ of $g(Y)$ to zero. At the same time we define the new cubic polynomial $h(Z)$ by putting

$$h(Z) := g(Y) = g(Z + (e_3 + M)) = Z(Z - e_1')(Z - e_2'),$$

where

$$e_1' := e_1 - e_3, \quad e_2' := e_2 - e_3.$$

Now the integral becomes

$$\int_y^\infty \frac{d\eta}{\sqrt{g(\eta)}} = \int_z^\infty \frac{d\zeta}{\sqrt{h(\zeta)}} = \int_{x-e_3}^\infty \frac{d\zeta}{\sqrt{\zeta(\zeta - e_1')(\zeta - e_2')}}.$$

We distinguish between two cases.

Case 1: $e_1, e_2 \in \mathbb{R}$, or $e_1, e_2 \in \mathbb{C} \setminus \mathbb{R}$, but $e_1 + e_2 < 2e_3$. We recall that in the latter case

$$e_1 + e_2 = e_1 + \overline{e}_1 = 2\mathrm{Re}(e_1).$$

Then

$$e_1' < 0 \text{ and } e_2' < 0 \quad \text{if } e_1, e_2 \in \mathbb{R}$$

and

$$e_1' + e_2' = (e_1 + e_2) - 2e_3 < 0 \quad \text{if } e_1, e_2 \in \mathbb{C} \setminus \mathbb{R}.$$

Under both conditions,

$$h(z) > z^3 \quad \text{for } z > 0,$$

because we were assuming that $e_3 \geq 0$. Of course,

$$z > 0 \Leftrightarrow y > e_3 + M \Leftrightarrow x > e_3.$$

Concerning the integral, we conclude that for $z > 0$

$$\int_z^\infty \frac{d\zeta}{\sqrt{h(\zeta)}} < \int_{x-e_3}^\infty \frac{d\zeta}{\sqrt{\zeta^3}} = \frac{2}{\sqrt{x - e_3}}.$$

However, since in this case always

$$\frac{e_1 + e_2}{2} \leq e_3,$$

we have from the hypothesis of the lemma

$$x > 2 \max\{0, e_3\} + M$$

so that

$$x - e_3 = \frac{x}{2} + \frac{x}{2} - e_3 > \frac{x}{2} + \frac{M}{2}$$

showing that

$$\frac{2}{\sqrt{x - e_3}} < \frac{2\sqrt{2}}{\sqrt{x}},$$

where we kept in mind that $M \geq 0$.

Case 2: $e_1, e_2 \in \mathbb{C} \setminus \mathbb{R}$ and $e_1 + e_2 \geq 2e_3$. In this case we always have

$$e_2 = \overline{e}_1.$$

The assumption implies

$$e'_1 + e'_2 = (e_1 + e_2) - 2e_3 \geq 0.$$

In addition,

$$e'_1 - e'_2 = e_1 - e_2$$

is purely imaginary. Therefore,

$$(Z - e'_1)(Z - e'_2) = Z^2 - (e'_1 + e'_2)Z + e'_1 e'_2$$

$$= \left(Z - \left(\frac{e'_1 + e'_2}{2}\right)\right)^2 - \left(\frac{e'_1 - e'_2}{2}\right)^2 > 0 \quad \text{with} \quad \left(\frac{e'_1 - e'_2}{2}\right)^2 < 0$$

Thus we have

$$\begin{aligned} \frac{h(Z)}{Z} &= (Z - e'_1)(Z - e'_2) \\ &= \left(Z - \frac{e'_1 + e'_2}{2}\right)^2 - \left(\frac{e'_1 - e'_2}{2}\right)^2 \\ &> \left(Z - \frac{e'_1 + e'_2}{2}\right)^2. \end{aligned}$$

Since also

$$Z \geq Z - \frac{e'_1 + e'_2}{2},$$

we obtain altogether

$$h(Z) > \left(Z - \frac{e_1' + e_2'}{2}\right)^3.$$

Now suppose that

$$z > 0 \Leftrightarrow y > e_3 + M \Leftrightarrow x > e_3.$$

For the integral we then get

$$\int_z^\infty \frac{d\zeta}{\sqrt{h(\zeta)}} < \int_{x-e_3}^\infty \frac{d\zeta}{\sqrt{\left(\zeta - \frac{e_1'+e_2'}{2}\right)^3}} = \frac{2}{\sqrt{(x-e_3) - \frac{e_1'+e_2'}{2}}}$$

The hypothesis of the lemma

$$x > 2\max\left\{\frac{e_1+e_2}{2}, e_3\right\} + M$$

implies in this case

$$\begin{aligned} x - e_3 &= \frac{x}{2} + \left(\frac{x}{2} - e_3\right) \\ &> \frac{x}{2} + \left(\frac{e_1+e_2}{2} - e_3\right) + \frac{M}{2} \\ &= \frac{x}{2} + \frac{e_1'+e_2'}{2} + \frac{M}{2}. \end{aligned}$$

This yields

$$\frac{2}{\sqrt{(x-e_3) - \frac{e_1'+e_2'}{2}}} < \frac{2\sqrt{2}}{\sqrt{x+M}} \leq \frac{2\sqrt{2}}{\sqrt{x}}$$

where we again referred to $M \geq 0$. The lemma is proved in both subcases. □

$p \neq \infty$. We denote by $\mathbb{Q}_p$ as usual the completion of $\mathbb{Q}$ with respect to p, and its ring of integers by $\mathbb{Z}_p$. As the prime p is fixed, we may assume that

$$\mathbb{Z} \subseteq \mathbb{Z}_p \text{ and } \mathbb{Q} \subseteq \mathbb{Q}_p.$$

Then, we have for our elliptic curve $E/\mathbb{Q}$,

$$E(\mathbb{Z}) \subseteq E(\mathbb{Z}_p) \text{ and } E(\mathbb{Q}) \subseteq E(\mathbb{Q}_p),$$

and we designate, again as usual, by $E_0(\mathbb{Q}_p)$ the subgroup of points P in the group $E(\mathbb{Q}_p)$ whose reduction $\tilde{P} = (P \bmod p) \in \widetilde{E}(\mathbb{F}_p)$ is nonsingular, and by $E_1(\mathbb{Q}_p)$ the subgroup of points $P \in E(\mathbb{Q}_p)$ which reduce to the zero element $\tilde{\mathcal{O}}$ of the nonsingular part of $\tilde{E}(\mathbb{F}_p)$:

$$\tilde{P} = \tilde{\mathcal{O}}$$

(see Chapter 4, Section 4.2).

Clearly,

$$E_1(\mathbb{Q}_p) \subseteq E_0(\mathbb{Q}_p) \subseteq E(\mathbb{Q}_p).$$

It is well-known that the index

$$[E(\mathbb{Q}_p) : E_0(\mathbb{Q}_p)] = c_p$$

is finite for a p-minimal elliptic curve E over $\mathbb{Q}$. In fact, c_p is the *Tamagawa number* of $E/\mathbb{Q}$ at p (see Theorem 4.11).

If we put

$$Z = -\frac{X}{Y}, \quad W = -\frac{1}{Y},$$

the long Weierstraß equation

$$Y^2 + (a_1X + a_3)Y = X^3 + a_2X^2 + a_4X + a_6$$

of the elliptic curve E over $\mathbb{Q}$ is turned into

$$W = Z^3 + (a_1Z + a_2Z^2)W + (a_3 + a_4Z)W^2 + a_6W^3.$$

We write the right hand side for short temporarily as $f(Z, W)$. Then

$$W = f(Z, W).$$

The W in this relation depends on Z:

$$W = W(Z).$$

Indeed W is the unique power series in Z fulfilling that relation. A recursive procedure repeatedly replacing W, W^2 and W^3 on the right by means of the relation $W = f(Z, W)$ finally leads to the power series in Z (see Silverman [204], Chapter IV)

$$\begin{aligned} W = {} & Z^3 + a_1Z^4 + (a_1^2 + a_2)Z^5 + (a_1^3 + 2a_1a_2 + a_3)Z^6 \\ & + (a_1^4 + 3a_1^2a_2 + 3a_1a_3 + a_2^2 + a_4)Z^7 \\ & + (a_1^5 + 4a_1^3a_2 + 6a_1^2a_3 + 3a_1a_2^2 + 3a_1a_4 + 3a_2a_3)Z^8 + \cdots \\ & \in \mathbb{Z}[a_1, a_2, a_3, a_4, a_6][[Z]]. \end{aligned}$$

Of course, in our situation, where the elliptic curve E is defined over $\mathbb{Q}$, we have $a_i \in \mathbb{Z}$ for $i = 1, 2, 3, 4, 6$. However, this is in general no longer true if $\mathbb{Q}$ is replaced by a number field $\mathbb{K}$. We note that if one assigns to the coefficient a_i the weight i:

$$\mathrm{wt}(a_i) = i \quad (i = 1, 2, 3, 4, 6),$$

then the coefficients of Z^n all have weight $n-3$. Since W depends on Z, so do X and Y. In fact, we have

$$\begin{aligned} X(Z) &= \frac{1}{Z^2} - \frac{a_1}{Z} - a_2 - a_3 Z - (a_1 a_3 + a_4) Z^2 \\ &\quad - (a_1^2 a_3 + a_1 a_4 + a_2 a_3) Z^3 - \cdots, \\ Y(Z) &= -\frac{1}{Z^3} + \frac{a_1}{Z^2} + \frac{a_2}{Z} + a_3 + (a_1 a_3 + a_4) Z \\ &\quad + (a_1^2 a_3 + a_1 a_4 + a_2 a_3) Z^2 + \cdots \end{aligned}$$

This each is the Laurent expansion of X, Y in terms of Z so that

$$X, Y \in \mathbb{Z}[a_1, a_2, a_3, a_4, a_6]\{\{Z\}\}.$$

The invariant differential of the curve $E/\mathbb{Q}$ has the expansion

$$\begin{aligned} \omega(Z) &= (1 + a_1 Z + (a_1^2 + a_2) Z^2 + (a_1^3 + 2a_1 a_2 + a_3) Z^3 \\ &\quad + (a_1^4 + 3a_1^2 a_2 + 6a_1 a_3 + a_2^2 + 2a_4) Z^4 \\ &\quad + (a_1^5 + 4a_1^3 a_2 + 12a_1^2 a_3 + 3a_1 a_2^2 + 6a_1 a_4 + 6a_2 a_3) Z^5 + \cdots) dZ. \end{aligned}$$

The above remark about the weights of the coefficients of Z holds correspondingly for $X(Z)$, $Y(Z)$ and $\omega(Z)$.

Now let $\mathcal{E}(p\mathbb{Z}_p)$ denote the formal group associated to the elliptic curve $E/\mathbb{Q}$ (see Silverman [204], Chapter IV). Then there is an isomorphism

$$\begin{aligned} \mathcal{E}(p\mathbb{Z}_p) &\longrightarrow E_1(\mathbb{Q}_p) \\ Z &\longmapsto \left\{ \begin{array}{ll} \mathcal{O} & \text{if } Z = 0 \\ \left(\frac{Z}{W(Z)}, -\frac{1}{W(Z)} \right) & \text{if } Z \neq 0 \end{array} \right\}. \end{aligned}$$

(Clearly, $Z \neq 0$ implies $W(Z) \neq 0$.)

This isomorphism can be extended by a homomorphism in the following way. Let $\hat{G}_a$ be the additive formal group so that $\hat{G}_a(p\mathbb{Z}_p)$ is simply $p\mathbb{Z}_p$ but furnished only with the usual addition on $p\mathbb{Z}_p$. The *p-adic elliptic logarithm* is the function

$$\begin{aligned} \log_{\hat{G}_a} Z &:= \int \omega(Z) \\ &= Z + \frac{a_1}{2} Z^2 + \frac{a_1^2 + a_2}{3} Z^3 + \frac{a_1^3 + 2a_1 a_2 + a_3}{4} Z^4 \\ &\quad + \frac{a_1^4 + 3a_1^2 a_2 + 6a_1 a_3 + a_2^2 + 2a_4}{5} Z^5 \\ &\quad + \frac{a_1^5 + 4a_1^3 a_2 + 12a_1^2 a_3 + 3a_1 a_2^2 + 6a_1 a_4 + 6a_2 a_3}{6} Z^6 + \cdots \\ &=: \sum_{i=1}^{\infty} \frac{d_i}{i} Z^i \in \mathbb{Q}[a_1, a_2, a_3, a_4, a_6][[Z]] \end{aligned}$$

with $d_i \in \mathbb{Z}[a_1, a_2, a_3, a_4, a_6]$, $d_1 = 1$.

For simplicity, we shall write $\log_p$ instead of $\log_{\hat{G}_a}$. The homomorphism mentioned is then the map

$$\begin{aligned} E_1(\mathbb{Q}_p) &\longrightarrow \hat{G}_a(p\mathbb{Z}_p) \\ \overline{P} = (\overline{x}, \overline{y}) &\longmapsto \log_p(\overline{P}) = \textstyle\int \omega(\overline{z}) = \sum_{i=1}^{\infty} \frac{d_i}{i}\overline{z}^i, \end{aligned}$$

where $X = \overline{x}$, $Y = \overline{y}$, $Z = \overline{z}$.

As usual, for points $\overline{P}, \overline{Q} \in E_1(\mathbb{Q}_p)$, the additive relation

$$\log_p(\overline{P} + \overline{Q}) = \log_p(\overline{P}) + \log_p(\overline{Q})$$

holds. Furthermore, we have for the multiplicative p-value

$$|\log_p(\overline{P})|_p = |\overline{z}|_p = \left| -\frac{\overline{x}}{\overline{y}} \right|_p$$

(because $d_1 = 1$).

9.3 S-integral points over $\mathbb{Q}$

Suppose now $\mathbb{K} = \mathbb{Q}$.

Let again E' be defined as the elliptic curve in short Weierstraß form into which E can be transformed:

$$E': \quad Y^2 = X^3 - 27c_4X - 54c_6.$$

The discriminant of $E'/\mathbb{Q}$ is

$$\Delta' = 2^6 \cdot 3^9(c_4^3 - c_6^2) = 2^{12} \cdot 3^{12} \cdot \Delta_0,$$

because

$$2^6 \cdot 3^3 \Delta_0 = c_4^3 - c_6^2.$$

Let t denote the order of the torsion subgroup $E(\mathbb{Q})_{\text{tors}}$ of $E(\mathbb{Q})$:

$$t := \#E(\mathbb{Q})_{\text{tors}}.$$

By a theorem of Mazur (see Theorem 6.8), since the basic number field is $\mathbb{K} = \mathbb{Q}$, we have

$$t \leq 12.$$

We assume now that a basis $P_1, \ldots, P_r$ of (the infinite part of) the Mordell–Weil group $E(\mathbb{Q})$ is known. Let the real number λ this time designate the smallest eigenvalue of the positive definite regulator matrix

$$\mathcal{R}_{E|\mathbb{K}} = (\langle P_i, P_j \rangle)_{1 \leq i, j \leq r},$$

where

$$\langle P_i, P_j \rangle := \hat{h}(P_i + P_j) - \hat{h}(P_i) - \hat{h}(P_j)$$

with the canonical height $\hat{h}$ on the group $E(\mathbb{Q})$ (see Definition 5.21).

We distinguish two cases.

In the *complex case*, where $p = \infty$, we consider the basis points P_i of $E/\mathbb{Q}$, call the elliptic logarithm $u_{i,\infty} = u(P_i)$ and put

$$\overline{u}_{i,\infty} := t \cdot \frac{u_{i,\infty}}{\omega_1},$$

where ω_1 is the real period of $E/\mathbb{Q}$ (see below or Chapter 2, Section 2.3).

In the *p-adic case* (in which, of course, $p \neq \infty$ is an ordinary prime of $\mathbb{Q}$), we proceed in a different manner. Suppose that the equation for $E/\mathbb{Q}$ has integral coefficients and is p-minimal. Let c_p be the Tamagawa number and let $\widetilde{E}/\mathbb{F}_p$ denote the reduction mod p of the curve $E/\mathbb{Q}_p$ with the number of points

$$\mathcal{N}_p := \#\widetilde{E}_{ns}(\mathbb{F}_p),$$

where $\widetilde{E}_{ns}$ is the nonsingular part of the reduced curve $\widetilde{E}/\mathbb{F}_p$. We put

$$m_p := \operatorname{lcm}(t, c_p \mathcal{N}_p).$$

Then, for the primes $p \in M_{\mathbb{Q}} \setminus S$, we define the normalized p-adic logarithms of the basis points P_i of $E/\mathbb{Q}$ by

$$\overline{u}_{i,p} := \log_p(m_p P_i),$$

where $\log_p(m_p P_i)$ is the ordinary p-adic logarithm of the point $m_p P_i$.

We introduce the following constants (with $|\ldots| = |\ldots|_\infty$).

$$\begin{aligned} \tilde{k}_1 &:= \frac{32}{3}\sqrt{|\Delta'|}\Big(8 + \frac{1}{2}\log|\Delta'|\Big)^4, \\ \tilde{k}_2 &:= 20^4 \max\Big\{3^6 c_4^2, 16\sqrt{|\Delta'|}^3\Big\}, \end{aligned}$$

and a fortiori

$$\begin{aligned} k_1 &:= \log\max\Big\{|b_2|, |b_4|^{\frac{1}{2}}, |b_6|^{\frac{1}{3}}, |b_8|^{\frac{1}{4}}\Big\}, \\ k_2 &:= 7 \cdot 10^{38s+49} s^{20s+15} \overline{p}^{24} (\log^* \overline{p})^{4s-2} \cdot \tilde{k}_1 (\log \tilde{k}_1)^2 ((20s-19)\tilde{k}_1 \\ &\quad + \log(e \cdot \tilde{k}_2)) + 2\log(|b_2| + 6) + \log 2, \end{aligned}$$

where e is the basis of the natural logarithm,

$$S := \{p_1, \ldots, p_{s-1}, p_s := \infty\}$$

is a finite set of places of $\mathbb{Q}$ including infinity, and $\overline{p}$ is the maximal prime of S excluding infinity ($\overline{p} = 1$ if $s - 1 = 0$, that is, if $S = \{\infty\}$):

$$\overline{p} := \max\{p_1, \ldots, p_{s-1}, 1\}.$$

As usual, for $z \in \mathbb{R}$, $z > 0$,

$$\log^* z := \max\{1, \log z\}.$$

Finally, we introduce the constant

$$k_3 := \left\{ \begin{array}{cl} \dfrac{2t}{3\omega_1} & \text{for } p = \infty \\ 1 & \text{for } p \neq \infty \end{array} \right\},$$

where ω_1 again is the real period of $E/\mathbb{Q}$. Note that

$$k_1 = -\mu_v \geq 0 \quad \forall\, v \in M_{\mathbb{K}},$$

in particular for $v = \infty$, i.e. for the infinite place $p_s = \infty$ (see Definition 5.9).

We designate by $\mathbb{Z}_S$ the integral domain of all S-integral elements in the field $\mathbb{Q}$.

We remark that, since S is finite, there exists a model for $E/\mathbb{Q}$ which is simultaneously p-minimal for every finite place in S.

Theorem 9.4. *Let $S := \{p_1, \ldots, p_{s-1}, p_s = \infty\}$ be a finite set of places of $\mathbb{Q}$ including infinity. Suppose that the elliptic curve $E/\mathbb{Q}$ is p-minimal for each (finite) prime p in S. Let $\{P_1, \ldots, P_r\}$ be a basis of the group $E(\mathbb{Q})$. Then, for an S-integral point*

$$P = (x, y) \in E(\mathbb{Z}_S)$$

and its representation

$$P = n_1 P_1 + \cdots + n_r P_r + T \quad (n_i \in \mathbb{Z},\ T \in E(\mathbb{Q})_{\text{tors}}),$$

the maximum of the absolute values of the coefficients of the P_i,

$$N := \max\{|n_1|, \ldots, |n_r|\},$$

satisfies the estimate

$$N \leq N_1 := \sqrt{\frac{k_1 + k_2}{\lambda}},$$

where $\lambda > 0$ is the minimal eigenvalue of the positive definite regulator matrix $\mathcal{R}_{E|\mathbb{Q}}$ of $E/\mathbb{Q}$. Moreover if either

$$\max\{|x|_{p_1}, \ldots, |x|_{p_s}\} > |x|_\infty$$

or

$$\max\{|x|_{p_1}, \ldots, |x|_{p_s}\} = |x|_\infty \quad \textit{and} \quad |x|_\infty > \frac{1}{36}(\max\{0, e_0\} + 6|b_2|),$$

there is a place p in S such that

$$|n_1\overline{u}_{1,p} + \cdots + n_r\overline{u}_{r,p} + n_{r+1}|_p \leq k_3 \exp\left(-\frac{\lambda}{2s}N^2 + \frac{k_1 + \log 2}{2s}\right),$$

where

$$n_{r+1} \in \left\{ \begin{array}{ll} \mathbb{Z}, & \textit{if } p = \infty \\ \{0\}, & \textit{if } p \neq \infty \end{array} \right\} \quad \textit{for } p \in S.$$

Before we are going to prove this theorem, we make a few remarks on the constants appearing therein.

If $E/\mathbb{Q}$ has rank $r \leq 2$, Rémond and Urfels [175] derived an explicit lower bound for linear forms in p-adic elliptic logarithms which leads to an explicit upper bound for N. This upper bound can be much better than the N_1 in the theorem (cf. [175]). But it seems to be difficult to generalize this bound for N to elliptic curves $E/\mathbb{Q}$ of arbitrary rank r. To obtain an explicit lower bound for linear forms in elliptic logarithms, if an elliptic curve $E/\mathbb{Q}$ has arbitrary rank r, is already rather tedious in the classical case of complex elliptic logarithms (see David [45]).

If the elliptic curve $E/\mathbb{Q}$ itself is given in short Weierstraß form

$$E: \quad Y^2 = X^3 + AX + B \quad (A, B \in \mathbb{Z})$$

it has discriminant

$$\Delta = -16\Delta_0$$

with

$$\Delta_0 = 4A^3 + 27B^2.$$

Then in the constants $\tilde{k}_1, \tilde{k}_2$, the discriminant Δ' can be replaced by Δ_0. More precisely, instead of $\tilde{k}_1, \tilde{k}_2$ one obtains the smaller constants

$$\begin{aligned} \tilde{k}_1^0 &:= \frac{32}{3}\sqrt{|\Delta_0|}\left(8 + \frac{1}{2}\log|\Delta_0|\right)^4, \\ \tilde{k}_2^0 &:= 10^4 \max\left\{16A^2, 256\sqrt{|\Delta_0|}^3\right\}. \end{aligned}$$

Moreover, if one then wants to determine only the ordinary integral points on $E/\mathbb{Q}$ (case $s = 1$ in the set S), instead of k_1 one can also take the smaller constants

$$\begin{aligned} k_1^0 &:= \log\max\left\{|2A|^{\frac{1}{2}}, |4B|^{\frac{1}{3}}\right\}, \\ k_2^0 &:= 5 \cdot 10^{64}\tilde{k}_1^0 \log\left(\tilde{k}_1^0(\tilde{k}_1^0 + \log\tilde{k}_2^0)\right). \end{aligned}$$

One then obtains the stronger estimate (for $r \geq 1$)

$$N \leq N_0 := \sqrt{\frac{k_1^0 + k_2^0}{\lambda}}.$$

It should be also noted in this connection that the short Weierstraß equation with $A, B \in \mathbb{Z}$ can be chosen p-minimal for every prime $p \neq 2, 3$.

The case of ordinary integral points on $E/\mathbb{Q}$, when $S = \{\infty\}$, was treated independently by Stroeker and Tzanakis [217] and by Gebel, Pethő and the second author [77]. In this case one needs only complex elliptic logarithms. Therefore, the work of David [45] can be applied. David requires that the elliptic curve $E/\mathbb{Q}$ is given in short Weierstraß form. Then one gets the estimate

$$N \leq N_2 := 2^{r+3}\sqrt{k_4 k_5}\log^{\frac{r+3}{2}}(k_5(r+3)^{r+3}),$$

where (see the article of Gebel, Pethő and the second author [81])

$$k_4 := \max\left\{1, \frac{1}{\lambda}\log\left(\frac{2\sqrt{2}\sqrt[3]{4t}}{\omega_1}\right)\right\},$$

$$k_5 := \max\left\{10^9, \frac{C}{\lambda}\right\} \cdot \left(\frac{h}{2}\right)^{r+2} \prod_{i=0}^{r} \log V_i$$

with (see David [45])

$$C := 2.9 \cdot 10^{6(r+2)} 4^{2(r+1)^2} (r+2)^{2r^2+13r+23.3},$$
$$h := \log\max\{4|Aj_2|, 4|Bj_2|, |j_1|\}.$$

In the constant h, which is sort of the height of the elliptic curve $E|\mathbb{Q}$, we wrote the absolute invariant of the elliptic curve E over $\mathbb{Q}$ as a fraction

$$j = \frac{j_1}{j_2} \quad \text{with } j_1, j_2 \in \mathbb{Z}, \ \gcd(j_1, j_2) = 1.$$

The numbers $V_i \in \mathbb{R}$ are chosen in such a way that

$$\log V_i \geq \max\left\{\hat{h}(P_i), h, \frac{3\pi|u_{i,\infty}|^2}{\omega_1^2 \operatorname{Im}(\tau)}\right\}, \qquad i = 1, \ldots, r,$$

($|\ldots| = |\ldots|_\infty$ is again the ordinary absolute value on $\mathbb{C}$ and $u_{i,\infty}$ is the complex elliptic logarithm of the basis point P_i)

$$\log V_0 \geq \max\left\{h, \frac{3\pi}{\operatorname{Im}(\tau)}\right\}.$$

Of course, $\mathrm{Im}(\tau)$ is the imaginary part of the inhomogeneous period

$$\tau := \frac{\omega_1}{\omega_2}$$

with real period ω_1 and complex period ω_2 of $E/\mathbb{Q}$ normalized in such a way that the imaginary part is positive:

$$\mathrm{Im}(\tau) > 0.$$

Next we describe the algorithm to compute S-integral points over $\mathbb{Q}$. After computing the constants in Theorem 9.4 we have to solve for every $p \in S$ the diophantine approximation problems

$$|n_1\overline{u}_{1,p} + \cdots + n_r\overline{u}_{r,p} + n_{r+1}|_p \leq k_3 \exp\left(-\frac{\lambda}{2s}N^2 + \frac{k_1 + \log 2}{2s}\right),$$

where $N = \max\{|n_1|, \ldots, |n_r|\} < N_1$. Here we apply the LLL-reduction following de Weger [232] (see appendix, Section A.7) to reduce the upper bound N_1 for N. This reduction is done for every $p \in S$ and can be repeated several times until the upper bound for N cannot be reduced any further (see the article of Pethő, Zimmer, Gebel, Herrmann [164]).

Algorithm 9.5 (S-integral points).

INPUT: A global minimal model of an elliptic curve $E|\mathbb{Q}$ with integral coefficients and a set $S = \{p_1, \ldots, p_{s-1}, \infty\}$ of places of $\mathbb{Q}$

OUTPUT: All S-integral points of $E(\mathbb{Q})$ provided a basis of $E(\mathbb{Q})$ can be computed.

1. Calculate the coefficients $A, B \in \mathbb{Z}$ of the short Weierstraß normal form for $E|\mathbb{Q}$ and its discriminant $\Delta_0 = 4A^3 + 27B^2$.
2. Compute the constants $\tilde{k}_1$, $\tilde{k}_2$, k_1, $k_2 \in \mathbb{R}$ in Theorem 9.4.
3. Compute the rank, a basis, and the minimal eigenvalue λ of the regulator matrix of $E|\mathbb{Q}$. If a basis of $E(\mathbb{Q})$ can not be computed then return with a failure message.
4. Calculate the constant N_1 of Theorem 9.4.
5. Calculate the real period ω_1 and the number of torsion points of $E|\mathbb{Q}$.
6. Compute the constant $k_3 \in \mathbb{R}$.
7. Apply LLL-reduction:
8.   Perform the reduction for $p_s = \infty$ to obtain a new upper bound $N \leq M_\infty$.
9.   For $i = 1$ to $s - 1$ perform the reduction for p_i to obtain a new upper bound $N \leq M_i$
10. $N_0 \leftarrow N_1$, $N_1 \leftarrow \max\{M_1, \ldots, M_{s-1}, M_\infty\}$
11. while $N_1 < N_0$.

12. Use the sieving procedure described in the thesis of Gebel [79] to find all S-integral points in $E(\mathbb{Q})$ using the upper bound N_1 for the maximal coefficient N.
13. Check all $x \in \mathbb{Z}_S$ such that $\max\{|x|_{p_1}, \ldots, |x|_{p_s}\} \leq \frac{1}{36}(\max\{0, e_0\} + 6|b_2|)$ whether there exist $y \in \mathbb{Z}_S$ such that $(x, y) \in E(\mathbb{Q})$.
14. Return the S-integral points found in Step 12 and Step 13.

9.4 Proof of the theorem

The proof falls into three parts. At first an explicit upper bound for the ordinary height of S-integral points in the Mordell–Weil group $E(\mathbb{Q})$ is obtained. Secondly, a relation between $h(P)$ and N is proved for S-integral points $P \in E(\mathbb{Q})$. Then the inequality

$$N \leq N_1$$

can be established. At third, the last inequality of the theorem is derived on the basis of some fundamental properties of complex ($p = \infty$) and p-adic ($p \neq \infty$) elliptic logarithms.

1. Estimation of the height of an S-integral point. For each point $P = (x, y)$ of the Mordell–Weil group $E(\mathbb{Q})$ of an elliptic curve in short Weierstraß form

$$E: \quad Y^2 = X^3 + AX + B \quad (A, B \in \mathbb{Z}),$$

the coordinates can be uniquely expressed as (see, e.g., Lemma 5.4)

$$x = \frac{\xi}{\zeta^2}, \quad y = \frac{\eta}{\zeta^3} \quad \text{with } \xi, \eta, \zeta \in \mathbb{Z},\ \zeta > 0,\ \gcd(\xi, \zeta) = \gcd(\eta, \zeta) = 1.$$

This holds especially for S-integral points, i.e. points in $E(\mathbb{Z}_S)$.

We have (cf. the definition of k_2)

$$K_0 := 7 \cdot 10^{38s+49} s^{20s+15} \overline{p}^{24} (\log^* \overline{p})^{4s-2} \tilde{k}_1^0 (\log \tilde{k}_1^0)^2 (20s - 19)\tilde{k}_1^0 + \log(e \cdot \tilde{k}_2^0))$$

and define, more generally, the constant

$$K = 7 \cdot 10^{38s+49} s^{20s+15} \overline{p}^{24} (\log^* \overline{p})^{4s-2} \tilde{k}_1 (\log \tilde{k}_1)^2 ((20s - 19)\tilde{k}_1 + \log(e \cdot \tilde{k}_2))$$

so that the conditions $\tilde{k}_1^0 \leq \tilde{k}_1$, $\tilde{k}_2^0 \leq \tilde{k}_2$ imply the inequality

$$K_0 \leq K.$$

Notice that

$$k_2 = K + 2\log(|b_2| + 6) + \log 2.$$

Then Hajdu and Herendi [90] prove:

Proposition 9.6. *The coordinates of an S-integral point* $P = \left(\frac{\xi}{\zeta^2}, \frac{\eta}{\zeta^3}\right)$ *with* $\xi, \eta, \zeta \in \mathbb{Z}$, $\zeta > 0$, $\gcd(\xi, \zeta) = \gcd(\eta, \zeta) = 1$, *of an elliptic curve* $E/\mathbb{Q}$ *in short Weierstraß form satisfy the inequality*

$$\log \max\{|\xi|, |\eta|, |\zeta|^3\} \leq K_0 = K_0(A, B).$$

For the proof of this proposition, we refer to Theorem 2 in the article of Hajdu and Herendi [90].

2. Proof of the inequality $N \leq N_1$. By the sum formula for the absolute values $v \in M_{\mathbb{K}}$, where $\mathbb{K}$, as always, is an algebraic number field, it follows from the (easy) right hand inequality in Theorem 5.35 c) that, with

$$k_1 = \log \max \left\{|b_2|, |b_4|^{\frac{1}{2}}, |b_6|^{\frac{1}{3}}, |b_8|^{\frac{1}{4}}\right\}$$

as before (see Theorem 4 of the article of the second author [256]), the estimate

$$h(P) \geq \hat{h}(P) - \frac{1}{2}(k_1 + \log 2) \quad \text{for } P \in E(\mathbb{K})$$

holds. This is true especially in the case $\mathbb{K} = \mathbb{Q}$ under consideration.

Now let

$$\mathcal{R} = \mathcal{R}_{E/\mathbb{Q}} = (\langle P_i, P_j \rangle)_{1 \leq i, j \leq r}$$

denote the regulator matrix of E over $\mathbb{Q}$ and

$$I = (\delta_{ij})_{1 \leq i, j \leq r}$$

be the $r \times r$ identity matrix. As $\mathcal{R}$ is a real symmetric matrix, it has only real eigenvalues, i.e. solutions of the characteristic polynomial

$$\det(\mathcal{R} - X \cdot I) \in \mathbb{R}[X].$$

Furthermore, as $\mathcal{R}$ is positive definite, the eigenvalues of $\mathcal{R}$ are all positive. Let λ be the smallest of them. Then (see Gantmacher [76], Chapter X §7, Theorem 10)

$$\min_{0 \neq \boldsymbol{X} \in \mathbb{R}^r} \frac{\mathcal{R}(\boldsymbol{X}, \boldsymbol{X})}{I(\boldsymbol{X}, \boldsymbol{X})} = \lambda,$$

where

$$\mathcal{R}(\boldsymbol{X}, \boldsymbol{X}) := \sum_{i,j=1}^{r} \langle P_i, P_j \rangle X_i X_j$$

and

$$I(\boldsymbol{X}, \boldsymbol{X}) := \sum_{i=1}^{r} X_i^2.$$

For a point

$$P = \sum_{i=1}^{r} n_i P_i + T \in E(\mathbb{Q})$$

with the basis points P_i and a torsion point T of $E(\mathbb{Q})$, we have therefore

$$2\hat{h}(P) \geq \lambda N^2.$$

This inequality is valid because

$$\hat{h}(P) = \hat{h}\Big(\sum_{i=1}^{r} n_i P_i + T\Big) = \frac{1}{2} \sum_{i,j=1}^{r} \langle P_i, P_j \rangle n_i n_j = \frac{1}{2} \mathcal{R}(\boldsymbol{n}, \boldsymbol{n})$$

(see Proposition 5.22), with the row vector

$$\boldsymbol{n} = (n_1, \ldots, n_r),$$

and with, as before,

$$N = \max\{|n_i| : i = 1, \ldots, r\}.$$

Altogether, we thus end up with the inequality

$$h(P) \geq \frac{1}{2}(\lambda N^2 - (k_1 + \log 2)).$$

On the other hand, we apply the proposition with $A := -27c_4$ and $B := -54c_6$. We know from the birational transformation over $\mathbb{Q}$

$$\varphi' : E' \longrightarrow E$$

of the short Weierstraß curve

$$E' : \quad Y^2 = X^3 - 27c_4 X - 54c_6$$

to the long Weierstraß curve

$$E : \quad Y^2 + a_1 XY + a_3 Y = X^3 + a_2 X^2 + a_4 X + a_6$$

that, for a point $P = (x, y) = \varphi'(P') \in E(\mathbb{Q})$ with $P' = (x', y') \in E'(\mathbb{Q})$, the first coordinate of P can be written in the form

$$x := \frac{\xi}{\zeta^2} = \frac{\xi' - 3b_2 \zeta'^2}{36 \zeta'^2},$$

since

$$x = \frac{x' - 3b_2}{36}.$$

Here

$$x' = \frac{\xi'}{\zeta'^2} \quad \text{with } \xi', \zeta' \in \mathbb{Z},\ \zeta > 0,\ \gcd(\xi', \zeta') = 1.$$

Thus

$$\begin{aligned}
&\log\max\{|\xi|, |\zeta|^2\} - \log\max\{|\xi'|, |\zeta'|^2\} \\
&\quad = \log\left(\frac{\max\{|\xi|, |\zeta|^2\}}{\max\{|\xi'|, |\zeta'|^2\}}\right) \\
&\quad = \log\left(\frac{\max\{|\xi' - 3b_2\zeta'^2|,\ 36|\zeta'|^2\}}{\max\{|\xi'|, |\zeta'|^2\}}\right) \\
&\quad = \log\left(\frac{\max\left\{\left|\frac{\xi'}{\zeta'^2}\right| + 3|b_2|,\ 36\right\}}{\max\left\{\left|\frac{\xi'}{\zeta'^2}\right|, 1\right\}}\right) \\
&\quad \le \log\max\{1 + 3|b_2|, 36\} \\
&\quad \le 2\log(|b_2| + 6).
\end{aligned}$$

The latter inequality arising from

$$\max\{1 + 3|b_2|, 36\} \le (|b_2| + 6)^2.$$

In the case of the rational number field $\mathbb{K} = \mathbb{Q}$, the ordinary height of $P \in E(\mathbb{Q})$ is simply

$$h(P) = \frac{1}{2}\log\max\{|\xi|, |\zeta|^2\}.$$

This holds by the sum formula for the absolute values of $\mathbb{Q}$ and by the fact that $\gcd(\xi, \zeta) = 1$.

Since Proposition 9.6 yields

$$\log\max\{|\xi'|, |\zeta'|^2\} \le K_0(-27c_4, -54c_6) = K_0,$$

we get from the former inequality

$$h(P) \le \frac{1}{2}K_0 + \log(|b_2| + 6) \le \frac{1}{2}K + \log(|b_2| + 6).$$

But

$$h(P) \ge \frac{1}{2}(\lambda N^2 - (k_1 + \log 2)),$$

so that in sum

$$\lambda N^2 - (k_1 + \log 2) \le K + 2\log(|b_2| + 6).$$

Hence

$$N \leq \sqrt{\frac{k_1 + K + 2\log(|b_2| + 6) + \log 2}{\lambda}}$$

$$\leq \sqrt{\frac{k_1 + k_2}{\lambda}} = N_1.$$

3. Elliptic logarithms and proof of the remaining inequality. Let $P = (x, y) \in E(\mathbb{Z}_S)$ be an S-integral point on the elliptic curve E in general Weierstraß form over the field $\mathbb{Q}$ of rational numbers. Take a $p \in S = \{p_1, \ldots, p_{s-1}, \infty\}$ such that

$$|x|_p = \max\{|x|_{p_1}, \ldots, |x|_{p_{s-1}}, |x|_\infty\}.$$

Then, since by the product formula for the absolute values of $\mathbb{Q}$,

$$|x|_p^{-1} = \prod_{p \neq p' \in M_{\mathbb{Q}}} |x|_{p'}$$

or the sum formula

$$\log |x|_p = \sum_{p \neq p' \in M_{\mathbb{Q}}} (-\log |x|_{p'})$$

(remember that $v_{p'}(x) = -\log |x|_{p'}$ for $p' \in M_{\mathbb{Q}}$), it cannot happen for S-integral points $P = (x, y) \in E(\mathbb{Q})$ that

$$|x|_p < 1 \Leftrightarrow |x|_p^{-1} > 1.$$

This is true because the inequalities

$$|x|_{p'} \leq 1 \quad \text{for } p' \in M_{\mathbb{Q}} \setminus S$$

and the assumption $|x|_p < 1$ (and hence $|x|_{p'} < 1$ for all $p' \in S$) lead to a contradiction to the product formula. Hence we always have

$$|x|_p \geq 1$$

so that always

$$\max\{1, |x|_p\} = |x|_p.$$

Therefore, for an S-integral point $P = (x, y) \in E(\mathbb{Q})$, the multiplicative height is

$$H(P) = \prod_{i=1}^{s} \max\{1, |x|_{p_i}\}$$

$$\leq \prod_{i=1}^{s} |x|_p = |x|_p^s$$

and thus additively

$$h(P) \leq \frac{s}{2} \log |x|_p.$$

We have seen already that

$$h(P) \geq \frac{1}{2}(\lambda N^2 - (k_1 + \log 2)).$$

Together this leads to

$$\log |x|_p^{\frac{1}{2}} \geq \frac{1}{2}\left(\frac{\lambda}{s} N^2 - \frac{k_1 + \log 2}{s}\right)$$

or multiplicatively to

$$\frac{1}{|x|_p^{\frac{1}{2}}} \leq \exp\left(\frac{k_1 + \log 2}{2s}\right) \exp\left(-\frac{\lambda}{2s} N^2\right).$$

With the constants

$$k_6 := \exp\left(\frac{k_1 + \log 2}{2s}\right) \quad \text{and} \quad k_7 := \frac{\lambda}{2s},$$

the latter inequality can be written as

$$\frac{1}{|x|_p^{\frac{1}{2}}} \leq k_6 \exp(-k_7 N^2).$$

Next we transform the upper bound for the inverse $|x|_p^{-1}$ into an upper bound for the p-value of the elliptic logarithm

$$\sum_{i=1}^{r} n_i \overline{u}_{i,p} + n_{r+1}.$$

To this end we must study elliptic logarithms in both cases, the classical case, where $p = \infty$, and in the p-adic case, in which p is a (finite) prime of $\mathbb{Q}$. Of course, if one wants to determine only the ordinary integral points on the curve E over $\mathbb{Q}$, that is, if $S = \{\infty\}$, it suffices to consider classical complex elliptic logarithms.

$p = \infty \in S$. Lemma 9.3 gives an estimate of the complex elliptic logarithm only for $x' \in \mathbb{R}$ with $x' > \max\{0, e_0\}$. We remark first that if $x' < 0$, an extra search for S-integral points is needed. However, since

$$y'^2 = x'^3 - 27c_4 x' - 54c_6 =: f(x')$$

is a square, $x' < 0$ is bounded. In addition, if $0 \leq x' \leq e_0$, the variable x' is also bounded.

To apply the lemma, let $P = (x, y) \in E(\mathbb{Q})$ be an S-integral point with elliptic logarithm $u \in \mathbb{R}$ normalized according to $0 < |u|_\infty < \frac{\omega_1}{2}$.

Suppose further that the maximum $\max\{|x|_{p_1}, \ldots, |x|_{p_{s-1}}, |x|_{p_s}\}$ is attained at the place $p_s = \infty$:

$$|x|_\infty = \max\{|x|_{p_1}, \ldots, |x|_{p_{s-1}}, |x|_{p_s}\}.$$

Assume finally that the first coordinate of the point is sufficiently large, viz. satisfies the inequality

$$|x|_\infty > \frac{1}{36}(\max\{0, e_0\} + 6|b_2|).$$

(Observe that $x > 0 \Leftarrow x = |x|_\infty$.) Then, Lemma 9.3 yields the result that modulo Λ

$$u \equiv \int_{36x+3b_2}^{\infty} \frac{d\xi}{\sqrt{\xi^3 - 27c_4\xi - 54c_6}} < \frac{2\sqrt{2}}{\sqrt{36x + 3b_2}}.$$

Since

$$\frac{2\sqrt{2}}{\sqrt{36x + 3b_2}} = \frac{2}{\sqrt{18x + \frac{3}{2}b_2}} = \frac{2}{3\sqrt{2x + \frac{1}{6}b_2}}$$

with

$$2x + \frac{1}{6}b_2 \geq x \;\Leftrightarrow\; x \geq -\frac{1}{6}b_2 \;\Leftarrow\; x \geq \frac{1}{6}|b_2|_\infty \geq 0$$

(the latter inequality following from the hypothesis $x > 0$) we thus obtain

$$\frac{2\sqrt{2}}{\sqrt{36x + 3b_2}} \leq \frac{2}{3\sqrt{|x|_\infty}}$$

(keeping in mind that $x > 0 \Rightarrow x = |x|_\infty$). Hence, for the absolute value of the elliptic logarithm $u \in \mathbb{R}$ of the S-integral point $P \in E(\mathbb{Q})$ we have then the upper estimate

$$|u|_\infty < \frac{2}{3\sqrt{|x|_\infty}}.$$

Notice that the assumption on x implies the assumption on x':

$$x' = 36x + 3b_2 > \max\{0, e_0\} + 6|b_2| + 3b_2 \geq \max\{0, e_0\}$$

so that the hypothesis of the lemma is indeed satisfied.

But we know already that

$$\frac{1}{\sqrt{|x|_\infty}} \leq k_6 \exp(-k_7 N^2).$$

Therefore

$$|u|_\infty < \frac{2}{3} k_6 \exp(-k_7 N^2).$$

The S-integral point $P \in E(\mathbb{Q})$ is represented in the form

$$P = \sum_{i=1}^{r} n_i P_i + T \quad (n_i \in \mathbb{Z})$$

through the base points P_i of the free part of the group $E(\mathbb{Q})$ and a torsion point $T \in E(\mathbb{Q})_{\mathrm{tors}}$. The number N is always given as the maximum

$$N = \max_{1\le i\le r} \{|n_i|\}.$$

If the order of the torsion group is $t = \#E(\mathbb{Q})_{\mathrm{tors}}$, we have

$$tP = \sum_{i=1}^{r} n_i(tP_i).$$

For the complex elliptic logarithms, this means

$$tu = \sum_{i=1}^{r} n_i(tu_{i,\infty}).$$

Hence,

$$|tu|_\infty \le \frac{2}{3} t k_6 \exp(-k_7 N^2).$$

We take as fundamental parallelogram of $\mathbb{C}/\Lambda$

$$\mathcal{F} := \Big\{ -\frac{\omega_1+\omega_2}{2} + d_1\omega_1 + d_2\omega_2 : 0 \le d_i < 1\Big\}.$$

There exists thus an integer $n_{r+1} \in \mathbb{Z}$ such that

$$\Big|\sum_{i=1}^{r} (n_i t) u_{i,\infty} + n_{r+1}\omega_1\Big|_\infty \le \frac{2}{3} t k_6 \exp(-k_7 N^2).$$

Hence we have proved that

$$\Big|\sum_{i=1}^{r} n_i \overline{u}_{i,\infty} + n_{r+1}\Big|_\infty \le \frac{2}{3}\frac{t}{\omega_1} k_6 \exp(-k_7 N^2)$$

for a suitable integer n_{r+1}. This finishes the proof of the theorem in the case of $p = \infty$ since

$$k_6 = \exp\left(\frac{k_1 + \log 2}{2s}\right), \quad k_7 = \frac{\lambda}{2s},$$

and also, for $p = \infty$,

$$k_3 = \frac{2t}{3\omega_1}, \quad \overline{u}_{i,\infty} = \frac{t}{\omega_1} \cdot u_{i,\infty} \quad (i = 1, \ldots, r),$$

where $u_{i,\infty}$ is the elliptic logarithm of the basis point $P_i \in E(\mathbb{Q})$ as before.

$\boldsymbol{p \neq \infty, p \in S.}$ In this case, $p = p_i$ $(1 \leq i < s)$ is a true prime number and one is thus dealing with p-adic elliptic logarithms which come into play if one wants to determine all S-integral points instead of only ordinary integral points on an elliptic curve E over the rationals $\mathbb{Q}$, i.e. when $S \supsetneq \{\infty\}$.

We recall that

$$m = m_p = \operatorname{lcm}(t, c_p \mathcal{N}_p) \quad \text{with } \mathcal{N}_p = \#\widetilde{E}_{ns}(\mathbb{F}_p),$$

where $\widetilde{E}/\mathbb{F}_p$ is the modulo p reduced curve $E/\mathbb{Q}$. Then, for the basis points $P_1, \ldots, P_r$ of $E(\mathbb{Q})$,

$$mP_i =: \overline{P}_i \in E_1(\mathbb{Q}_p) \quad (i = 1, \ldots, r),$$

since

$$[E(\mathbb{Q}_p) : E_0(\mathbb{Q}_p)] = c_p,$$
$$E_0(\mathbb{Q}_p)/E_1(\mathbb{Q}_p) \cong \widetilde{E}_{ns}(\mathbb{F}_p).$$

and, for the torsion point $T \in E(\mathbb{Q})$,

$$mT = \mathcal{O}.$$

The S-integral point $P \in E(\mathbb{Q})$ had a representation

$$P = \sum_{i=1}^{r} n_i P_i + T \quad (n_i \in \mathbb{Z}),$$

and this inherits a representation of its m-multiple

$$\overline{P} = mP = \sum_{i=1}^{r} \overline{n}_i P_i \quad (\overline{n}_i = mn_i \in \mathbb{Z}).$$

For $\overline{P} = (\overline{x}, \overline{y}) \in E_1(\mathbb{Q}_p)$ (even for $\overline{P} \in E_1(\mathbb{Q})$), we have the Laurent expansion

$$\overline{x} = \frac{\overline{z}}{w(\overline{z})} = \frac{1}{\overline{z}^2} - \frac{a_1}{\overline{z}} - a_2 - a_3\overline{z} - (a_4 + a_1a_3)\overline{z}^2 - \cdots,$$

where

$$\overline{z} = -\frac{\overline{x}}{\overline{y}}, \quad \overline{w} = w(\overline{z}) = -\frac{1}{\overline{y}}.$$

The inequality

$$v_p(\overline{z}) > 0 \iff |\overline{z}|_p < 1$$

for $\overline{P} = (\overline{x}, \overline{y})$ associated with the S-integral point $P = (x, y)$ by the relation $\overline{P} = mP$ shows therefore that, since $\overline{z} = \overline{u}$,

$$|\overline{x}|_p \leq \frac{1}{|\overline{z}|_p^2} = \frac{1}{|\overline{u}|_p^2},$$

where

$$\overline{u} = \sum_{i=1}^{r} \overline{n}_i u_{i,p} = \sum_{i=1}^{r} n_i \overline{u}_{i,p}$$

is the p-adic elliptic logarithm of $\overline{P}$ so that

$$\overline{u} = \overline{u}_p = \log_p(\overline{P}).$$

Now

$$v_p(\overline{x}) \leq v_p(x) \Leftrightarrow |\overline{x}|_p \geq |x|_p$$

(see the article of Pethő, Zimmer, Gebel, Herrmann [164] and the article of Smart [211]) leads to the estimates

$$|\overline{u}|_p \leq \frac{1}{|\overline{x}|_p^{1/2}} \leq \frac{1}{|x|_p^{1/2}}.$$

However, for an S-integral point $P = (x, y) \in E(\mathbb{Q})$, we have seen earlier that

$$\frac{1}{\sqrt{|x|_p}} \leq k_6 \exp(-k_7 N^2).$$

Therefore

$$\left| \sum_{i=1}^{r} n_i \overline{u}_{i,p} + n_{r+1} \right|_p \leq k_6 \exp(-k_7 N^2)$$

with $n_{r+1} = 0$ and again

$$k_6 = \exp\left(\frac{k_1 + \log 2}{2s}\right), \quad k_7 = \frac{\lambda}{2s},$$

($k_3 = 1$ in this case). This completely proves the second inequality of the theorem.

9.5 Example

As an example we consider the elliptic curve

$$E: Y^2 + 67Y = X^3 - 21X^2 - 10X + 30$$

over $\mathbb{Q}$. The following tabular contains all $\{2, 3, 5, 7, \infty\}$-integral points of $E(\mathbb{Q})$. We give the coordinates ξ, η, ζ of the points

$$\left(\frac{\xi}{\zeta^2}, \frac{\eta}{\zeta^3}\right).$$

ξ	η	ζ
−27632	−32330033	3^4
4685	47235	$2 \cdot 7$
−1409	−153811	$3 \cdot 5$
363169	−290490423	$2^2 \cdot 7^2$
7097617	−15551454599	$2^4 \cdot 3^3$
5961	−464584	5^2
505	4677	2^2
1598	−50863	7
−22991	−5085377	$2^2 \cdot 3 \cdot 5$
1889	−63729	2^3
223039829191	56598084474118811	$3^7 \cdot 5 \cdot 7$
85	−1045	3
−6	−18	1
601	−20163	2^3
1436	39259	5
23	−78	1
61	353	1
2773	−268057	$2 \cdot 3^2$
−180	−1455	7
11246	−1780094	$5 \cdot 7$
75	518	1
193	−1817	3
−11	−15	2
8	−49	1
−61	−162	5
204	2726	1
685	−10637	$2 \cdot 3$
72979	18952417	$3 \cdot 5$
912	−15534	7
1549	59019	2
39280513	246100479911	$2^2 \cdot 3^2$
1265233	1423154832	1
296999095	3491224286418	7^4
2725	−273095	$2 \cdot 3^2$
3599	195343	5
−467	−3291	$2 \cdot 7$
209	2832	1
19	−49	1
240	3518	1
97	361	$2^2 \cdot 3$

ξ	η	ζ
6	-57	1
1	0	1
2510	125190	1
289	-2377	2^2
4374	288551	1
3607	209969	3
1345	-242847	2^4
2	-1	1
5	-60	1
42421	8734985	1
18	-34	1
18	-33	1
42421	-8735052	1
5	-7	1
2	-66	1
1345	-31585	2^4
3607	-211778	3
4374	-288618	1
289	-1911	2^2
2510	-125257	1
1	-67	1
6	-10	1
97	-116137	$2^2 \cdot 3$
240	-3585	1
19	-18	1
209	-2899	1
-467	-180557	$2 \cdot 7$
3599	-203718	5
2725	-117649	$2 \cdot 3^2$
296999095	-4418590528885	7^4
1265233	-1423154899	1
39280513	-246103605863	$2^2 \cdot 3^2$
1549	-59555	2
912	-7447	7
72979	-19178542	$3 \cdot 5$
685	-3835	$2 \cdot 3$
204	-2793	1
-61	-8213	5
8	-18	1
-11	-521	2
193	8	3

ξ	η	ζ
75	−585	1
11246	−1092531	$5 \cdot 7$
−180	−21526	7
2773	−122687	$2 \cdot 3^2$
61	−420	1
23	11	1
1436	−47634	5
601	−14141	2^3
−6	−49	1
85	−764	3
223039829191	−86646756593886686	$3^7 \cdot 5 \cdot 7$
1889	29425	2^3
−22991	−9386623	$2^2 \cdot 3 \cdot 5$
1598	27882	7
505	−8965	2^2
5961	−582291	5^2
7097617	10149809543	$2^4 \cdot 3^3$
363169	−213988489	$2^2 \cdot 7^2$
−1409	−72314	$3 \cdot 5$
4685	−231083	$2 \cdot 7$
−27632	−3276514	3^4
−5322567	−25548906669	$2^7 \cdot 7$
115423127	−791223680206	7^4
51244	10189486	$3 \cdot 7$
4426	287189	3
17553	2301039	2^2
454	−4917	5
1553	60752	1
190	−21526	7
57	−229	2^2
40286	−8037266	5
681	−266475	2^4
629	−15543	1
2055825	−2894083783	2^6
984496	−893269369	$3 \cdot 5 \cdot 7$
170703201	2215102092399	$2^6 \cdot 5$
13204	−563275	3^3
1503	−57894	1
1229	−29931	$2 \cdot 7$
4324	−207418	5^2

ξ	η	ζ
-3	-539	2
64	-379	3
-2589	-7908930	7^2
127	-1341	1
20	-10	1
461	-9219	2
-99	-7724	5
703	-15877	3^2
2877865	-4272809493	$2^5 \cdot 7$
1019	-1455	7
88	-754	1
2109	-2173	$2 \cdot 5$
-386	-42497	3^2
9	-24	1
-5	-57	1
401	1617	2^2
46	-265	1
1257	10088	7
6166651	-12910544351	$3 \cdot 5^3$
2553	-107027	2^3
250	1205	3
-166	-4202	5
617	-17301	2^3
28933	2760787	3^3
-249135	-193364977	$2^2 \cdot 7^2$
2234	-89566	7
1649	15465	2^3
-59	-1061	3
241	-4239	5
40	143	1
27	-106	1
3385	115885	3^2
-276	-4950	7
37	-327	2
-136	-1613	5
18244	-1811051	3^3
-19	-69	2
436	-15007	7
55281	10829696	5^2
187	-1691	3

ξ	η	ζ
96	798	1
−809	41311	$3^2 \cdot 5$
156529	60718071	2^4
4635481	−20930839957	$2^2 \cdot 3^3 \cdot 7$
−1727	−8673	2^5
8466	750214	5
42817	−4975201	$2^4 \cdot 3$
281689	146714777	$3 \cdot 7$
13225	1499277	2^2
2879725	−4881240595	$2 \cdot 3^2$
205171	−92711982	7
26876	−4366949	5
505624465	−3170009770777	$2^3 \cdot 3^7$
3517321454	−208601430525642	5^2
5959	−583198	5^2
160504	−46815634	3^4
4421	−205673	$2 \cdot 7$
4480423921	223622288059456	$3^5 \cdot 5 \cdot 7$
7809	−426777	$2^2 \cdot 5$
7942	648801	7
62009	14618771	2^4
466	−42130	3^2
1377	−9327	2^5
4559	307082	1
63604	−205161011	$3 \cdot 7^2$
10359488	−33343178529	1
37473	−7252003	1
827469	−732604547	$2 \cdot 5^2$
7402504	−19317737242	$3^3 \cdot 7$
68385	491665	$2^3 \cdot 7$

9.6 Exercises

1) Consider the example in Section 9.5. Compute the constants of Section 9.3 for this example, especially the constant N_1 of Theorem 9.4. (A basis for this curve is given in Exercise 1, Section 8.5.) Check for some points of the example, that they satisfy Theorem 9.4.

2) Compute the coefficient of Z^4 in the development of the coordinate $X(Z)$, the coefficient of Z^3 in the development of the coordinate $Y(Z)$ and the coefficient of Z^6 in the

development of the invariant differential $\omega(Z)$.

3) According to Proposition 9.6

$$\log\max\{|\xi|, |\eta|, |\zeta|^3\} \le K_0 = K_0(A, B)$$

for an elliptic curve

$$Y^2 = X^3 + AX + B \quad (A, B \in \mathbb{Q})$$

with an S-integral point

$$P = \left(\frac{\xi}{\zeta^2}, \frac{\eta}{\zeta^3}\right) \in E(\mathbb{Q}),$$

where $S = \{2, 3, \infty\}$ and

$$\xi, \eta, \zeta \in \mathbb{Z}, \quad \zeta > 0, \ \gcd(\xi, \zeta) = \gcd(\eta, \zeta) = 1.$$

Compute the bound K_0 explicitly for the Mordell-curve

$$Y^2 = X^3 + k \quad (k \in \mathbb{Q})$$

with $k = 1$.

4) Generalise the theorem of Hajdu and Herendi to quadratic number fields (instead of $\mathbb{Q}$).

5) Develop the details for Step 13 of Algorithm 9.5.

Appendix A
Algorithmic theory of diophantine equations

In this appendix we start with Hilbert's tenth problem which was solved negatively by Matijasevich. Then we turn to positive aspects of the algorithmic theory of diophantine equations. Here a distinguished role is played by Baker's theory on the lower bounds for linear forms in logarithms of algebraic numbers. We give an overview of the most important equations, where Baker's theory can be applied. Finally, we concentrate on numerical methods, which make it possible to solve concrete equations.

A.1 Hilbert's 10th problem

At the II. International Congress of Mathematicians, held in Paris in 1900 David Hilbert posed 23 problems, which he considered most important for the 20th century mathematics. His tenth problem was the following [99]:

Determination of the Solvability of a Diophantine Equation. *Given a diophantine equation with any number of unknown quantities and with numerical rational integral coefficients: To devise a process according to which it can be determined by a finite number of operations whether the equation is solvable in rational integers.*

Notice that Hilbert did not use the word algorithm.Moreover, his question is positive. He said "To derive a process..." and not "Whether there exists a process...", which would be more natural nowadays. The reason is that the exact notion of algorithm was defined 30 years later by Gödel.

After important preliminary works, of Davis, Putnam and Robinson, Matijasevich proved in 1970 [137], that this problem of Hilbert is not solvable. More precisely he proved

Theorem A.1 ([137]). *There exists a polynomial*

$$P(a_1, \dots, a_m, X_1, \dots, X_n) \in \mathbb{Z}[a_1, \dots, a_m, X_1, \dots, X_n]$$

for which the solvability of the diophantine equation

$$P(a_1, \dots, a_m, X_1, \dots, X_n) = 0$$

for any values of the parameters $a_1, \dots, a_m \in \mathbb{Z}$ is algorithmically unsolvable.

Later Matijasevich and Robinson [138] proved that one can take $n \leq 14$.

On the other hand, it is clear that a diophantine equation in one unknown is algorithmically solvable. The status of the question is not yet decided even for $n = 3$. Indeed, we do not know any algorithm, which decides for any given $a, b \in \mathbb{Q}$ whether the set of rational points of the elliptic curve

$$Y^2 = X^3 + aX + b$$

is empty. Writing $X = \frac{x}{z^2}$ and $Y = \frac{y}{z^3}$ with $x, y, z \in \mathbb{Z}$, $z > 0$, and $\gcd(x, z) = \gcd(y, z) = 1$ one obtains the diophantine equation

$$x^3 + axz^4 + bz^6 - y^2 = 0$$

in three unknowns.

For further details about this topics we refer to the book of Matijasevich [139].

A.2 Introduction to Baker's method

Although there exist no general algorithms for the solution of diophantine equations, there exist wide classes of such equations, which can be solved by algorithms. We mentioned already the equations with one unknown. There exist also well known algorithms for linear or quadratic equations. However in the sequel we will concentrate on a very powerful and general method, which is called *Baker's method*. It originates from the seventh Hilbert problem and combines algebraic number theoretical and diophantine approximation tools. As our space is limited we will give here only a brief overview. For details we refer to the books Baker [11], Shorey and Tijdeman [200], Gaál [75] and Smart [213] and to the survey paper Evertse, Győry, Stewart and Tijdeman [61].

Before turning to Baker's method, we have to introduce measures for algebraic numbers. Let α be an algebraic number with defining polynomial $a_d X^d + \cdots + a_0 \in \mathbb{Z}[X]$ and with absolute conjugates $\alpha^{(1)}, \ldots, \alpha^{(d)}$. To measure α, several height concepts are available, like

$$H(\alpha) = \max_{0 \le j \le d} \{|a_j|\},$$

$$|\alpha| = \max_{1 \le j \le d} \{|\alpha^{(j)}|\}$$

and

$$h(\alpha) = \frac{1}{d} \log \left(|a_d| \prod_{j=1}^{d} \max \left\{1, |\alpha^{(j)}|\right\} \right).$$

All of these have the common feature that there exist only finitely many algebraic numbers with bounded degree and height. In the sequel we will use mainly the absolute logarithmic height $h(\alpha)$.

Let $\mathbb{K}$ be a number field of degree $d_{\mathbb{K}}$ over $\mathbb{Q}$ and denote by $\mathbb{Z}_{\mathbb{K}}$ the ring of integers of $\mathbb{K}$. Let S be a finite set of places of $\mathbb{K}$ including all archimedean (or infinite) ones. Denote by $\mathbb{Z}_{\mathbb{K},S}$ the ring of S-integers of $\mathbb{K}$, i.e. the set of those $\alpha \in \mathbb{K}$ with $|\alpha|_{\mathfrak{p}} \leq 1$ for all $\mathfrak{p} \notin S$. If S contains no finite places, then $\mathbb{Z}_{\mathbb{K},S} = \mathbb{Z}_{\mathbb{K}}$. The elements $\alpha \in \mathbb{K}$ with $|\alpha|_{\mathfrak{p}} = 1$ for all $\mathfrak{p} \notin S$ are called S-units. Their set will be denoted by $U_{\mathbb{K},S}$. By the generalization of Dirichlet's unit theorem (see Hasse [92] and Lang[114]), $U_{\mathbb{K},S}$ is a finitely generated group.

Let $F(X_1, \ldots, X_n) \in \mathbb{Z}_{\mathbb{K}}[X_1, \ldots, X_n]$ and $m \in \mathbb{Z}_{\mathbb{K}}$. To solve the diophantine equation

$$F(X_1, \ldots, X_n) = m \tag{A.1}$$

in $(X_1, \ldots, X_n) \in \mathbb{Z}^n_{\mathbb{K},S}$, it is enough (at least in principle) to find a computable constant c, which depends only on the height of m, on the degree of P, on the heights of the coefficients of P and on the maximum of the rational primes lying below the finite places of S, such that the inequality

$$x = \max\{h(x_1), \ldots, h(x_n)\} \leq c$$

holds for any solution $(x_1, \ldots, x_n) \in \mathbb{Z}^n_{\mathbb{K},S}$ of (A.1).

Having such a bound, one can enumerate all elements β of $\mathbb{K}$ with $h(\beta) \leq c$; then testing all n-tuples whether they satisfy (A.1), we are able to find all solutions of it algorithmically. Observe that this procedure can be very lengthy if c is large, which is typical in this theory. We will come back to this problem later.

Now we concentrate on the core of Baker's method; how to find the upper bound c.

The first step is the transformation of (A.1) into finitely many S-unit equations. More precisely, one has to find a finite extension $\mathbb{L}$ of $\mathbb{K}$ and a finite set $\mathcal{A}$ of $\mathbb{L}$ such that for any solution $(x_1, \ldots, x_n) \in \mathbb{Z}^n_{\mathbb{K},S}$ of (A.1), there exist $\alpha_1, \alpha_2 \in \mathcal{A}$ and $E_1, E_2 \in U_{\mathbb{L},\tilde{S}}$ such that

$$\alpha_1 E_1 + \alpha_2 E_2 = 1 \tag{A.2}$$

holds. Here $\tilde{S}$ denotes the set of extensions to $\mathbb{L}$ of the elements of S. Hence $\tilde{S}$ includes all archimedean places of $\mathbb{L}$.[1]

Of course this step depends strongly on the structure of the set of candidates for solutions of (A.1). Sometimes there are several different ways for the transformation of (A.1) to the form (A.2). The different methods lead to upper bounds for x, where the sizes are usually similar. From a computational point of view however the methods can be very different. You find a comparison of four methods for the solution of elliptic equations in Pethő and Zimmer [165].

In other applications – like finding common values of second order linear recursive sequences or perfect powers in such sequences – one can skip this step landing directly at Step 3 below.

---

[1]There are interesting diophantine problems, which can be transformed into unit equations with more then two unknowns. Then there exist upper bounds for the number of solutions.

N.B. The number it is finite, but there is no known general upper bound for the height of the solutions.

In the next steps one establishes a computable upper bound for the solutions of (A.2). We discuss this in detail in Section A.3, but for the presentation of the general strategy we give some explanation.

We mentioned above that $U_{\mathbb{L},\tilde{S}}$ is finitely generated. As $U_{\mathbb{L},\tilde{S}}$ is often larger as necessary, it is better to choose the smallest possible subgroups U_1 and U_2 of $U_{\mathbb{L},\tilde{S}}$ such that $E_i \in U_i$ $i = 1, 2$ hold for all units E_1, E_2, which can occur in (A.2). Of course these groups are finitely generated as well.

Let $\varepsilon_{i1}, \dots, \varepsilon_{ir_i}$, $i = 1, 2$, be a basis of U_i. Then there exist integers $a_{i1}, \dots, a_{ir_i} \in \mathbb{Z}$ such that

$$E_i = \varepsilon_{i1}^{a_{i1}} \dots \varepsilon_{ir_i}^{a_{ir_i}}, \quad i = 1, 2 \tag{A.3}$$

hold. Put $A_i = \max\{|a_{ij}| : 1 \le j \le r_i\}$.

In this way we associate to any solution $(x_1, \dots, x_n) \in \mathbb{Z}^n_{\mathbb{K},S}$ of the polynomial diophantine equation (A.1) a pair of solutions $(a_{11}, \dots, a_{1r_1}) \in \mathbb{Z}^{r_1}$, $(a_{21}, \dots, a_{2r_2}) \in \mathbb{Z}^{r_2}$ of the exponential diophantine equation (A.2). Examining the transformation

$$(x_1, \dots, x_n) \mapsto (a_{11}, \dots, a_{1r_1}; a_{21}, \dots, a_{2r_2})$$

one has to find a function f such that $x \le f(\max\{A_1, A_2\})$. This completes Step 2.

In the third step we prove that the solutions of (A.2) satisfy a very strong approximation property. To be more precise, assume that $A_2 \le A_1$ and that A_1 is large enough. Then there exists a place $\mathfrak{P} \in \tilde{S}$ such that the inequality

$$|\Lambda|_{\mathfrak{P}} = \left| \log \alpha_2 + \sum_{\ell=1}^{r_2} a_{2\ell} \log \varepsilon_{2\ell} + a_{2,0} \log(-1) \right|_{\mathfrak{P}} \le c_1 \exp(-c_2 A_1) \tag{A.4}$$

holds for any solution of (A.2), where c_1, c_2 are constants and the integer $a_{2,0} = 0$ provided $\mathfrak{P}$ is not a complex place.

This inequality has important consequences. The first, and most important, is that one can use an appropriate lower bound for linear forms in logarithms of algebraic numbers. We give a collection of such results in Section A.6.

These have the following form: If $\Lambda \neq 0$, then

$$|\Lambda|_{\mathfrak{P}} \ge \exp(-c_3 g(A_2))$$

with an appropriate function g, which typically is the logarithm function. Comparing this inequality with the upper bound, we obtain that if $\Lambda \neq 0$, then

$$-c_3 g(A_2) \le -c_2 A_1 + \log c_1.$$

If $g(X)$ is increasing and $g(X) = o(X)$, which is always valid in the actual applications, we have

$$c_2 A_1 - \log c_1 \le c_3 g(A_2) \le c_3 g(A_1),$$

which can hold only if $A_2 \le A_1 \le c_4$. As in the above investigation the role of A_1 and A_2 is symmetric. We obtain $\max\{A_1, A_2\} \le c_4$.

Finally using that $x \le f(\max\{A_1, A_2\})$ we have $x \le c = f(c_4)$.

Note that f is also usually the exponential function and c_4 depends exponentially on the heights of $\alpha_i, \varepsilon_{i\ell}, \ell = 1, \dots, r_i, i = 1, 2$, and on the degree of $\mathbb{L}$, thus c depends doubly exponentially on most of the parameters. (You will find some explicit result in Section A.5.) Hence c_4 and c are huge numbers, and a direct search for the solutions, mentioned at the beginning of this section, is hopeless.

On the other hand (A.4) makes it possible to reduce c_4 considerably provided the coefficients of Λ are given. About this numerical reduction process we will report in Section A.8. Sometimes the reduced bound for $\max\{A_1, A_2\}$ is already so small, that one can find the solutions of (A.1) by a direct search. In other cases one has to work out special methods, e.g. sieving procedures, to find the solutions below the reduced bound. For details we refer to Gaál [75] and Smart [213].

A.3 S-unit equations

Let $\mathbb{K}$ be an algebraic number field, $\alpha_1, \alpha_2 \in \mathbb{K}$ and U_1, U_2 be finitely generated (multiplicative) subgroups of $\mathbb{K}$. Our aim in this section is to show that (A.2) has only finitely many computable solutions. As a byproduct we prove inequality (A.4) too.

It was shown by Siegel [201] that unit equations (S contains only the infinite places) have only finitely many solutions. Fifty years later Győry [88] gave an effectively computable upper bound for the size of the solutions.

Before turning to the proof, we need some preparation. Because the (multiplicative) torsion subgroup of $\mathbb{K}$ is finite, if we incorporate the torsion elements of U_i in $\alpha_i, i = 1, 2$, we obtain finitely many equations of the shape (A.2). Hence we may assume without loss of generality that U_1 and U_2 are torsion-free. Put r_i for the rank of $U_i, i = 1, 2$.

Let $M_{\mathbb{K}}$ again denote the set of all inequivalent places of $\mathbb{K}$ and denote by $r + 1$ the number of inequivalent archimedean or infinite places of $\mathbb{K}$. (Notice that r is the rank of the unit groups $U_{\mathbb{K}}$ of $\mathbb{Z}_{\mathbb{K}}$.) Let

$$S_i = \{\mathfrak{p} \in M_{\mathbb{K}} : |\alpha|_{\mathfrak{p}} \ne 1 \text{ for some } \alpha \in U_i\}, \quad i = 1, 2,$$

and put $s_i = |S_i| - 1$.

One can define a lattice in $\mathbb{R}^{s_i}$ associated to the group U_i via the logarithmic map

$$\mathrm{Log}_i : \begin{cases} U_i & \to \mathbb{R}^{s_i} \\ \alpha & \to (\log|\alpha|_{\mathfrak{p}_1}, \dots, \log|\alpha|_{\mathfrak{p}_{s_i}})^T, \end{cases} \qquad \text{where } S_i = \{\mathfrak{p}_1, \dots, \mathfrak{p}_{s_i+1}\}.$$

In the definition of this map, one element of S_i is omitted. This is the one corresponding to an embedding of $\mathbb{K}$ into the field $\mathbb{C}$ of the complex numbers.

The image of Log_i is a lattice in $\mathbb{R}^{s_i}$ of rank r_i. Let again $\varepsilon_{i1}, \dots, \varepsilon_{ir_i}, i = 1, 2$, be a basis of U_i.

From this stage on we are studying U_1 to obtain an upper bound for A_1 provided that $A_2 \leq A_1$. As the role of U_1 and U_2 are symmetrical, this causes no loss of generality. Let $\tilde{R} = (\log|\varepsilon_{1i}|_{\mathfrak{p}_j}) \in \mathbb{R}^{s_1 \times r_1}$ be the matrix whose columns are $\mathrm{Log}_1\varepsilon_{1i}$. This matrix has a $r_1 \times r_1$ non-singular minor, say R, because U_1 is a group of rank r_1. Set $c_1 = \|R^{-1}\|_\infty$,[2] where $\|.\|_\infty$ denotes the row sum norm of a matrix. We choose $c_2 \in \mathbb{R}$ to be a constant such that

$$0 < c_2 < 1/c_1 s_1.$$

Let $\mathfrak{p}_j \in S_1$ be such that $A_1 = |a_{1k}|$ for some $1 \leq k \leq r_1$ and

$$|E_1|_{\mathfrak{p}_j} = \min\{|E_1|_{\mathfrak{p}} : \mathfrak{p} \in S_1\}.$$

We are now in the position to state our first result.

Lemma A.2. *In the above notation, we have*

$$|E_1|_{\mathfrak{p}_j} \leq \exp(-c_2 A_1).$$

Proof. Equation (A.3) and the definition of the Log_1 mapping implies

$$\mathrm{Log}_1(E_1) = \tilde{R}(a_{11}, \ldots, a_{1r_1})^T.$$

Let $V = \{v_1, \ldots, v_{r_1}\}$ denote the choice of rows of $\tilde{R}$ which make up the matrix R. Then we obtain

$$(a_{11}, \ldots, a_{1r_1})^T = R^{-1}(\log|E_1|_{\mathfrak{p}_{v_1}}, \ldots, \log|E_1|_{\mathfrak{p}_{v_{r_1}}})^T,$$

which implies

$$A_1 = \max_{1\leq i\leq r_1}\{|a_{1i}|\} \leq \left\|R^{-1}\right\|_\infty \max_{v\in V}\{|\log|E_1|_{\mathfrak{p}_v}|\} \leq c_1 \max_{v\in S_1}\{|\log|E_1|_{\mathfrak{p}_v}|\}.$$

Let $1 \leq J \leq |S_1|$ be such that

$$\log|E_1|_{\mathfrak{p}_J} = \max\{\log|E_1|_{\mathfrak{p}} : \mathfrak{p} \in S_1\}.$$

As $\prod_{\mathfrak{p}\in S_1} |E_1|_{\mathfrak{p}} = 1$ we have

$$\sum_{\mathfrak{p}\in S_1} \log|E_1|_{\mathfrak{p}} = 0.$$

Hence $\log|E_1|_{\mathfrak{p}_J}$ is non-negative, moreover

$$\log|E_1|_{\mathfrak{p}_J} \geq A_1/c_1.$$

[2] Notice that the enumeration of the constants is independent of their enumeration in the preceding section.

This implies that $\log|E_1|_{\mathfrak{p}_j} \le -\frac{1}{s_1}\log|E_1|_{\mathfrak{p}_J}$, i.e

$$\log|E_1|_{\mathfrak{p}_j} \le -\frac{A_1}{c_1 s_1}.$$

□

Put $c_3 = |\alpha_1|_{\mathfrak{p}_j}$ and $L = \alpha_2 E_2$. Then by (A.2) and by Lemma A.2, we have (with the above constant c_2)

$$|1 - L|_{\mathfrak{p}_j} \le c_3 \exp(-c_2 A_1). \tag{A.5}$$

We distinguish two cases according as $\mathfrak{p}_j$ is a finite or an infinite place.

Case I: $\mathfrak{p}_j$ is an infinite place. Let $\mathbb{K}^{(j)}$ denote the conjugate of $\mathbb{K}$ corresponding to the place $\mathfrak{p}_j$. Consequently, let $\alpha^{(j)}$ denote the image of $\alpha \in \mathbb{K}$ under the isomorphism $\mathbb{K} \to \mathbb{K}^{(j)}$. Then $|\alpha|_{\mathfrak{p}_j} = |\alpha^{(j)}|$, where $|\cdot|$ denotes the (usual) absolute value of a complex number. Using this notation we can rewrite (A.5) as

$$\left|1 - L^{(j)}\right| \le c_3 \exp(-c_2 A_1).$$

If A_1 is large enough, then $|1 - L^{(j)}| < 1/2$ and using well known properties of the complex logarithm function the inequality implies

$$\left|\log L^{(j)}\right| \le 2c_3 \exp(-c_2 A_1).$$

Put $\Lambda^{(j)} = \log L^{(j)}$. Then there exists an integer a_{20} such that

$$\Lambda^{(j)} = \log \alpha_2^{(j)} + \sum_{\ell=1}^{r_2} a_{2\ell} \log \varepsilon_{2\ell}^{(j)} + a_{20} 2\pi i$$

and

$$\left|\Lambda^{(j)}\right| \le 2c_3 \exp(-c_2 A_1). \tag{A.6}$$

As $|a_{2\ell}| \le A_2 \le A_1$, $1 \le \ell \le r_2$, and

$$\begin{aligned}|a_{20} 2\pi i| &\le 2c_3 \exp(-c_2 A_1) + \left|\log \alpha_2^{(j)} + \sum_{\ell=1}^{r_2} a_{2\ell} \log \varepsilon_{2\ell}^{(j)}\right| \\ &\le c_4 A_2 \le c_4 A_1,\end{aligned}$$

we obtain that the linear form on the left hand side of (A.6) is very well approximable.

On the other hand we can apply Theorem A.10 (see Section A.6) of Baker and Wüstholz to $\Lambda^{(j)}$, because it is obviously a non zero linear form in logarithms of algebraic numbers. Hence we obtain

$$\left|\Lambda^{(j)}\right| \ge \exp(-c_5 \log(c_6 A_1)),$$

with suitable constants c_5, c_6.

Comparing the lower and upper bounds for $|\Lambda^{(j)}|$, we get

$$-c_5 \log(c_6 A_1) \le -c_2 A_1 + \log(2c_3),$$

which can hold only if $A_2 \le A_1 \le c_7$. Thus (A.2) can have only finitely many solutions if $A_2 \le A_1$ and such that $\mathfrak{p}_j$ is an infinite place.

Case II: $\mathfrak{p}_j$ is a finite place. In this case Lemma A.2 implies

$$|\alpha_2 E_2 - 1|_{\mathfrak{p}_j} = \left|\alpha_2 \varepsilon_{21}^{a_{21}} \cdot \dots \cdot \varepsilon_{2r_2}^{a_{2r_2}} - 1\right|_{\mathfrak{p}_j} \le c_3 \exp(-c_2 A_1). \tag{A.7}$$

As $\alpha_2 E_2 - 1 = -\alpha_1 E_1 \ne 0$ we may apply Theorem A.11 (see Section A.6) of Yu and obtain

$$|\alpha_2 E_2 - 1|_{\mathfrak{p}_j} \ge \exp(-c_8 \log(k A_2)) \ge \exp(-c_8 \log(n A_1)),$$

where n again denotes the degree of $\mathbb{K}$ over $\mathbb{Q}$ and c_8 is a suitable constant which depends in this case on the rational prime p_j lying under $\mathfrak{p}_j$ too. Comparing again the lower and upper bounds we get $A_2 \le A_1 \le c_9$.

Thus (A.2) can have only finitely many solutions also if $A_2 \le A_1$ and $\mathfrak{p}_j$ is a finite place. Interchanging the roles of A_1 and A_2 we conclude that the S-unit equation (A.2) has only finitely many solutions.

Unfortunately (A.7) is not a linear form, and is not directly applicable for the reduction of the huge upper bound. Moreover one cannot transform (A.7) directly into such a form because some of $\alpha_2, \varepsilon_{2\ell}, 1 \le \ell \le r_2$ may not be $\mathfrak{p}_j$-adic units. However, after changing these elements appropriately we can always get a suitable form.

To do this let p_j be the rational prime lying under $\mathfrak{p}_j$. Denote by f_j the residue degree and by e_j the ramification index of $\mathfrak{p}_j$.

From the inequality

$$|\alpha_1 E_1|_{\mathfrak{p}_j} \le c_3 \exp(-c_2 A_1)$$

we get

$$-\frac{1}{e_j} \operatorname{ord}_{\mathfrak{p}_j}(\alpha_1 E_1) \log(p_j) \le -c_2 A_1 + \log(c_3).$$

Hence

$$\operatorname{ord}_{\mathfrak{p}_j}(\alpha_1 E_1) \ge (c_2 A_1 - \log(c_3)) / \frac{1}{e_j} \log(p_j) = c_9 A_1 - c_{10},$$

and $\operatorname{ord}_{\mathfrak{p}_j}(\alpha_1 E_1) > 0$, whenever $A_1 > c_{10}/c_9$. Thus $\operatorname{ord}_{\mathfrak{p}_j}(\alpha_2 E_2) = 0$.

The next lemma shows that in this situation, one can find other generators of U_2, which are already $\mathfrak{p}_j$-adic units.

Lemma A.3. *If* $\operatorname{ord}_{\mathfrak{p}_j}(\alpha_2 E_2) = 0$*, then there exist elements* $\beta_i \in \mathbb{K}$*,* $i = 0, \dots, t_2$*, such that*

1. $\operatorname{ord}_{\mathfrak{p}_j}(\beta_i) = 0, \ i = 0, \dots, t_2 = r_2 - 1.$

2. *$t_2 = r_2$, if $\mathfrak{p}_j \notin S_2$, otherwise $t_2 = r_2 - 1$.*

3. *There are integers b_{2i}, with $|b_{2i}| \le |a_{2i}| \le A_2$, such that*

$$\alpha_2 E_2 = \beta_0 \prod_{i=1}^{t_2} \beta_i^{b_{2i}}.$$

Proof. If $\mathfrak{p}_j \notin S_2$, we must have $\mathrm{ord}_{\mathfrak{p}_j}(\varepsilon_{2i}) = 0, i = 1, \ldots, t_2$. Hence $\mathrm{ord}_{\mathfrak{p}_j}(\alpha_2) = 0$ holds, too. Thus we can take $\beta_0 = \alpha_2$ and $\beta_i = \varepsilon_{2i}, i = 1, \ldots, r_2$.

Hence, in the sequel we may assume $\mathfrak{p}_j \in S_2$. For $i = 1, \ldots, r_2$ set $n_i = \mathrm{ord}_{\mathfrak{p}_j}(\varepsilon_{2i})$, and set $n_0 = \mathrm{ord}_{\mathfrak{p}_j}(\alpha_2)$. Define $k \in \{1, \ldots, r_2\}$ to be the index which satisfies

$$n_k = \min\{|n_i| : 1 \le i \le r_2, n_i \ne 0\}.$$

Such a k exists because of the definition of S_2. By relabelling the ε_{2i} we may assume that $k = r_2$. The relation $\mathrm{ord}_{\mathfrak{p}_j}(\alpha_2 E_2) = 0$ implies the linear equation

$$n_0 + \sum_{i=1}^{r_2} a_{2i} n_i = 0.$$

We define b_{2i} and q_i by the relation $a_{2i} = n_{r_2} b_{2i} + q_i$ with $0 \le q_i < |n_{r_2}|$. Therefore, $|b_{2i}| \le |a_{2i}| \le A_2$. We further define

$$\beta_i = \varepsilon_{2i}^{n_{r2}} \varepsilon_{2r_2}^{-n_i}, \quad i = 1, \ldots, r_2 - 1.$$

Then $\mathrm{ord}_{\mathfrak{p}_j}(\beta_i) = n_{r_2} n_i - n_i n_{r_2} = 0$.

Furthermore, $\sigma = -\left(n_0 + \sum_{i=1}^{r_2-1} n_i q_i\right)$. Then

$$\begin{aligned}\sigma &= -\left(n_0 + \sum_{i=1}^{r_2-1} n_i (a_{2i} - n_{r_2} b_{2i})\right)\\ &= n_{r_2} a_{2r_2} + n_{r_2} \sum_{i=1}^{r_2-1} b_{2i}\\ &\equiv 0 \pmod{n_{r_2}}.\end{aligned}$$

Putting

$$\beta_0 = \alpha_2 \varepsilon_{2r_2}^{\sigma/n_{r2}} \prod_{i=1}^{r_2-1} \varepsilon_{2i}^{q_i},$$

we see that $\beta_0 \in \mathbb{K}$ and

$$\mathrm{ord}_{\mathfrak{p}_j}(\beta_0) = n_0 + \sigma + \sum_{i=1}^{r_2-1} q_i n_i = 0.$$

Finally, a simple computation shows that the relation in Part 3 of the lemma is true too. □

Combining Lemma A.3 with (A.7) we get

$$\left|\beta_0 \prod_{i=1}^{t_2} \beta_i^{b_{2i}} - 1\right|_{\mathfrak{p}_j} \leq c_3 \exp(-c_2 A_1).$$

On the other hand,

$$\mathrm{ord}_{\mathfrak{p}_j}\Big(\beta_0 \prod_{i=1}^{t_2} \beta_i^{b_{2i}} - 1\Big) = \mathrm{ord}_{\mathfrak{p}_j}(\alpha_1 E_1) \geq c_9 A_1 - c_{10} \geq \frac{1}{p_j - 1}$$

whenever $A_1 \geq \frac{c_{10}+1}{c_9}$. Using well known properties of the p-adic logarithm, this implies

$$\begin{aligned}\mathrm{ord}_{\mathfrak{p}_j}\Big(\beta_0 \prod_{i=1}^{t_2} \beta_i^{b_{2i}} - 1\Big) &= \mathrm{ord}_{\mathfrak{p}_j}\Big(\log_{\mathfrak{p}_j}(\beta_0) + \sum_{i=1}^{t_2} b_{2i} \log_{\mathfrak{p}_j}(\beta_i)\Big)\\ &\geq c_9 A_1 - c_{10}.\end{aligned}$$

This is already an inequality for linear forms with p-adic coefficients, which can be used to reduce the upper bound for A_2. We will discuss the reduction procedure in Section A.8.

A.4 Thue equations

Having described the general strategy of the application of Baker's method, we give here a concrete example. We concentrate on the transformation of the polynomial equation (A.1) into the exponential equation (A.2).

Let $F(X_1, X_2) \in \mathbb{Z}_{\mathbb{K}}[X_1, X_2]$ be homogenous, irreducible, and of degree $k \geq 3$. Let $0 \neq m \in \mathbb{Z}_{\mathbb{K}}$. Then the diophantine equation

$$F(X_1, X_2) = m \tag{A.8}$$

is called a *Thue equation*. For $\mathbb{K} = \mathbb{Q}$ Thue [223] proved that (A.8) has only finitely many solutions. His result was made effective by Baker [7]. The first effective finiteness result in the general case was proved by Baker and Coates [10].

In order to solve (A.8) we observe first, that multiplying it by a suitable element of $\mathbb{K}$ and performing a linear transformation of the variables we can achieve that the leading coefficient with respect to $X_1$of F is one. We will assume this in the sequel.

Let $\alpha = \alpha^{(1)}, \ldots, \alpha^{(k)}$ denote the roots of $F(X_1, 1)$. Put $\mathbb{L} = \mathbb{K}(\alpha)$ and denote by $U_{\mathbb{L}}^*$ the group of non-torsion units of $\mathbb{Z}_{\mathbb{L}}$. Then (A.8) can be rewritten as

$$N_{\mathbb{L}/\mathbb{K}}(X_1 - \alpha X_2) = \prod_{j=1}^{k} \left(X_1 - \alpha^{(j)} X_2\right) = m.$$

There is, by Dirichlet's unit theorem, a finite set $\mathcal{A} \subset \mathbb{Z}_{\mathbb{L}}$ such that, for any $\beta \in \mathbb{Z}_{\mathbb{L}}$ with $N_{\mathbb{L}/\mathbb{K}}(\beta) = m$, there exist a $\mu \in \mathcal{A}$ and an $\varepsilon \in U_{\mathbb{L}}^*$ satisfying $\beta = \mu \cdot \varepsilon$. Thus, if $x_1, x_2 \in \mathbb{Z}_{\mathbb{K}}$ is a solution of (A.8), then

$$\beta^{(j)} := x_1 - \alpha^{(j)} x_2 = \mu^{(j)} \varepsilon^{(j)}, \quad 1 \le j \le k \tag{A.9}$$

hold with suitably chosen elements $\mu \in \mathcal{A}$ and $\varepsilon \in U_{\mathbb{L}}^*$.

As $k \ge 3$, the number of equations in (A.9) is at least three, but we have only two unknowns X_1 and X_2. Hence X_1 and X_2 can be eliminated and one obtains a relation

$$\left(\alpha^{(j)} - \alpha^{(h)}\right)\mu^{(t)}\varepsilon^{(t)} + \left(\alpha^{(t)} - \alpha^{(j)}\right)\mu^{(h)}\varepsilon^{(h)} + \left(\alpha^{(h)} - \alpha^{(t)}\right)\mu^{(j)}\varepsilon^{(j)} = 0$$

for any $1 \le j < h < t \le k$. Dividing here by $(\alpha^{(t)} - \alpha^{(h)})\mu^{(j)}\varepsilon^{(j)} \ne 0$ we get (see (A.2))

$$a_1 E_1 + a_2 E_2 = 1,$$

where

$$a_1 = \frac{\alpha^{(j)} - \alpha^{(h)}}{\alpha^{(t)} - \alpha^{(h)}} \cdot \frac{\mu^{(t)}}{\mu^{(j)}}, \qquad E_1 = \frac{\varepsilon^{(t)}}{\varepsilon^{(j)}}$$

and

$$a_2 = \frac{\alpha^{(t)} - \alpha^{(j)}}{\alpha^{(t)} - \alpha^{(h)}} \cdot \frac{\mu^{(h)}}{\mu^{(j)}}, \qquad E_2 = \frac{\varepsilon^{(h)}}{\varepsilon^{(j)}}.$$

This is already a unit equation, and it is clear that there are only finitely many possibilities for a_1 and a_2.

Let $\varepsilon_1, \ldots, \varepsilon_r$ denote a basis of $U_{\mathbb{L}}^*$. Then there exist integers $b_1, \ldots, b_r$ such that $\varepsilon = \varepsilon_1^{b_1} \cdots \varepsilon_r^{b_r}$.

Inserting this into (A.9) yields

$$\beta^{(j)} = \mu^{(j)} \varepsilon_1^{(j)b_1} \ldots \varepsilon_r^{(j)b_r}, \quad 1 \le j \le k.$$

Put $A = \max\{|b_1|, \ldots, |b_r|\}$. Then

$$\left|\beta^{(j)}\right| \le c_7 \exp(c_8 A), \quad 1 \le j \le k.$$

Choosing $1 \le j < h \le k$ such that $\mathbb{K}^{(j)} = \mathbb{K}^{(h)}$, but $\mathbb{L}^{(j)} \ne \mathbb{L}^{(h)}$, so that $\alpha^{(j)} \ne \alpha^{(h)}$, we obtain

$$x_1^{(j)} = \frac{\alpha^{(h)}\beta^{(j)} - \alpha^{(j)}\beta^{(h)}}{\alpha^{(h)} - \alpha^{(j)}} \quad \text{and} \quad x_2^{(j)} = \frac{\beta^{(j)} - \beta^{(h)}}{\alpha^{(h)} - \alpha^{(j)}}.$$

This, together with the upper bound for $\left|\beta^{(j)}\right|$, implies the inequality

$$\max\{h(x_1),\ h(x_2)\} \le c_8 A + c_9.$$

Therefore the general method for the solution of unit equations from the last section can be applied.

Working out the details and using ideas developed in the last thirty years, Bugeaud and Győry [23] obtained the best known bound for the solutions of (A.8).

Theorem A.4. *Let* $n = [\mathbb{K} : \mathbb{Q}]$ *and* $n_{\mathbb{L}} = [\mathbb{L} : \mathbb{Q}]$. *Let* R *be the regulator and* r *be the unit rank of* $\mathbb{L}$. *Let further* $M \ge \exp(h(m))$, $M^* = \max\{e, |N_{\mathbb{K}/\mathbb{Q}}(m)|\}$ *and* $A = \exp(\max\{1, h(\alpha)\})$. *Then all solutions* $x_1, x_2 \in \mathbb{Z}_{\mathbb{K}}$ *of Equation* (A.8) *satisfy*

$$\max\{h(x_1), h(x_2)\} \le B^{1/n_{\mathbb{L}}} \exp\left(c_7 R(\log^* R)(R + \log AM^*)\right),$$

where $c_7 = 3^{r+26}(r+1)^{7r+19} n^{4r+2} n_{\mathbb{L}}^{2(n_{\mathbb{L}}+r+6)}$.

If $\mathbb{K} = \mathbb{Q}$ then Bugeaud and Győry proved a more explicit bound. Let $H \ge 3$ be an upper bound for the maximum of the absolute values of the coefficients of F. Then we have the following theorem.

Theorem A.5. *If* $\mathbb{K} = \mathbb{Q}$ *then all solutions of Equation* (A.8) *satisfy*

$$\max\{|x_1|, |x_2|\} < \exp\left(3^{3(k+9)} k^{18(k+1)} H^{2k-2} (\log H)^{2k-1} \log^* |m|\right),$$

where $\log^* |m| = \max\{1, \log |m|\}$.

A.5 Small collection of other results

There are many other families of classical equations for which Baker's method can be applied. In this section we present some typical results. For the proofs we refer to the literature.

Let $\mathbb{K}$ be a number field of degree n and with discriminant $d_{\mathbb{K}}$. Let S be a finite set of places on $\mathbb{K}$, including the set S_∞ of infinite places. Denote by t the number of finite places of S. Let p be the largest rational prime lying below the finite places of S, with the convention that $p = 1$ if $S = S_\infty$. Let $\mathbb{Z}_{\mathbb{K},S}$ denote the ring of S-integers in $\mathbb{K}$.

Take $f(X) \in \mathbb{Z}_{\mathbb{K}}[X]$ to be monic of degree k and assume that its height is bounded by $H \ge e^e$. Let Δ denote the discriminant of f/f', where f' is the derivative of f. Finally let $0 \ne a \in \mathbb{Z}_{\mathbb{K}}$ and $A \ge \max\{|N_{\mathbb{K}/\mathbb{Q}}(a)|, e\}$.

Hyperelliptic equations. If $k \ge 3$ then

$$aY^2 = f(X) \quad \text{with } x, y \in \mathbb{Z}_{\mathbb{K},S} \tag{A.10}$$

is called a *hyperelliptic equation*. In the special case of $k = 3$, we obtain an elliptic equation, which was studied in Chapter 9. Baker [8] proved the first effective upper bound for $\max\{|x|, |y|\}$ of a solution $x, y \in \mathbb{Z}_{\mathbb{K},S}$ provided that $\mathbb{K} = \mathbb{Q}$ and $S = S_\infty$. Quantitative improvements and generalizations of Baker's theorem were later obtained by several authors. Presently the best known bound is due to Bugeaud [24]. We state it here in a bit simpler form.

Theorem A.6. *Assume that $f(X)$ has at least three simple roots. Then all solutions $x, y \in \mathbb{Z}_{\mathbb{K},S}$ of* (A.10) *satisfy*

$$\begin{aligned} h(x) \le H^2 \exp\big(&c_1(n,k,t) p^{4k^3n} (\log^* p)^{4k^2nt} \\ &\times |d_{\mathbb{K}}|^{15k^2/2} A^{3k^2} |N_{\mathbb{K}/\mathbb{Q}}(\Delta)|^{12k} \\ &\times (\log |A d_{\mathbb{K}} N_{\mathbb{K}/\mathbb{Q}}(\Delta)|)^{6k^2n} \log\log H\big), \end{aligned}$$

where $c_1(n,k,t)$ is an effectively computable constant.

Superelliptic equations. If $m \ge 3$ and $k \ge 2$, then

$$aY^m = f(X) \quad \text{with solutions in } x, y \in \mathbb{Z}_{\mathbb{K},S} \tag{A.11}$$

is called a *superelliptic equation*. Again Baker [8] was the first who proved an effective upper bound for $\max\{|x|, |y|\}$ provided $\mathbb{K} = \mathbb{Q}$ and $S = S_\infty$. We state here the presently best known bound due to Bugeaud [24].

Theorem A.7. *Suppose $m \ge 3$ and that $f(X)$ has at least two simple roots. If m is not a power of 2, let q be the smallest odd prime dividing it, otherwise put $q = 4$. Then all solutions of* (A.11) *satisfy*

$$\begin{aligned} h(x) \le H^{q+1} \exp\big(&c_3(k,n,m,t) P^{nk^2q^3} (\log^* p)^{tk^2q} \\ &\times |d_{\mathbb{K}}|^{5k^2q/2} |N_{\mathbb{K}/\mathbb{Q}}(\Delta)|^{5kq} A^{k^2q} \\ &\times (\log |A d_{\mathbb{K}} N_{\mathbb{K}/\mathbb{Q}}(\Delta)|)^{2nk^2q}\big), \end{aligned}$$

where $c_3(k,n,m,t)$ is an effectively computable constant.

We remark that one can use Baker's method also if m in (A.11) is regarded as a variable. Tijdeman [224] proved that if $\mathbb{K} = \mathbb{Q}$, $S = S_\infty$ and $f(X)$ has at least two simple rational roots and if $|y| > 1$, $m \ge 2$, then m is bounded by an effectively computable constant. For generalizations and related questions we refer to the book [200].

Discriminant and index form equations. Let $\mathbb{K}$ be a number field of degree $n \ge 2$ and with discriminant $d_{\mathbb{K}}$. Denote $\alpha = \alpha^{(1)}, \dots, \alpha^{(n)}$ the conjugates of $\alpha \in \mathbb{K}$. Let

$\alpha_1, \dots, \alpha_m \in \mathbb{Z}_{\mathbb{K}}$ be $\mathbb{Q}$-linearly independent, and let $L(\boldsymbol{X}) = \alpha_1 X_1 + \dots + \alpha_m X_m$. Assume that $\mathbb{K} = \mathbb{Q}(\alpha_1, \dots, \alpha_m)$ and define

$$L^{(i)}(\boldsymbol{X}) = \alpha_1^{(i)} X_1 + \dots + \alpha_m^{(i)} X_m, \quad 1 \le i \le n.$$

Then

$$d_{\mathbb{K}/\mathbb{Q}}(L(\boldsymbol{X})) = \prod_{1 \le i < j \le n} \left(L^{(i)}(\boldsymbol{X}) - L^{(j)}(\boldsymbol{X})\right)^2$$

is a polynomial with rational integer coefficients of degree $n(n-1)$. It is called a *discriminant form*. If $0 \neq d \in \mathbb{Z}$, then

$$d_{\mathbb{K}/\mathbb{Q}}(x_1\alpha_1 + \dots + x_m\alpha_m) = d \quad \text{for } x_1, \dots, x_m \in \mathbb{Z} \tag{A.12}$$

is a *discriminant form equation*.

If $1, \alpha_1, \dots, \alpha_m$ is a $\mathbb{Z}$-basis of $\mathbb{Z}_{\mathbb{K}}$, then $m = n - 1$ and

$$d_{\mathbb{K}/\mathbb{Q}}(L(\boldsymbol{X})) = (I(\boldsymbol{X}))^2 d_{\mathbb{K}}$$

holds, where $I(\boldsymbol{X}) \in \mathbb{Z}[\boldsymbol{X}]$ is of degree $\frac{n(n-1)}{2}$. If $0 \neq I \in \mathbb{Z}$, then

$$I(x_1, \dots, x_{n-1}) = \pm I \quad \text{for } x_1, \dots, x_{n-1} \in \mathbb{Z} \tag{A.13}$$

is called an *index form equation*.

It was proved by Győry [87] that (A.12) and (A.13) have only finitely many effectively computable solutions. Improving that bounds he proved in [89] the following theorems.

Theorem A.8. *Let $A \ge \max\{|\alpha_i| : 1 \le i \le m\}$ and $n_2 = \frac{n(n-1)}{2}$. Then any solution $(x_1, \dots, x_m) \in \mathbb{Z}^m$ of* (A.12) *satisfies*

$$\max_{1 \le i \le m} \{|x_i|\} < A^{m-1} \exp\left(c|d_{\mathbb{K}}|^{n-1}(\log|d_{\mathbb{K}}|)^{2n_2-1}(|d_{\mathbb{K}}|^{n-1} + \log|d|)\right),$$

where $c = n^5 n_2^{8n_2+25}$.

Theorem A.9. *Any solution $(x_1, \dots, x_{n-1}) \in \mathbb{Z}^{n-1}$ of* (A.13) *satisfies*

$$\max_{1 \le i \le n-1} \{|x_i|\} < A^{n-2} \exp\left(2c|d_{\mathbb{K}}|^{n-1}(\log|d_{\mathbb{K}}|)^{2n_2-1}(|d_{\mathbb{K}}|^{n-1} + \log|I|)\right),$$

where A, n_2 and c are specified in Theorem A.8.

For proofs, applications and further results we refer to the book of Gaál [75].

A.6 Lower bounds for linear forms in logarithms

In the preceding sections we have shown that lower bounds for linear forms in logarithms are very useful to prove algorithmic solvability of wide classes of diophantine equations. The first general results were proved by Baker [5]. Since then several generalizations and improvements appeared in the literature.

Complex logarithms. Let $\alpha_1, \dots, \alpha_k$, $k \geq 2$ be algebraic numbers not equal to 0 or 1. Let $\mathbb{K} = \mathbb{Q}(\alpha_1, \dots, \alpha_k)$ and set $n = [\mathbb{K} : \mathbb{Q}]$. Define the modified height of the algebraic number α by the formula

$$h_m(\alpha) = \max\left\{h(\alpha), \frac{|\log\alpha|}{n}, \frac{1}{n}\right\}.$$

The following theorem is due to Baker and Wüstholz [12].

Theorem A.10. *Let $b_1, \dots, b_k$ be integers such that*

$$\Lambda = b_1 \log\alpha_1 + \dots + b_k \log\alpha_k$$

is non-zero. Then if $B = \max\{|b_1|, \dots, |b_k|, 3\}$ we have the inequality

$$\log|\Lambda| > -c_1 h_m(\alpha_1) \dots h_m(\alpha_k) \log B$$

with

$$c_1 = 18(k+1)!\, k^{k+1} (32n)^{k+2} \log(2kn).$$

There are results for linear forms in two or three logarithms, which have much better dependence on k and n. They are very useful for the solutions of parametric families of diophantine equations. (See e.g. [122, 97].)

p-adic logarithms. Let $v = v_\wp$ be a finite place on $\mathbb{K}$, corresponding to the prime ideal $\wp$ of $\mathbb{Z}_\mathbb{K}$. Let p denote the rational prime lying below $\wp$, and denote by $|\cdot|_v$ the corresponding normalized non-archimedean valuation. Assume that $A_1, \dots, A_k$ are positive real numbers such that

$$\log A_i \geq \max\left\{\log h(\alpha_i), \frac{|\log\alpha_i|}{10n}, \log p\right\}, \quad i = 1, \dots, k.$$

The next Theorem is a simple consequence of the main result of Yu [240].

Theorem A.11. *Let $b_1, \dots, b_k$ be integers such that*

$$\Lambda = \alpha_1^{b_1} \dots \alpha_k^{b_k} - 1$$

is non-zero. Then if $B = \max\{|b_1|, \dots, |b_k|, 3\}$ we have the inequality

$$|\Lambda|_v \geq \exp\left(n(\log p)\Phi \log(nB)\right),$$

where

$$\Phi = c_2(k) \left(\frac{n}{\sqrt{\log p}} \right)^{2(k+1)} p^n \log A_1 \dots \log A_k \log(10kn \log A),$$

where $c_2(k) = 22000(9.5(k+1))^{2(k+1)}$ *and* $A = \max\{A_1, \dots, A_k, e\}$.

A.7 LLL-algorithm

To explain the reduction procedure mentioned in Section A.2 we have to introduce the LLL-algorithm which, among other things, is very useful in solving numerical diophantine approximation problems. It was designed by Lenstra, Lenstra and Lovász [128]. We describe here only the for us most important features of the LLL-algorithm. For more details we refer to [128] and [213].

Let the lattice $\mathcal{L} \subset \mathbb{Z}^r$ be given by the basis $\boldsymbol{b}_1, \dots, \boldsymbol{b}_r \in \mathbb{Z}^r$. Let $d(\mathcal{L})$ denote the volume of the fundamental parallelotop of $\mathcal{L}$, i.e. of $\{\alpha_1 \boldsymbol{b}_1 + \cdots + \alpha_r \boldsymbol{b}_r : 0 \leq \alpha_i \leq 1,\ 1 \leq i \leq r\}$ and let $\lambda(\mathcal{L})$ denote the length of the shortest non-zero element of $\mathcal{L}$. For $\boldsymbol{x} = (x_1, \dots, x_r)^T \in \mathbb{R}^r$, let $|\boldsymbol{x}|$ designate the Euclidean length of $\boldsymbol{x}$, i.e. $|\boldsymbol{x}| = \left(\sum_{i=1}^r x_i^2\right)^{1/2}$.

In the above notation *the* LLL-*algorithm computes from the basis* $\boldsymbol{b}_1, \dots, \boldsymbol{b}_r$ *of* $\mathcal{L}$ *an* LLL-*reduced basis* $\boldsymbol{a}_1, \dots \boldsymbol{a}_r$ *of* $\mathcal{L}$ *for which – among other conditions – the inequalities*

$$|\boldsymbol{a}_1| \leq 2^{\frac{r(r-1)}{4}} d(\mathcal{L})^{1/r} \tag{A.14}$$

and

$$|\boldsymbol{a}_1| \leq 2^{(r-1)/2} \lambda(\mathcal{L}) \tag{A.15}$$

hold. Moreover the algorithm is polynomial in the maximum of $|\boldsymbol{b}_1|, \dots, |\boldsymbol{b}_r|$.

Observe that not only the theoretical complexity of the LLL-algorithm is good, but it also works very well in the practice.

We intend to apply the LLL-algorithm to solve a simultaneous approximation problem and its dual problem, i.e. to find small values of linear forms.

Concerning the first problem we prove

Proposition A.12. *Let* $\vartheta_1, \dots, \vartheta_r \in \mathbb{Q}$ *be non-zero elements and* $Q > 2^{r(r+1)^2/4}$ *be an integer. Then using the* LLL-*algorithm one can compute integers* $1 \leq q \leq Q$, $p_1, \dots, p_r$ *such that*

$$|q\vartheta_i - p_i| \leq 2^{\frac{(r+1)^2}{4}} Q^{-1/r}, \qquad i = 1, \dots, r. \tag{A.16}$$

Proof. Let $\boldsymbol{b}_i = \boldsymbol{e}_i, i = 1, \ldots, r$, where $\boldsymbol{e}_i$ denotes the i-th $(n+1)$-dimensional unit vector. Set further $\boldsymbol{b}_{r+1} = \left(\vartheta_1, \ldots, \vartheta_r, 2^{\frac{(r+1)^2}{4}} Q^{-\frac{r+1}{r}}\right)^T$ and consider the lattice $\mathcal{L} \subseteq \mathbb{R}^{r+1}$ generated by $\boldsymbol{b}_1, \ldots, \boldsymbol{b}_{r+1}$. Then $d(\mathcal{L}) = 2^{\frac{(r+1)}{4}} Q^{-\frac{r+1}{r}}$. Applying the LLL-algorithm to the basis $\boldsymbol{b}_1, \ldots, \boldsymbol{b}_{r+1}$ of $\mathcal{L}$ we obtain an LLL-reduced basis $\boldsymbol{a}_1, \ldots, \boldsymbol{a}_{r+1}$ of $\mathcal{L}$ for which by (A.14)

$$|\boldsymbol{a}_1| < 2^{\frac{(r+1)r}{4}} \left(2^{\frac{(r+1)^2}{4}} Q^{-\frac{r+1}{r}}\right)^{1/(r+1)} = 2^{\frac{(r+1)^2}{4}} Q^{-1/r}.$$

On the other hand $\boldsymbol{a}_1$ is an integral linear combination of $\boldsymbol{b}_1, \ldots, \boldsymbol{b}_{r+1}$. Let $-p_1, \ldots, -p_r$ and q denote the coefficients. Then

$$\boldsymbol{a}_1 = \left(q\vartheta_1 - p_1, \ldots, q\vartheta_r - p_r, q2^{\frac{(r+1)^2}{4}} Q^{-\frac{r+1}{r}}\right)^T.$$

As the absolute values of all the coordinates of $\boldsymbol{a}_1$ are not greater than the Euclidean length of $\boldsymbol{a}_1$, we obtain immediately the inequalities for $|q\vartheta_i - p_i|$. Finally $q2^{\frac{(r+1)^2}{4}} Q^{-\frac{r+1}{r}} \le 2^{\frac{(r+1)^2}{4}} Q^{-1/r}$ implies $q \le Q$. □

As a rational lattice can always be imbedded into an integer lattice, one can apply the LLL-algorithm to $\mathcal{L}$.

Notice that by Proposition A.12 the LLL-algorithm solves Dirichlet's local simultaneous approximation problem, up to a constant factor.

Now we turn to the dual problem.

Proposition A.13. *Let $\vartheta_1, \ldots, \vartheta_r \in \mathbb{Q}$ be non-zero elements and $Q > 2^{r(r+1)/4}$ be an integer. Using the* LLL*-algorithm, one can compute integers $p_1, \ldots, p_r, q$ such that not all of $p_1, \ldots, p_r$ are* 0 *and*

$$|p_i| \le Q, \quad 1 \le i \le r, \tag{A.17}$$

$$\left|\sum_{i=1}^{r} p_i\vartheta_i + q\right| \le 2^{r^2(r+1)/4} Q^{-n}. \tag{A.18}$$

Proof. Put $C = 2^{-r(r+1)^2/4} Q^{r+1}$, $\boldsymbol{b}_i = \boldsymbol{e}_i + (0, \ldots, 0, C\vartheta_i)^T$, $i = 1, \ldots, r$, and $\boldsymbol{b}_{r+1} = C\boldsymbol{e}_{r+1}$. Then applying the LLL-algorithm to the basis $\boldsymbol{b}_1, \ldots, \boldsymbol{b}_{r+1}$ we can prove the assertion of Proposition A.13. □

The LLL-algorithm can be applied to find small values of rational linear forms with respect not only to the archimedean but also to the non-archimedean valuations.

Proposition A.14. *Let $\vartheta_1, \ldots, \vartheta_r \in \mathbb{Q}$ be non-zero elements. Let p be a prime and $u \in \mathbb{Z}$ be such that $p^u > 2^{(r+1)^2/4}\left(\sum_{i=1}^r |\vartheta_i| + 1\right)^{(r+1)/r}$. Then, using the* LLL*-algorithm, one can compute integers $p_1, \ldots, p_r, q$ such that not all of $p_1, \ldots, p_r$ are* *zero,*

$$|p_i| \le 2^{r(r+1)/4} p^{u/(r+1)}, \quad i = 1, \dots, r$$

and either $q = 0$ *or*

$$\Big|\sum_{i=1}^{r} p_i \vartheta_i\Big|_p \le p^{-u}.$$

Proof. In this case the LLL-algorithm should be applied to the collection of linearly independent vectors $\boldsymbol{b}_i = \boldsymbol{e}_i + (0, \dots, 0, \vartheta_i)^T$, $i = 1, \dots, r$, $\boldsymbol{b}_{r+1} = p^u \boldsymbol{e}_{r+1}$. Then the first member of the LLL-reduced basis, $\boldsymbol{a}_1$, has the form

$$\boldsymbol{a}_1 = \Big(p_1, \dots, p_r, \sum_{i=1}^{r} p_i \vartheta_i + q p^u\Big)^T.$$

As the lattice determinant is $d(\mathcal{L}) = p^u$, one obtains immediately the upper bound for p_i, $i = 1, \dots, r$. To prove the second inequality, one has to notice that if $q \neq 0$, the last coordinate of $\boldsymbol{a}_1$ must be zero. □

A.8 Reduction of the large bound

We showed in Section A.2 how to prove an effective upper bound for the solutions of unit equations. As a byproduct we obtained the result that if A_2 is large enough, the inequalities

$$|\Lambda| \le c_3 \exp(-c_2 A_1) \quad \text{and} \quad A_1 \le c_6$$

are simultaneously true. Analogous inequalities were proved in Chapter 9 for linear forms in complex and p-adic elliptic logarithms. A common setting of these problems is the following:

The complex case. *Let* $0 \neq \vartheta_1, \dots, \vartheta_r \in \mathbb{C}$, $\vartheta_{r+1} \in \mathbb{C}$, $c_1, c_2, B_0 \in \mathbb{R}$. *Find all* $b_1, \dots, b_r, b_{r+1} \in \mathbb{Z}$ *such that simultaneously*

$$B = \max_{1 \le j \le r+1} \{|b_j|\} \le B_0 \tag{A.19}$$

and

$$\Big|\sum_{j=1}^{r} b_j \vartheta_j + \vartheta_{r+1} + b_{r+1}\Big| < c_2 \exp(-c_1 B). \tag{A.20}$$

The p-adic case. *Let* $0 \neq \vartheta_1, \dots, \vartheta_r \in \mathbb{Q}_p$, $\vartheta_{r+1} \in \mathbb{Q}_p$, $c_1, c_2, B_0 \in \mathbb{R}$. *Find all* $b_1, \dots, b_r, b_{r+1} \in \mathbb{Z}$ *such that simultaneously*

$$B = \max_{1 \le j \le r+1} \{|b_j|\} \le B_0 \tag{A.21}$$

and

$$\left|\sum_{j=1}^{r} b_j\vartheta_j + \vartheta_{r+1} + b_{r+1}\right|_p < c_2\exp(-c_1 B). \tag{A.22}$$

Let $\varepsilon > 0$. By Khinchin's theorem the inequality[3]

$$\left|\sum_{j=1}^{r} b_j\vartheta_j + \vartheta_{r+1} + b_{r+1}\right| < B^{-r-1-\varepsilon},$$

where $B = \max_{1\le j\le r+1}|b_j|$, has, for almost all $(\vartheta_1,\dots,\vartheta_{r+1}) \in \mathbb{R}^{r+1}$, only finitely many solutions in $(b_1,\dots,b_{r+1}) \in \mathbb{Z}^{r+1}$. Thus we may expect that (A.19)–(A.20) and (A.21)–(A.22) have only few solutions. The first method to reduce B_0 by using numerical diophantine approximation techniques was used by Baker and Davenport [9]. Later there appeared several variants and improvements, cf. Ellison [59], Pethő and Schulenberg [161] and de Weger [232].

Let us first consider the inequalities (A.19) and (A.20) restricting ourself to the case $(\vartheta_1,\dots,\vartheta_{r+1}) \in \mathbb{R}^{r+1}$. A detailed study of the general problem can be found in the book of Smart [213].

Case I: $r = 1$, $\vartheta_1 \in \mathbb{R}$, $\vartheta_2 = 0$. Compute the convergents $\frac{p_0}{q_0}, \frac{p_1}{q_1}, \dots, \frac{p_n}{q_n}$ in the continued fraction expansion of ϑ_1; up to one we have $q_n > B_0$. Then $|b_1|, |b_2| \le B_0 < q_n$ and we have

$$|p_n - q_n\vartheta_1| < |b_1\vartheta_1 + b_2| < c_2\exp(-c_1 B)$$

by the extremality property of the convergents and by (A.20). This implies

$$B < \frac{1}{c_1}\log\left(c_2|p_n - q_n\vartheta_1|^{-1}\right).$$

We can expect that if B_0 is large, so that $|p_n - q_n\vartheta_1| \approx 1/q_n \approx 1/B_0$. Hence the new bound for B will be about $\frac{1}{c_1}\log(c_2 B_0)$.

Case II: $r \ge 1$, $\vartheta_{r+1} \ne 0$ ([9], [59], [161]). In the next lemma $\|x\|$ denotes the distance of $x \in \mathbb{R}$ to the nearest integer.

Lemma A.15. *Let $C > B_0^r$. Assume that there exist $D \in \mathbb{R}$, $q, p_1, \dots, p_r \in \mathbb{Z}$ such that*

$$1 \le q \le DC, \tag{A.23}$$

$$|q\vartheta_j - p_j| < \frac{1}{DC^{1/r}}, \quad j = 1,\dots,r \tag{A.24}$$

$$\|q\vartheta_{r+1}\| \ge \frac{2r}{D}. \tag{A.25}$$

[3] See W. Schmidt, Diophantine Approximation, Lecture Notes in Math. 785, Springer-Verlag, Heidelberg, New York 1980.

Then we have

$$B \leq \frac{1}{c_1} \log \frac{D^2 C c_2}{r}$$

for all solutions $(b_1, \ldots, b_{r+1})^T \in \mathbb{Z}^{r+1}$ *of* (A.19) *and* (A.20).

Proof. Multiplying (A.20) by q and using (A.23), we obtain

$$\Big| \sum_{j=1}^{r} q b_j \vartheta_j + q b_{r+1} + q \vartheta_{r+1} \Big| < q c_2 \exp(-c_1 B) \leq C D c_2 \exp(-c_1 B).$$

On the other hand (A.24), together with the assumption $C > B_0^r$, implies

$$\Big| \sum_{j=1}^{r} b_j (q \vartheta_j - p_j) \Big| \leq \sum_{j=1}^{r} |b_j|\, |q \vartheta_j - p_j| < \frac{r}{D}.$$

Hence

$$\begin{aligned} &\Big| \sum_{j=1}^{r} b_j q \vartheta_j + q b_{r+1} + q \vartheta_{r+1} \Big| \\ &\quad = \Big| \sum_{j=1}^{r} b_j (q \vartheta_j - p_j) + \sum_{j=1}^{r} b_j p_j + q b_{r+1} + q \vartheta_{r+1} \Big| \\ &\quad \geq \Big| \sum_{j=1}^{r} b_j p_j + q b_{r+1} + q \vartheta_{r+1} \Big| - \Big| \sum_{j=1}^{r} b_j (q \vartheta_j - p_j) \Big| \\ &\quad \geq \| q \vartheta_{r+1} \| - \frac{r}{D} \\ &\quad \geq \frac{r}{D}. \end{aligned}$$

Combining this inequality with the first one of this proof we obtain immediately the asserted upper bound for B. □

To find integers $q, p_1, \ldots, p_r$ satisfying (A.23) and (A.24) one can use, for $r = 1$, the continued fraction expansion of ϑ_1. For $r > 1$, the LLL-algorithm solves the problem by Proposition A.12 of Section A.7. Having q one can easily check (A.25). If it holds, then one obtains a new bound for B which has the expected size constant times $\log B_0$, provided B_0 was large enough. Otherwise one starts the process again, enlarging D.

One can apply the reduction based on the lemma both in the complex (more precisely the real) and in the p-adic cases. To do this one has to compute rational approximations of $\vartheta_1, \ldots, \vartheta_{r+1}$ with a suitable precision. Nevertheless, one has to be careful to control the necessary precision.

Case III: $r \geq 1$, $\vartheta_{r+1} = 0$. In this case the lemma originally due to Baker and Davenport [9] cannot be applied. On the other hand, it is very important to have an analogous procedure, for example for computing S-integral points on elliptic curves using elliptic logarithm in which case one has always homogenous linear forms. Indeed, in this case de Weger [232] worked out a reduction process essentially based on the Propositions A.13 and A.14 of Section A.7.

Consider for example the system (A.19) and (A.20), where ϑ_j, $j = 1, \dots, r$, already denote a suitable rational approximation of the occurring real numbers. Consider further the lattice $\mathcal{L}$ defined in Proposition A.13 with $Q = DB_0$, where $D \geq 1$ has to be chosen properly. It is clear that if $0 \neq (b_1, \dots, b_{r+1})^T \in \mathbb{Z}^{r+1}$ is a solution of (A.19) and (A.20), then

$$\boldsymbol{b} = \Bigl(b_1, \dots, b_r, \sum_{j=1}^{r} b_j C\vartheta_j + Cb_{r+1}\Bigr)^T$$

is a non-zero element of $\mathcal{L}$.

Computing the LLL-basis of $\mathcal{L}$ we obtain a lower bound for the shortest non-zero element of $\mathcal{L}$ by (A.15). This is actually

$$\lambda(\mathcal{L}) \geq 2^{-r/2}|\boldsymbol{a}_1|.$$

On the other hand

$$\begin{aligned}\lambda(\mathcal{L})^2 \leq |\boldsymbol{b}|^2 &\leq rB_0^2 + C^2\Bigl(\sum_{j=1}^{r} b_j\vartheta_j + b_{r+1}\Bigr)^2 \\ &\leq rB_0^2 + C^2c_2^2\exp(-2c_1B).\end{aligned}$$

If $2^{-r}|\boldsymbol{a}_1|^2 - rB_0^2 > 0$, which can be achieved by choosing D large enough, then

$$B \leq \frac{1}{c_1}\log(Cc_2(2^{-r}|\boldsymbol{a}_1|^2 - rB_0^2)^{-1/2}).$$

Observe that if D is too large, then the right hand side can be greater then B_0.

So far we considered the reduction process only for linear forms with complex coefficients. Finally we will study the homogenous case for linear forms with p-adic coefficients. Therefore consider the system (A.21) and (A.22) and assume for simplicity that $\vartheta_{r+1} = b_{r+1} = 0$. Multiplying $\vartheta_1, \dots, \vartheta_r$ by a suitable power of p we can achieve that the new numbers already are p-adic integers. Thus we may assume that $\vartheta_1, \dots, \vartheta_r$ are p-adic integers.

Choose an integer $D \geq 1$ and the integer u such that $p^u > DB_0^{r+1}$. Compute $\tilde{\vartheta}_i \in \mathbb{Z}, i = 1, \dots, r$ such that

$$|\vartheta_i - \tilde{\vartheta}_i|_p \leq p^{-u-1}, \quad i = 1, \dots, r.$$

Consider the lattice $\mathcal{L}$ generated by the vectors $\boldsymbol{b}_i = \boldsymbol{e}_i + (0, \ldots, 0, \vartheta_i)^T$, $i = 1, \ldots, r$ and $\boldsymbol{b}_{r+1} = p^u \boldsymbol{e}_{r+1}$, i.e. the lattice defined in the proof of Proposition A.14. Let $A \in \mathbb{Z}^{(r+1)\times(r+1)}$ be the matrix whose columns are the $\boldsymbol{b}_i$, $i = 1, \ldots, r+1$. Then $\mathcal{L} = A\mathbb{Z}^{r+1}$.

We apply the LLL-algorithm to $\mathcal{L}$ and assume that $\boldsymbol{a}_1 = (p_1, \ldots, p_r, 0)^T$. If this does not hold then enlarge u and start the process again. In the lucky case we obtain the (numerical) lower bound

$$\lambda(\mathcal{L}) \geq 2^{-r/2}|\boldsymbol{a}_1|.$$

On the other hand, let $b_1, \ldots, b_r \in \mathbb{Z}$ be a non-trivial solution of (A.21) and (A.22). Then

$$\Big|\sum_{j=1}^{r} b_j \tilde{\vartheta}_j\Big|_p \leq \max\left\{\Big|\sum_{j=1}^{r} b_j \vartheta_j\Big|_p, p^{-u-1}\right\}$$

and, if $c_2 \exp(-c_1 B) \leq p^{-u-1}$, then

$$\Big|\sum_{j=1}^{r} b_j \tilde{\vartheta}_j\Big|_p = p^v \leq p^{-u-1}. \tag{A.26}$$

As $\sum_{j=1}^{r} b_j \tilde{\vartheta}_j$ is an integer, there exists an $f \in \mathbb{Z}$ such that

$$\sum_{j=1}^{r} b_j \tilde{\vartheta}_j = f p^v.$$

Therefore,

$$A(b_1, \ldots, b_r, -f p^{v-u})^T = (b_1, \ldots, b_r, 0),$$

hence $\mathbf{0} \neq \boldsymbol{b} = (b_1, \ldots, b_r, 0) \in \mathcal{L}$ and we get

$$\lambda(\mathcal{L}) \leq |\boldsymbol{b}| \leq \sqrt{r} B_0.$$

Now we compare the lower and upper bound for $\lambda(\mathcal{L})$. If we find that the lower bound is larger then the upper bound, we get a contradiction, i.e. inequality (A.26) is false. This implies

$$B < \frac{(r+1)\log B_0}{c_1} + \frac{\log(pDc_2)}{c_1}.$$

De Weger's reduction can be applied for inhomogeneous complex and p-adic linear forms, too. The key idea is that we know $|\boldsymbol{a}_1|$ and have not only a lower bound for $\lambda(\mathcal{L})$ but also for the distance of any given vector from the lattice. For details we refer to [213, 232].

Appendix B
Multiquadratic number fields

In the book we presented the basic facts about elliptic curves over number fields. The interplay between the theory of elliptic curves and the arithmetic of algebraic number fields is a decisive feature of arithmetic algebraic geometry. If the number fields $\mathbb{K}$ are restricted to have degree at most four,

$$[\mathbb{K} : \mathbb{Q}] \leq 4,$$

one gets e.g. some special results about torsion groups of elliptic curves over such fields $\mathbb{K}$ (see Sections 6.1 and 6.2, Abel-Hollinger and Zimmer [1], Fung et al. [74], Hollinger [100], Müller et al. [149], Stein [215], Vessis [227], and Weis [236]).

The composite $\mathbb{K}$ of quadratic fields plays a crucial role already in the papers [1], [100], [215]. More precisely, the composite of *two* quadratic fields, that is biquadratic fields, were taken herein as basic fields of elliptic curves.

However, the arithmetic of multiquadratic fields is similar to that of biquadratic fields. This similarity was already used by Hollinger (see [1], [100] and Laska–Lorenz [121]).

We give therefore here a brief introduction into the arithmetic of multiquadratic fields.

B.1 Multiquadratic fields and Galois groups

Let $a_1, \ldots, a_r \in \mathbb{Z}$ such that for all $i = 2, \ldots, r$

$$\sqrt{a_i} \notin \mathbb{Q}\left(\sqrt{a_1}, \ldots, \sqrt{a_{i-1}}\right).$$

Then $\mathbb{K} = \mathbb{Q}\left(\sqrt{a_1}, \ldots, \sqrt{a_r}\right)$ is called a *multiquadratic number field.* The degree of a multiquadratic field is

$$[\mathbb{K} : \mathbb{Q}] = 2^r.$$

Clearly, $\mathbb{K}_i := \mathbb{Q}\left(\sqrt{a_1}, \ldots, \sqrt{a_i}\right)$ is the composite of

$$\mathbb{K}_{i-1} := \mathbb{Q}\left(\sqrt{a_1}, \ldots, \sqrt{a_{i-1}}\right) \quad \text{and} \quad \mathbb{k}_i := \mathbb{Q}\left(\sqrt{a_i}\right), \quad i = 2, \ldots, r.$$

Since all the quadratic subfields $\mathbb{k}_i = \mathbb{Q}\left(\sqrt{a_i}\right)$ are Galois over $\mathbb{Q}$ with Galois group $G_i = \langle \sigma_i \rangle$, where

$$\sigma_i : \sqrt{a_i} \mapsto -\sqrt{a_i},$$
$$\sigma_i|_{\mathbb{Q}} = \mathrm{id}_{\mathbb{Q}},$$

and since

$$\mathbb{K}_{i-1} \cap \mathbb{k}_i = \mathbb{Q} \quad \text{for } i = 2, \ldots, r$$

$\mathbb{K}|\mathbb{Q}$ is also Galois with Galois group isomorphic to the direct product of the groups G_i:

$$G \cong G_1 \times \ldots \times G_r.$$

We note that, for $i = 2, \ldots, r$, the fields $\mathbb{K}_{i-1}$ and $\mathbb{k}_i$ are linearly disjoint over $\mathbb{Q}$.

Moreover, each intermediate field subextension $\mathbb{L}|\mathbb{k}$,

$$\mathbb{Q} \leq \mathbb{k} < \mathbb{L} \leq \mathbb{K},$$

is also Galois.

We extend the automorphisms σ_i to

$$\sigma_i : \sqrt{a_i} \mapsto -\sqrt{a_i}, \quad \sigma_i|_{\hat{\mathbb{K}}_i} = \mathrm{id}_{\hat{\mathbb{K}}_i},$$

(the notation differs from that used in the biquadratic case, where $r = 2$) with the fields ($a_0 := 0$)

$$\begin{aligned} \hat{\mathbb{K}}_i &:= \mathbb{Q}\big(\sqrt{a_1}, \ldots, \sqrt{a_{i-1}}, \sqrt{a_{i+1}}, \ldots, \sqrt{a_r}\big) \\ &= \mathbb{Q}\big(\sqrt{a_1}, \ldots, \widehat{\sqrt{a_i}}, \ldots, \sqrt{a_r}\big) \quad (i = 1, \ldots, r). \end{aligned}$$

Then

$$G = \langle \sigma_1, \ldots, \sigma_i, \ldots \sigma_r \rangle.$$

The subfields $\hat{\mathbb{K}}_i$ of $\mathbb{K}$ have Galois groups

$$\hat{G}_i = \langle \sigma_1, \ldots, \widehat{\sigma_i}, \ldots, \sigma_r \rangle \cong G/G_i \quad (i = 1, \ldots, r).$$

The corresponding Hasse diagrams with (antiton) Galois groups are

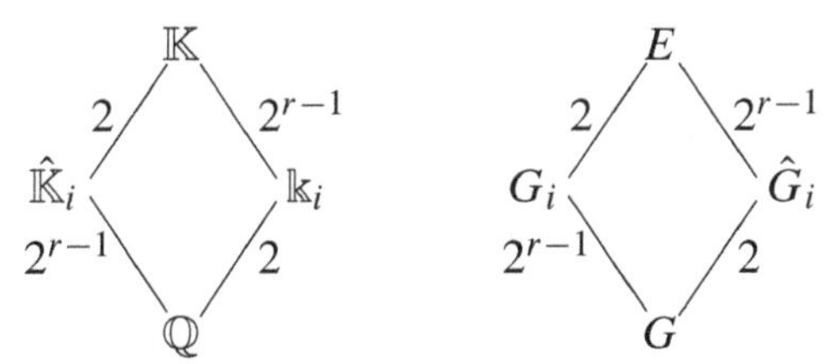

$(1 \leq i \leq r)$

B.2 Discriminants

The numbers $a_i \in \mathbb{Z}$ can be normalized in such a way that

$$a_i = \pm \prod_{\mu=1}^{m_i} p_{i\mu} \quad \text{with primes } p_{i\mu} \in \mathbb{P},\ p_{i\mu} \neq p_{i\nu} \text{ for } \mu \neq \nu\ (i = 1, \ldots, r) \tag{B.1}$$

and, for any prechosen prime $p \in \mathbb{P}$,

$$p \nmid a_i \quad \text{for } i = 2, \ldots, r. \tag{B.2}$$

To see that the latter assertion is true, we argue as follows (see Schmal [187]). If $p \nmid a_i$ for all $i = 1, \ldots, r$ nothing is to be proved.

Otherwise, we may arrange the numbering of the a_i in such a way that $p \mid a_1$. Suppose then that additionally $p \mid a_i$ for some $i \geq 2$. In this case we have $p \nmid \frac{1}{p^2} a_1 a_i =: \widetilde{a_i} \in \mathbb{Z}$ and replace $\sqrt{a_i}$ by $\frac{1}{p}\sqrt{a_1 a_i}$. We repeat this process for other i's if need be.

Of course, if we put

$$\widetilde{a_i} := \left\{ \begin{array}{lll} \frac{1}{p^2} a_1 a_i, & \text{if } p \mid a_i & \text{for } i = 2, \ldots, r \\ a_i, & \text{if } p \nmid a_i & \text{for } i = 2, \ldots, r \\ a_1 & & \text{for } i = 1 \end{array} \right\},$$

we have

$$\mathbb{K} = \mathbb{Q}\left(\sqrt{a_1}, \ldots, \sqrt{a_r}\right) = \mathbb{Q}\left(\sqrt{\widetilde{a_1}}, \ldots, \sqrt{\widetilde{a_r}}\right).$$

We can achieve an additional normalization (see Schmal [187]):

Lemma B.1. *For* $\mathbb{K} = \mathbb{Q}\left(\sqrt{a_1}, \ldots, \sqrt{a_r}\right)$ *with* (B.1), (B.2), *if we take* $p = 2$, *the* a_i*'s can be arranged in such a way that*

$$a_i \equiv 1 \pmod 4 \quad \textit{for} \left\{ \begin{array}{ll} 2 \leq i \leq r & \textit{if } 2 \nmid a_1 \\ 3 \leq i \leq r & \textit{if } 2 \mid a_1 \end{array} \right\}. \tag{B.3}$$

Proof. If $a_i \equiv 1 \pmod 4$ for all indices $1 \leq i \leq r$, nothing is to be proved. Assume therefore that $a_1 \not\equiv 1 \pmod 4$.

In case $2 \nmid a_1$, suppose that $a_i \not\equiv 1 \pmod 4$ for some index $i \geq 2$. Then both a_1 and a_i are odd by (B.2) for $p = 2$, that is,

$$a_1 = 1 + 2k_1, \quad a_i = 1 + 2k_i \quad \text{for some integers } k_1, k_i \in \mathbb{Z}.$$

The incongruences $a_1 \not\equiv 1 \pmod 4$, $a_i \not\equiv 1 \pmod 4$ imply that both k_1 and k_i are odd. Hence

$$a_1 a_i = (1 + 2k_1)(1 + 2k_i) \equiv 1 + 2(k_1 + k_i) \equiv 1 \pmod 4.$$

Then we replace a_i by $\widetilde{a_i} := a_1 a_i$ to obtain $\widetilde{a_i} \equiv 1 \pmod 4$. This process has to be repeated if need be.

In case $2 \mid a_1$ and $a_i \equiv 1 \pmod 4$ for all $2 \leq i \leq r$ nothing is to be proved. This holds also in the case that just one $a_i \not\equiv 1 \pmod 4$, $3 \leq i \leq r$, because we can then interchange a_2 and a_i keeping in mind that $a_2 \equiv 1 \pmod 4$. Let therefore $a_2 \not\equiv 1 \pmod 4$ and $a_i \not\equiv 1 \pmod 4$ for some $3 \leq i \leq r$. We then perform the same process as above in the case $2 \nmid a_1$ but with a_2 instead of a_1 and with the same $a_i \not\equiv 1 \pmod 4$. This process has to be repeated if need be. Again, altogether, $\mathbb{Q}\left(\sqrt{a_1}, \ldots, \sqrt{a_r}\right) = \mathbb{Q}\left(\sqrt{\widetilde{a_1}}, \ldots, \sqrt{\widetilde{a_r}}\right)$ for

$$\widetilde{a_i} = \left\{ \begin{array}{ll} a_1 a_i & \text{if } 2 \nmid a_1 \text{ and } a_i \not\equiv 1 \pmod 4,\ i \geq 2, \\ a_2 a_i & \text{if } 2 \mid a_1 \text{ and } a_i \not\equiv 1 \pmod 4,\ i \geq 3, \\ a_i & \text{if } a_i \equiv 1 \pmod 4) \text{ or } i = 1, \end{array} \right\} \quad (i = 1, \ldots, r).$$

□

The discriminant of a quadratic field

$$\mathbb{k}_i = \mathbb{Q}\left(\sqrt{a_i}\right) \quad (1 \leq i \leq r)$$

with a_i satisfying (B.1) is well-known:

$$d_{\mathbb{k}_i} = \left\{ \begin{array}{ll} a_i & \text{if } a_i \equiv 1 \pmod 4 \\ 4a_i & \text{if } a_i \equiv 2, 3 \pmod 4 \end{array} \right\} \quad (1 \leq i \leq r).$$

(see Hasse [92], but also Hollinger [100] and Stein [215]).

Let therefore

$$\mathbb{K} = \mathbb{Q}\left(\sqrt{a_1}, \ldots, \sqrt{a_r}\right)$$

be a multiquadratic number field of degree

$$[\mathbb{K} : \mathbb{Q}] = 2^r$$

with integers $a_1, \ldots, a_r \in \mathbb{Z}$ satisfying the conditions (B.1), (B.2), (B.3), and assume that $r \geq 2$. The conditions (B.1), (B.2), (B.3) amount to

$$\begin{array}{c} (a_1, a_2) \equiv (1,1), (2,1), (3,1) \text{ or } (2,3) \pmod 4 \\ \text{and} \\ a_i \equiv 1 \pmod 4 \quad \text{for each } 3 \leq i \leq r \ \text{(at least)} \end{array} \tag{B.4}$$

(see Schmal [186], [187]). Under these conditions on the $a_i \in \mathbb{Z}$, we prove

Theorem B.2. *The multiquadratic number field*

$$\mathbb{K} = \mathbb{Q}\left(\sqrt{a_1}, \ldots, \sqrt{a_r}\right)$$

of degree

$$[\mathbb{K} : \mathbb{Q}] = 2^r, \quad r \geq 2,$$

with rational integers $a_i \in \mathbb{Z}$ satisfying (B.1)–(B.4) *and with the product decomposition*

$$\prod_{i=1}^{r} a_i =: \pm \prod_{j=1}^{s} p_j^{n_j} \quad (n_j \in \mathbb{N},\ p_j \in \mathbb{P}, 2 \leq p_1 < \ldots < p_s)$$

has discriminant

$$d_{\mathbb{K}} = (2^{\rho} p_1 \ldots p_s)^{2^{r-1}},$$

where

$$\rho := \begin{Bmatrix} 0 \\ 2 \\ 3 \end{Bmatrix} \text{ in case } (a_1, a_2) \equiv \begin{Bmatrix} (1,1) \\ (2,1) \text{ or } (3,1) \\ (2,3) \end{Bmatrix} \pmod 4.$$

We note in this connection that the cases

$$(a_1, a_2) \equiv (2,2),\ (3,3) \pmod 4$$

cannot occur by the normalization requirement (B.2).

Proof of Theorem B.2. Observe first that $p_1 = 2$ is possible, namely if

$$(a_1, a_2) \equiv (2,1) \text{ or } (2,3) \pmod 4.$$

We apply induction on r. (The case of $r = 1$, that is, of quadratic fields, is covered by Theorem B.2 if one excludes the case $\rho = 3$.)

The case $r = 2$ has been taken care of by Williams [238]. However, Williams proves the assertion in terms of an integral basis of biquadratic fields (see also Section B.5) referring to the defining formula

$$d_{\mathbb{K}} = |\mathrm{Tr}(\mu_i, \mu_j)|,$$

where $\mu_1, \ldots, \mu_n$ is an integral basis of the number field $\mathbb{K}$ and Tr is the trace of $\mathbb{K}$ over $\mathbb{Q}$. In the general situation, it is easier to use the composition formula for discriminants (see Hasse [92]).

Let $r \geq 3$. By induction hypothesis, the multiquadratic field

$$\mathbb{K}' := \mathbb{K}_{r-1} = \mathbb{Q}\left(\sqrt{a_1}, \ldots, \sqrt{a_{r-1}}\right)$$

has discriminant

$$d_{\mathbb{K}'} = (2^\rho p'_1 \ldots p'_{s'})^{2^{r-2}},$$

where

$$\rho := \begin{Bmatrix} 0 \\ 2 \\ 3 \end{Bmatrix} \text{ in case } (a_1, a_2) \equiv \begin{Bmatrix} (1,1) \\ (2,1) \text{ or } (3,1) \\ (2,3) \end{Bmatrix} \pmod 4$$

and

$$\prod_{i'=1}^{r-1} a_{i'} = \prod_{j'=1}^{s'} p'^{\,n'_{j'}}_{j'} \quad (n'_{j'} \in \mathbb{N},\ p'_{j'} \in \mathbb{P},\ 2 \leq p'_1 < \ldots < p'_{s'}).$$

The quadratic field

$$\mathbb{k}_r := \mathbb{Q}\left(\sqrt{a_r}\right)$$

has discriminant

$$d_{\mathbb{k}_r} = \begin{Bmatrix} a_r & \text{if } a_r \equiv 1 \pmod 4 \\ 4a_r & \text{if } a_r \equiv 2, 3 \pmod 4 \end{Bmatrix}.$$

By the composition theorem for discriminants (see Hasse [92]) the composite

$$\mathbb{K} := \mathbb{K}_r = \mathbb{K}'\mathbb{k}_r = \mathbb{Q}\left(\sqrt{a_1}, \ldots, \sqrt{a_{r-1}}, \sqrt{a_r}\right)$$

has discriminant

$$d_{\mathbb{K}} = \mathcal{N}_{\mathbb{K}'|\mathbb{Q}}(\vartheta_{\mathbb{K}|\mathbb{K}'})d_{\mathbb{K}'}^2$$

Hence

$$\begin{aligned} d_{\mathbb{K}} &= {a'}_r^{2^{r-1}} \left((2^\rho p'_1 \ldots p'_{s'})\right)^{2^{r-1}} \\ &= (2^\rho p_1 \ldots p_s)^{2^{r-1}}, \end{aligned}$$

where $a'_r := \prod_{\substack{p'/a_r, p' \neq p'_i \\ i=1,\ldots,s'}} p'$ and we have the prime decompositions

$$\prod_{\nu=1}^{r} a_\nu = \left(\prod_{j'=1}^{s'} {p'}_{j'}^{n'_{j'}}\right) a_r = \prod_{j=1}^{s} p_j^{n_j} \quad (n_j \in \mathbb{N},\, p_j \in \mathbb{P},\, 2 \leq p_1 < \ldots < p_s).$$

Of course,

$$s' \leq s$$

(and the $p'_{j'}$ occur among the p_j). The relative discriminant $\vartheta_{\mathbb{K}/\mathbb{K}'}$ of $\mathbb{K}$ over $\mathbb{K}'$ is associated to the element $a'_r \in \mathbb{Z}$, that is,

$$\vartheta_{\mathbb{K}/\mathbb{K}'} \cong a'_r.$$

Thus Theorem B.2 is proved by induction. □

B.3 Integral Bases

In case $r = 1$,

$$\left\{\hat{y}_1 = 1,\ \hat{y}_2 = \frac{1}{2^{\delta_2}}(\sqrt{a_i} - a_i)\right\}$$

with

$$\delta_2 = \left\{\begin{array}{ll} 1 & \text{if } a_i \equiv 1 \pmod 4 \\ 0 & \text{if } a_i \equiv 2, 3 \pmod 4 \end{array}\right\}$$

is an integral basis of the quadratic field

$$\mathbb{k}_i := \mathbb{Q}\left(\sqrt{a_i}\right) \quad (1 \leq i \leq r).$$

In case $r = 2$, Williams [238] has given an integral basis of the biquadratic field

$$\mathbb{K} := \mathbb{Q}\left(\sqrt{a_1}, \sqrt{a_2}\right)$$

which we shall relate to another integral basis of $\mathbb{K}$ given by Schmal [186], [187] (see Section B.5).

Assume therefore $r \geq 3$ so that the normalization of the a_i yields (see (B.3))

$$a_i \equiv 1 \pmod 4 \quad \text{for each } 3 \leq i \leq r.$$

We put again $\mathbb{K}' = \mathbb{K}_{r-1} = \mathbb{Q}\left(\sqrt{a_1}, \ldots, \sqrt{a_{r-1}}\right)$ and $\mathbb{k}_r = \mathbb{Q}\left(\sqrt{a_r}\right)$ so that the multiquadratic field

$$\mathbb{K} = \mathbb{Q}\left(\sqrt{a_1}, \ldots, \sqrt{a_{r-1}}, \sqrt{a_r}\right)$$

is the composite of $\mathbb{K}'$ and $\mathbb{k}_r$:

$$\mathbb{K} = \mathbb{K}'\mathbb{k}_r.$$

It is obvious that $\mathbb{K}'$ and $\mathbb{k}_r$ are linearly disjoint over $\mathbb{Q}$ (as mentioned already).

Now let

$$j - 1 = j_r 2^{r-1} + \cdots + j_2 2 + j_1 \quad (j_i \in \mathbb{Z},\ 0 \leq j_i \leq 1)$$

be the 2-adic representation for $j \in \{1, \ldots, 2^r\}$ and put

$$b_j := a_1^{j_1} \ldots a_{r-1}^{j_{r-1}} a_r^{j_r}$$

and

$$\omega_j := \sqrt{b_j} = \sqrt{a_1}^{j_1} \ldots \sqrt{a_{r-1}}^{j_{r-1}} \sqrt{a_r}^{j_r} \quad (j = 1, \ldots, 2^r).$$

Let furthermore $g_j \in \mathbb{Z}$ such that

$$0 \leq v_p(g_j^{-2} b_j) \leq 1 \quad \text{for each prime } p \in \mathbb{P} \quad (j = 1, \ldots, 2^r),$$

v_p being the normalized additive p-adic valuation of $\mathbb{Q}$.

Put $x_j := \dfrac{1}{g_j}\omega_j$ for $j \in \{1, \ldots, 2^r\}$. Then $\{x_1, \ldots, x_{2^r}\}$ is a basis of $\mathbb{K}$ over $\mathbb{Q}$. We obtain an integral basis of $\mathbb{K}|\mathbb{Q}$ in the following manner (see Schmal [186], [187]).

Theorem B.3. *Let*

$$\omega_j := \prod_{i=1}^{r} \sqrt{a_i}^{\alpha_{ij}} \quad (\alpha_{ij} \in \mathbb{Z}, 0 \leq \alpha_{ij} \leq 1)$$

for $j = 1, \ldots, 2^r$ as above. Then $\{y_1, \ldots, y_{2^r}\}$ with

$$y_j := \frac{1}{2^{\delta_j} g_j} \prod_{i=1}^{r} (\sqrt{a_i} - a_i)^{\alpha_{ij}}$$

is an integral basis of $\mathbb{K} = \mathbb{Q}\left(\sqrt{a_1}, \ldots, \sqrt{a_{r-1}}, \sqrt{a_r}\right)$ over $\mathbb{Q}$, where $g_j \in \mathbb{Z}$ as above and

$$\delta_1 := 0,$$

$$\delta_2 := \begin{Bmatrix} 1 & \textit{for } a_1 \equiv 1 \pmod 4 \\ 0 & \textit{for } a_1 \equiv 2, 3 \pmod 4 \end{Bmatrix},$$

and for $j > 2$,

$$\delta_j := \sum_{i=1}^{r} \alpha_{ij} - \beta_j \quad (2 < j \le 2^r) \tag{B.5}$$

with

$$\beta_j := \begin{Bmatrix} 1 & \textit{for } (a_1, a_2) \equiv (2, 1), (3, 1) \pmod 4,\ j_1 = 1, \\ 1 & \textit{for } (a_1, a_2) \equiv (2, 3) \pmod 4,\ j_1 = 1 \textit{ or } j_2 = 1, \\ 0 & \textit{else.} \end{Bmatrix}$$

Proof. The proof will be given only in the case that $\gcd(a_1, \ldots, a_{r-1}, a_r) = 1$ and thus $\gcd(d_{\mathbb{K}'}, d_{\mathbb{k}_r}) = 1$ (but see Schmal [187] for the general case). We apply induction on r.

In case $r = 1$, we obtain $\mathbb{K} = \mathbb{Q}\left(\sqrt{a_1}\right)$ is a quadratic field. Furthermore, as pointed out above,

$$\left\{1, \frac{1}{2^{\delta_2}}\left(\sqrt{a_1} - a_1\right)\right\} = \begin{Bmatrix} \left\{1, \frac{1}{2}\left(\sqrt{a_1} - a_1\right)\right\} & \text{if } a_1 \equiv 1 \pmod 4 \\ \left\{1, \left(\sqrt{a_1} - a_1\right)\right\} & \text{if } a_1 \equiv 2, 3 \pmod 4 \end{Bmatrix}$$

is an integral basis of $\mathbb{K}|\mathbb{Q}$ (cf. Hasse [92], Hollinger [100], Stein [215]).

In case $r = 2$, Schmal [186], [187] gives an integral basis of the biquadratic field $\mathbb{K} = \mathbb{Q}\left(\sqrt{a_1}, \sqrt{a_2}\right)$ over $\mathbb{Q}$ which differs from the one obtained by Williams [238] (see Section B.5).

Suppose now $r \ge 3$. Assume by induction that $\{y_1, \ldots, y_{2^{r-1}}\}$ is an integral basis of $\mathbb{K}'|\mathbb{Q}$. Since $a_r \equiv 1 \pmod 4$ by Lemma B.1, we put

$$y_{2^{r-1}+j} := \frac{1}{2} y_j \left(\sqrt{a_r} - a_r\right) \quad (1 \le j \le 2^{r-1}).$$

Then

$$\{y_1, \ldots, y_{2^{r-1}}, y_{2^{r-1}+1}, \ldots, y_{2^r}\}$$

is an integral basis of the multiquadratic field $\mathbb{K}$ itself (see, for example, Thome [222] Cor. 8.1). This is true because $\gcd(d_{\mathbb{K}'}, d_{\mathbb{k}_r}) = 1$, and

$$\left\{1, \frac{1}{2}\left(\sqrt{a_r} - a_r\right)\right\}$$

is an integral basis of the quadratic field $\mathbb{k}_r|\mathbb{Q}$,

$$\{y_1, \ldots, y_{2^{r-1}}\}$$

is an integral basis of $\mathbb{K}'|\mathbb{Q}$ and hence the products

$$\begin{aligned} &\left\{y_1 \cdot 1, \ldots, y_{2^{r-1}} \cdot 1, \tfrac{1}{2}\left(\sqrt{a_r} - a_r\right) \cdot y_1, \ldots, \tfrac{1}{2}\left(\sqrt{a_r} - a_r\right) \cdot y_{2^{r-1}}\right\} \\ &= \{y_1, \ldots, y_{2^{r-1}}, y_{2^{r-1}+1}, \ldots, y_{2^r}\} \end{aligned} \tag{B.6}$$

form an integral basis of $\mathbb{K}|\mathbb{Q}$. Remember that $\mathbb{K}'$ and $\mathbb{k}_r$ are linearly disjoint over $\mathbb{Q}$ and that $\mathbb{K} = \mathbb{K}'\mathbb{k}_r$. □

Note that the discriminant of $\mathbb{K}$ is also a product. We know that in case $\gcd(d_{\mathbb{K}'}, d_{\mathbb{k}_r}) = 1$,

$$d_{\mathbb{K}} = d_{\mathbb{K}'}^2 \cdot d_{\mathbb{k}_r}^{2^{r-1}}$$

(see e.g. Thome [222], Cor. 8.1).

The relation (B.6) also shows that

$$\delta_{2^{r-1}+j} = \delta_j + 1 \quad (j = 1, \ldots, 2^{r-1}).$$

We have furthermore

$$\beta_{2^{r-1}+j} = \beta_j \quad (j = 1, \ldots, 2^{r-1}),$$

$$\alpha_{r,2^{r-1}+j} = 1 \quad (j = 1, \ldots, 2^{r-1})$$

and

$$\alpha_{i,j} = \alpha_{i,2^{r-1}+j} \quad (i = 1, \ldots, r-1, \ j = 1, \ldots, 2^{r-1}).$$

In particular, the relation (B.5) is true.

B.4 Decomposition Law

In the sequel we shall need the following auxiliary result.

Lemma B.4. *Let $\mathbb{k}$ be a number field as well as $\mathbb{K}$ and $\mathbb{K}'$ be finite Galois extensions of $\mathbb{K}$. Denote by $\mathbb{L} = \mathbb{K}\mathbb{K}'$ their composite. Suppose furthermore that $\mathbb{K}$ and $\mathbb{K}'$ are linearly disjoint over $\mathbb{k}$. Let $\mathfrak{p}$ resp. $\mathfrak{p}'$ be prime divisors of $\mathbb{K}$ resp. $\mathbb{K}'$ lying over a prime divisor $\wp$ of $\mathbb{k}$:*

$$\mathfrak{p}|\wp \text{ in } \mathbb{K}, \quad \mathfrak{p}'|\wp \text{ in } \mathbb{K}',$$

and let $\mathfrak{P}$ denote a prime divisor of $\mathbb{L}$ lying over $\mathfrak{p}$ resp. $\mathfrak{p}'$:

$$\mathfrak{P}|\mathfrak{p}|\wp \quad \text{resp.} \quad \mathfrak{P}|\mathfrak{p}'|\wp \quad \text{in } \mathbb{L}.$$

Then the ramification orders of $\mathfrak{P}$ over $\mathfrak{p}$ and $\mathfrak{p}'$ over $\wp$ are equal:

$$e_{\mathfrak{P}|\mathfrak{p}} = e_{\mathfrak{p}'|\wp}.$$

Similarly,

$$e_{\mathfrak{P}|\mathfrak{p}'} = e_{\mathfrak{p}|\wp}.$$

The same identities hold for the residue class degrees:

$$f_{\mathfrak{P}|\mathfrak{p}} = f_{\mathfrak{p}'|\wp},$$

$$f_{\mathfrak{P}|\mathfrak{p}'} = f_{\mathfrak{p}|\wp}.$$

Proof. Denote the Galois group of $\mathbb{K}|\mathbb{k}$ by

$$G := \mathrm{Gal}(\mathbb{K}|\mathbb{k})$$

and the Galois group of $\mathbb{K}'|\mathbb{k}$ by

$$G' := \mathrm{Gal}(\mathbb{K}'|\mathbb{k}).$$

The extension $\mathbb{L}|\mathbb{K}$ is also Galois with Galois group

$$\mathcal{G} := \mathrm{Gal}(\mathbb{L}|\mathbb{k}) \cong G \times G'$$

(see Lang [113]).

Let $\mathbb{k}_\wp$, $\mathbb{K}_\mathfrak{p}$, $\mathbb{K}'_{\mathfrak{p}'}$, resp. $\mathbb{L}_\mathfrak{P}$ denote the completions of $\mathbb{k}$, $\mathbb{K}$, $\mathbb{K}'$, resp. $\mathbb{L}$ with respect to $\wp$, $\mathfrak{p}$, $\mathfrak{p}'$, resp. $\mathfrak{P}$. Then we obtain the diagrams

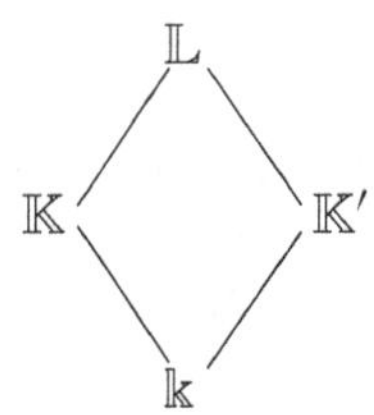

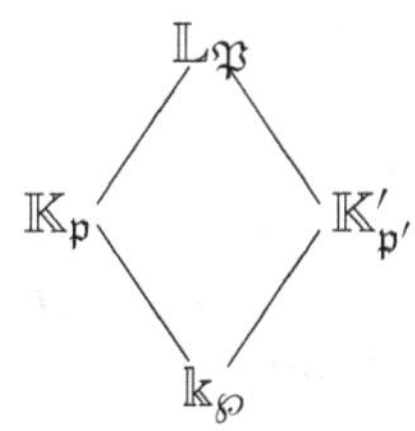

The local extensions $\mathbb{K}_\mathfrak{p}|\mathbb{k}_\wp$, $\mathbb{K}'_{\mathfrak{p}'}|\mathbb{k}_\wp$, $\mathbb{L}_\mathfrak{P}|\mathbb{k}_\wp$ are also Galois with Galois groups

$$\begin{aligned}
\mathrm{Gal}(\mathbb{K}_\mathfrak{p}|\mathbb{k}_\wp) &=: G_\mathfrak{p} \le G,\\
\mathrm{Gal}(\mathbb{K}'_{\mathfrak{p}'}|\mathbb{k}_\wp) &=: G'_{\mathfrak{p}'} \le G',\\
\mathrm{Gal}(\mathbb{L}_\mathfrak{P}|\mathbb{k}_\wp) &\cong G_\mathfrak{p} \times G'_{\mathfrak{p}'} \le G \times G' \cong \mathcal{G}.
\end{aligned}$$

By the ramification theory of Hilbert the subgroups $G_\mathfrak{p}$ of G, $G'_{\mathfrak{p}'}$ of G' and $G_\mathfrak{p} \times G'_{\mathfrak{p}'}$ of $\mathcal{G}$ (up to isomorphy) are the decomposition groups of $\mathfrak{p}$, $\mathfrak{p}'$ and $\mathfrak{P}$, each over $\mathbb{k}$ (see, e.g., Leutbecher [130] or Neukirch [157]).

Since $\mathbb{K}$ and $\mathbb{K}'$ are linearly disjoint over $\mathbb{K}$, the completions $\mathbb{K}_\mathfrak{p}$ and $\mathbb{K}'_{\mathfrak{p}'}$ are linearly disjoint over the completion $\mathbb{k}_\wp$.

We keep $\wp$ fixed and write for short

$$\begin{aligned}
e_\mathfrak{p} &:= e_{\mathfrak{p}|\wp}, & e_{\mathfrak{p}'} &:= e_{\mathfrak{p}'|\wp},\\
f_\mathfrak{p} &:= f_{\mathfrak{p}|\wp}, & f_{\mathfrak{p}'} &:= f_{\mathfrak{p}'|\wp},\\
e &:= e_{\mathfrak{P}|\mathfrak{p}}, & e' &:= e_{\mathfrak{P}|\mathfrak{p}'},\\
f &:= f_{\mathfrak{P}|\mathfrak{p}}, & f' &:= f_{\mathfrak{P}|\mathfrak{p}'}.
\end{aligned}$$

Locally, we have then

$$\begin{aligned}
&\mathfrak{p} = \mathfrak{P}^e, \quad \mathfrak{p}' = \mathfrak{P}^{e'} \quad \text{in } \mathbb{L}_\mathfrak{P},\\
&\wp = \mathfrak{p}^{e_\mathfrak{p}} \quad \text{in } \mathbb{K}_\mathfrak{p},\\
&\wp = \mathfrak{p}'^{e_{\mathfrak{p}'}} \quad \text{in } \mathbb{K}'_{\mathfrak{p}'}.
\end{aligned}$$

Let $\pi \in \mathbb{K}_{\mathfrak{p}}$ be a prime element for $\mathfrak{p}$ and $\pi' \in \mathbb{K}'_{\mathfrak{p}'}$ be a prime element for $\mathfrak{p}'$. Then $1, \pi, \dots, \pi^{e_{\mathfrak{p}}-1}$ resp. $1, \pi', \dots, \pi'^{e_{\mathfrak{p}'}-1}$ are linearly independent over $\mathbb{k}_{\wp}$. The elements $1, \pi, \dots, \pi^{e_{\mathfrak{p}}-1}$ resp. $1, \pi', \dots, \pi'^{e_{\mathfrak{p}'}-1}$ remain linearly independent over $\mathbb{K}'_{\mathfrak{p}'}$ resp. $\mathbb{K}_{\mathfrak{p}}$ (see Hasse [92]) because $\mathbb{K}_{\mathfrak{p}}$ and $\mathbb{K}'_{\mathfrak{p}'}$ are linearly disjoint over $\mathbb{k}_{\wp}$. Hence the ramification orders satisfy

$$\begin{aligned} e &= e_{\mathfrak{P}|\mathfrak{p}} = e_{\mathfrak{p}'|\wp} = e_{\mathfrak{p}'}, \\ e' &= e_{\mathfrak{P}|\mathfrak{p}'} = e_{\mathfrak{p}|\wp} = e_{\mathfrak{p}}, \end{aligned}$$

as asserted in the lemma. We get from this the same results for the residue degrees:

$$\begin{aligned} f &= f_{\mathfrak{P}|\mathfrak{p}} = f_{\mathfrak{p}'|\wp} = f_{\mathfrak{p}'}, \\ f' &= f_{\mathfrak{P}|\mathfrak{p}'} = f_{\mathfrak{p}|\wp} = f_{\mathfrak{p}}. \end{aligned}$$

The assumption that $\mathbb{K}|\mathbb{k}$ and $\mathbb{K}'|\mathbb{k}$ are Galois extensions can be dropped. □

The decomposition law for $r = 2$ was given by Cohn [36] and is contained also in Hollinger [1], [100]. However, it can be generalized to arbitrary multiquadratic fields $\mathbb{K} = \mathbb{Q}\left(\sqrt{a_1}, \dots, \sqrt{a_r}\right)$ in the following manner (see Hollinger [100]). Remember that $\mathbb{K}|\mathbb{Q}$ is a Galois extension.

Let $\left(\dfrac{a}{p}\right)$ for $a \in \mathbb{Z}$ and $p \in \mathbb{P}$, $p \neq 2$, denote the Legendre symbol, with the agreement that $\left(\dfrac{a}{p}\right) = 0$ for $p \mid a$.

The $a_i \in \mathbb{Z}$ can be arranged in such a way that it suffices to consider the cases treated in the subsequent decomposition theorem.

Theorem B.5. *The decomposition of primes in* $\mathbb{K} = \mathbb{Q}\left(\sqrt{a_1}, \dots, \sqrt{a_r}\right)$, *where* $[\mathbb{K} : \mathbb{Q}] = 2^r$ *is as follows.*

(a) *Let p be a (non-archimedean) prime number.*

(1) $p \in \mathbb{P}, p \neq 2$:

If $\left(\dfrac{a_i}{p}\right) = 1$ *for* $i = 1, \dots, r$,

$p \cong \mathfrak{p}_1 \dots \mathfrak{p}_{2^r}$ *splits completely.*

If $\left(\dfrac{a_1}{p}\right) = -1$ *and* $\left(\dfrac{a_i}{p}\right) = 1$ *for* $i = 2, \dots, r$,

$p \cong \mathfrak{p}_1 \dots \mathfrak{p}_{2^{r-1}}$ *is inert with residue degree* $f = 2$.

If $\left(\frac{a_1}{p}\right) = 0$ *and* $\left(\frac{a_i}{p}\right) = 1$ *for* $i = 2, \ldots, r$,

$$p \cong \mathfrak{p}_1^2 \ldots \mathfrak{p}_{2^{r-1}}^2 \quad \text{ramifies of order } e = 2 \text{ and residue degree } f = 1.$$

If $\left(\frac{a_1}{p}\right) = 0$, $\left(\frac{a_2}{p}\right) = -1$, *and* $\left(\frac{a_i}{p}\right) = 1$ *for* $i = 3, \ldots, r$,

$$p \cong \mathfrak{p}_1^2 \ldots \mathfrak{p}_{2^{r-2}}^2 \quad \text{ramifies of order } e = 2 \text{ and residue degree } f = 2.$$

(2) $p = 2$:

If $a_i \equiv 1 \pmod 8$ *for* $i = 1, \ldots, r$,

$$p \cong \mathfrak{p}_1 \ldots \mathfrak{p}_{2^r} \quad \text{splits completely.}$$

If $a_1 \equiv 5 \pmod 8$ *and* $a_i \equiv 1 \pmod 8$ *for* $i = 2, \ldots, r$,

$$p \cong \mathfrak{p}_1 \ldots \mathfrak{p}_{2^{r-1}} \quad \text{is inert with residue degree } f = 2.$$

If $a_1 \equiv 2, 3 \pmod 4$ *and* $a_i \equiv 1 \pmod 8$ *for* $i = 2, \ldots, r$,

$$p \cong \mathfrak{p}_1^2 \ldots \mathfrak{p}_{2^{r-1}}^2 \quad \text{ramifies of order } e = 2 \text{ and residue degree } f = 1.$$

If $a_1 \equiv 2, 3 \pmod 4$, $a_2 \equiv 5 \pmod 8$, *and* $a_i \equiv 1 \pmod 8$ *for* $i = 3, \ldots, r$,

$$p \cong \mathfrak{p}_1^2 \ldots \mathfrak{p}_{2^{r-2}}^2 \quad \text{ramifies of order } e = 2 \text{ and residue degree } f = 2.$$

If $a_1 \equiv 2 \pmod 4$, $a_2 \equiv 3 \pmod 4$, *and* $a_i \equiv 1 \pmod 8$ *for* $i = 3, \ldots, r$,

$$p \cong \mathfrak{p}_1^4 \ldots \mathfrak{p}_{2^{r-2}}^4 \quad \text{ramifies of order } e = 4 \text{ and residue degree } f = 1.$$

(*which occurs only if* $r \geq 2$).

If $a_1 \equiv 2 \pmod 4$, $a_2 \equiv 3 \pmod 4$, $a_3 \equiv 5 \pmod 8$, *and* $a_i \equiv 1 \pmod 8$ *for* $i = 4, \ldots, r$,

$$p \cong \mathfrak{p}_1^4 \ldots \mathfrak{p}_{2^{r-3}}^4 \quad \text{ramifies of order } e = 4 \text{ and residue degree } f = 2$$

(*which occurs only if* $r \geq 3$).

(b) $p = \infty$ *archimedean.*

(1) *If* $a_i > 0$ *for* $i = 1, \dots, r$,

$$p \cong \mathfrak{p}_1 \dots \mathfrak{p}_{2^r} \text{ splits completely.}$$

(2) *If* $a_1 < 0$ *and* $a_i > 0$ *for* $i = 2, \dots, r$,

$$p \cong \mathfrak{p}_1^2 \dots \mathfrak{p}_{2^{r-1}}^2 \quad \text{ramifies of order } e = 2 \text{ and residue degree } f = 1.$$

(by definition, see Hasse [92]*).*

Proof. If $r = 1$, $\mathbb{K} = \mathbb{Q}\left(\sqrt{a_1}\right)$ is a quadratic field. The first three cases for non-archimedean p's each give the well-known decomposition of primes in a quadratic field, and for the archimedean prime $p = \infty$, we get the corresponding decomposition law:

(a) Let p be a prime number.

(1) for $p \in \mathbb{P}$, $p \neq 2$:

$$p \cong \mathfrak{p}_1\mathfrak{p}_2 \quad \text{splits completely if } \left(\frac{a_1}{p}\right) = 1,$$

$$p \cong \mathfrak{p}_1 \quad \text{is inert if } \left(\frac{a_1}{p}\right) = -1,$$

$$p \cong \mathfrak{p}_1^2 \quad \text{ramifies if } \left(\frac{a_1}{p}\right) = 0,$$

(2) for $p = 2$:

$$\begin{array}{ll} p \cong \mathfrak{p}_1\mathfrak{p}_2 & \text{splits completely if } a_1 \equiv 1 \pmod 8, \\ p \cong \mathfrak{p}_1 & \text{is inert if } a_1 \equiv 5 \pmod 8, \\ p \cong \mathfrak{p}_1^2 & \text{ramifies if } a_1 \equiv 2, 3 \pmod 4, \end{array}$$

(b) for $p = \infty$:

$$\begin{array}{ll} p \cong \mathfrak{p}_1\mathfrak{p}_2 & \text{splits completely if } a_1 > 0, \\ p \cong \mathfrak{p}_1^2 & \text{ramifies if } a_1 < 0 \end{array}$$

(see Hasse [92]).

If $r = 2$, $\mathbb{K} = \mathbb{Q}\left(\sqrt{a_1}, \sqrt{a_2}\right)$ is biquadratic and we refer to Section B.5 for a direct proof.

Let $r \geq 2$.

(a) $\mathbb{K}|\mathbb{Q}$ is a Galois extension. It is therefore sufficient to establish the relation (see Neukirch [157])

$$[\mathbb{K} : \mathbb{Q}] = \left[\mathbb{Q}\left(\sqrt{a_1}, \ldots, \sqrt{a_r}\right) : \mathbb{Q}\right] = 2^r = efg. \tag{B.7}$$

We proceed by induction on r. Clearly, the relation (B.7) holds for $r = 1$. Let $r \geq 2$ and

$$\mathbb{K}' := \mathbb{Q}\left(\sqrt{a_1}, \ldots, \sqrt{a_{r-1}}\right) = \mathbb{K}_{r-1}.$$

We have the tower of fields

$$\mathbb{Q} \leq \mathbb{K}' \leq \mathbb{K}.$$

As above we put

$$\mathbb{k}_r = \mathbb{Q}\left(\sqrt{a_r}\right).$$

Assume the asserted relation (B.7) holds for $r - 1$. Then we have

$$[\mathbb{K}' : \mathbb{Q}] = 2^{r-1} = e' f' g'. \tag{B.8}$$

For the quadratic field $\mathbb{k}_r$, we have

$$[\mathbb{k}_r : \mathbb{Q}] = 2 = e_r f_r g_r. \tag{B.9}$$

Moreover, this relation remains true upon transition from $\mathbb{k}_r|\mathbb{Q}$ to $\mathbb{K}|\mathbb{K}'$ by Lemma B.4.

We know that the ramification order $e = e(\mathbb{K}|\mathbb{Q})$ is the product of the ramification orders $e_r = e(\mathbb{K}|\mathbb{K}')$ and $e' = e(\mathbb{K}'|\mathbb{Q})$:

$$e = e_r e'.$$

The corresponding relation holds also for the residue degree $f(\mathbb{K}|\mathbb{Q})$:

$$f = f_r f',$$

where $f_r = f(\mathbb{K}|\mathbb{K}')$ and $f' = f(\mathbb{K}'|\mathbb{Q})$ (see the exercise at the end of Appendix B and e.g. Leutbecher [130]).

Hence

$$2^r = 2 \cdot 2^{r-1} = e_r f_r g_r \cdot e' f' g' = efg_r g',$$

and we must have $g = g_r g'$. Thus (B.7) is true.

(b) For $p = \infty$ (archimedean) we proceed in a manner analogous to that in the proof of Lemma B.1. In case $a_i > 0$ for all indices $i = 1, \ldots, r$, nothing will be changed. Otherwise we may arrange the numbering of the a_i in such a way that $a_1 < 0$. Next we put

$$\widetilde{a_i} := \left\{ \begin{array}{ll} a_1 a_i & \text{if } a_i < 0 \text{ for } i \geq 2 \\ a_i & \text{if } a_i > 0 \text{ for } i \geq 2 \\ a_1 & \text{for } i = 1. \end{array} \right\}$$

and have again

$$\mathbb{K} = \mathbb{Q}\left(\sqrt{a_1}, \sqrt{a_2}, \dots, \sqrt{a_r}\right) = \mathbb{Q}\left(\sqrt{\widetilde{a_1}}, \sqrt{\widetilde{a_2}}, \dots, \sqrt{\widetilde{a_r}}\right).$$

The proof then follows by induction on r.

Altogether, Theorem B.5 is thus proved by induction. □

For multiquadratic fields $\mathbb{K}$ as ground fields of elliptic curves E it is possible to determine all abelian groups which are candidates for torsion groups $E(\mathbb{K})_{\text{tors}}$ (see Section 6.2, Abel–Hollinger and Zimmer [1], or Hollinger [100]). However, it is difficult to actually compute $E(\mathbb{K})_{\text{tors}}$ because these candidates may have high orders.

B.5 Biquadratic number fields

The case $r = 2$ deserves some extra consideration, especially because torsion groups of elliptic curves have been calculated over such fields.

A number field

$$\mathbb{K} := \mathbb{Q}(\sqrt{a}, \sqrt{b})$$

with $a = a_1, b = a_2 \in \mathbb{Z}$, both square-free and such that

$$\sqrt{b} \notin \mathbb{Q}(\sqrt{a})$$

implying that

$$[\mathbb{K} : \mathbb{Q}] = 2^2,$$

is called *biquadratic*.

Let $t := \gcd(a, b)$ and $a' := \dfrac{a}{t}$, $b' := \dfrac{b}{t}$. We assume that Condition (B.4) is fulfilled.

Proposition B.6. *The integral elements of the biquadratic field*

$$\mathbb{K} = \mathbb{Q}(\sqrt{a}, \sqrt{b})$$

are

(A)

$$\gamma = \frac{1}{4}\left(c_0 + c_1\sqrt{a} + c_2\sqrt{b} + c_3\sqrt{a'b'}\right) \quad (c_i \in \mathbb{Z})$$

in case $(a, b) \equiv (1, 1) \pmod 4$ *with* $(a', b') \equiv (1, 1)$ *or with* $(a', b') \equiv (3, 3) \pmod 4$, *in which case*

$$c_0 \equiv c_1 \equiv c_2 \equiv c_3 \pmod 2$$

and

$$c_0 - c_1 \pm c_2 - c_3 \equiv 0 \pmod 4.$$

(The minus sign in front of c_2 *occurs in case* $(a', b') \equiv (3, 3) \pmod 4$*.)*

(B)

$$\gamma = \frac{1}{2}\left(c_0 + c_1\sqrt{a} + c_2\sqrt{b} + c_3\sqrt{a'b'}\right) \quad (c_i \in \mathbb{Z})$$

in case $(a, b) \equiv (2, 1) \pmod 4$, $(a, b) \equiv (3, 1) \pmod 4$, *or* $(a, b) \equiv (2, 3) \pmod 4$, *in which cases*

$$c_0 \equiv c_2 \pmod 2 \quad \text{and} \quad c_1 \equiv c_3 \pmod 2.$$

Additionally, $c_0 \equiv c_2 \equiv 0 \pmod 2$ *and* $c_1 \equiv c_3 \pmod 2$ *in case* $(a, b) \equiv (2, 3) \pmod 4$.

Proof. We follow the paper [238] of Williams. Let

$$(a, b) \equiv (1, 1), (2, 1), (3, 1) \text{ or } (2, 3) \pmod 4 \tag{B.10}$$

(see Condition (B.4).

(The situation is symmetric in a and b so that a and b can be interchanged if need be. Even say b and the old ab may be interchanged to ensure (B.10).) Then, a number $\gamma \in \mathbb{K}$ can be represented in the form

$$\gamma = c_0 + c_1\sqrt{a} + c_2\sqrt{b} + c_3\sqrt{a'b'} \quad \text{with } c_i \in \mathbb{Q}.$$

Suppose that γ is an integer of $\mathbb{K}$. Then, so are its conjugates

$$\begin{aligned}
\gamma = \gamma^{(1)} &= c_0 + c_1\sqrt{a} + c_2\sqrt{b} + c_3\sqrt{a'b'},\\
\gamma^{\sigma_a} := \gamma^{(2)} &= c_0 + c_1\sqrt{a} - c_2\sqrt{b} - c_3\sqrt{a'b'},\\
\gamma^{\sigma_b} := \gamma^{(3)} &= c_0 - c_1\sqrt{a} + c_2\sqrt{b} - c_3\sqrt{a'b'},\\
\gamma^{\sigma_a \circ \sigma_b} := \gamma^{(4)} &= c_0 - c_1\sqrt{a} - c_2\sqrt{b} + c_3\sqrt{a'b'},
\end{aligned}$$

(with $\sigma_a := \sigma_2, \sigma_b := \sigma_1$) and their sums

$$\begin{aligned}
\gamma^{(1)} + \gamma^{(2)} &= 2c_0 + 2c_1\sqrt{a} \in \mathbb{Q}\left(\sqrt{a}\right),\\
\gamma^{(1)} + \gamma^{(3)} &= 2c_0 + 2c_2\sqrt{b} \in \mathbb{Q}(\sqrt{b}),\\
\gamma^{(1)} + \gamma^{(4)} &= 2c_0 + 2c_3\sqrt{a'b'} \in \mathbb{Q}(\sqrt{a'b'}).
\end{aligned}$$

Hence the sums are indeed integers already of the quadratic fields $\Bbbk_a := \Bbbk_1 = \mathbb{Q}(\sqrt{a})$, $\Bbbk_b := \Bbbk_2 = \mathbb{Q}(\sqrt{b})$ resp. $\Bbbk_{ab} := \mathbb{Q}(\sqrt{a'b'})$.

We consider the last three congruences of (B.10) leaving the case of the first congruence to the reader (but see Williams [238]).

Because of the last three congruences in (B.10), at least two of the three numbers $a, b, ab = t^2a'b'$ are $\not\equiv 1 \pmod 4$. Therefore, the well-known integral basis property for quadratic fields and the fact that the sum of the conjugates of γ listed above all contain the summand $2c_0$ show that

$$2c_0, 2c_1, 2c_2, 2c_3 \in \mathbb{Z}.$$

This implies that we can represent the integral element $\gamma \in \mathbb{K}$ (in new notation) in the form

$$\gamma = \frac{1}{2}\left(c_0 + c_1\sqrt{a} + c_2\sqrt{b} + c_3\sqrt{a'b'}\right)$$

with each $c_i \in \mathbb{Z}$. Now an elementary calculation (for which we used the SIMATH package) yields for γ the fourth degree equation

$$\gamma^4 - 2c_0\gamma^3 + \left(c + \frac{d}{2}\right)\gamma^2 + \frac{(c_3a'b'e - c_0d)}{2}\gamma + \frac{(d^2 - a'b'e^2)}{16} = 0,$$

where we took the short hand notation

$$\begin{aligned} c &:= c_0^2 - a'b'c_3^2, \\ d &:= c_0^2 - ac_1^2 - bc_2^2 + a'b'c_3^2, \\ e &:= 2(c_0c_3 - c_1c_2t). \end{aligned}$$

If γ is quadratic, especially if $\gamma \in \Bbbk_a$, $\gamma \in \Bbbk_b$ or $\gamma \in \Bbbk_{ab}$, the proposition is true. Otherwise, the equation for γ is a proper fourth degree equation. It has rational integral coefficients. In particular,

$$d^2 - a'b'e^2 \equiv 0 \pmod{16}. \tag{B.11}$$

The number e is, by definition, even. It follows from the congruence relation (B.11) that d must be even too.

Now consider the special case where $(a, b) \equiv (2, 1) \pmod 4$. In this case t must be odd so that we have

$$t \equiv 1 \pmod 2, \qquad \text{thus } a'b' \equiv 2 \pmod 4.$$

Then surely the congruence (B.11) is valid if and only if

$$d \equiv e \equiv 0 \pmod 4. \tag{B.12}$$

By definition of d and e, (B.12) and hence (B.11) is thus equivalent to the two congruences

$$\begin{aligned} c_0^2 - 2c_1^2 - c_2^2 + 2c_3^2 &\equiv 0 \pmod 4, \\ c_0c_3 - c_1c_2 &\equiv 0 \pmod 2. \end{aligned} \tag{B.13}$$

In case $c_0 \not\equiv c_2 \pmod 2$, the first congruence in (B.13) cannot be solved since then $c_0^2 - c_2^2 \equiv 1 \pmod 2$. Hence we must have $c_0 \equiv c_2 \pmod 2$ so that $c_0^2 - c_2^2 \equiv 0 \pmod 4$. The first congruence in (B.13) then implies that also $2(c_1^2 - c_3^2) \equiv 0 \pmod 4$ so that $c_1 \equiv c_3 \pmod 2$. The congruences (B.13) are then both soluble, and this proves the first case of Part (B) of the proposition.

The other cases are proved in a similar manner. It should be pointed out that the proofs of the other cases of the proposition are a little more involved. But altogether, the proof of Proposition B.6 is elementary. □

An integral basis of the field $\mathbb{K} = \mathbb{Q}(\sqrt{a}, \sqrt{b})$ is now clear.

Theorem B.7. *The biquadratic field $\mathbb{K} = \mathbb{Q}(\sqrt{a}, \sqrt{b})$ has integral basis*

(a)

$$\left\{1, \frac{1}{2}(1+\sqrt{a}), \frac{1}{2}(1+\sqrt{b}), \frac{1}{4}(1 \pm \sqrt{a} + \sqrt{b} + \sqrt{a'b'})\right\}$$

for (a, b) with $(a', b') \equiv (1, 1) \pmod 4$ or $(a, b) \equiv (1, 1)$ with $(a', b') \equiv (3, 3) \pmod 4$,

(The minus sign in front of $\sqrt{a}$ occurs in the latter case.)

(b)

$$\left\{1, \sqrt{a}, \frac{1}{2}(1+\sqrt{b}), \frac{1}{2}(\sqrt{a} + \sqrt{a'b'})\right\}$$

for $(a, b) \equiv (2, 1)$ or $(a, b) \equiv (3, 1) \pmod 4$,

(c)

$$\left\{1, \sqrt{a}, \sqrt{b}, \frac{1}{2}(\sqrt{a} + \sqrt{a'b'})\right\}$$

for $(a, b) \equiv (2, 3) \pmod 4$.

(Again we make use of the fact that a and b as well as b and the old ab may be interchanged.)

Proof. The proof is elementary too since we can rely on Proposition B.6 (see Williams [238]).

We confine ourselves to treating Cases (b) and (c).

Since $(a, b) \equiv (2, 1)$ or $(3, 1) \pmod 4$ in Case (b) and $(a, b) \equiv (2, 3) \pmod 4$ in Case (c), we know from the proof of Proposition B.6 that all integers of the field $\mathbb{K} = \mathbb{Q}(\sqrt{a}, \sqrt{b})$ have the form

$$\gamma = \frac{1}{2}\left(c_0 + c_1\sqrt{a} + c_2\sqrt{b} + c_3\sqrt{a'b'}\right) \tag{B.14}$$

with $c_i \in \mathbb{Z}$ and

$$c_0 \equiv c_2 \pmod 2, \quad c_1 \equiv c_3 \pmod 2.$$

We have in Case (b)

$$\gamma = c_0' + c_1'\sqrt{a} + \frac{1}{2}c_2'(1+\sqrt{b}) + \frac{1}{2}c_3'(\sqrt{a}+\sqrt{a'b'}) \quad (c_i' \in \mathbb{Q})$$

so that according to (B.14)

$$\begin{aligned} c_0' + \frac{1}{2}c_2' &= \frac{1}{2}c_0, \\ c_1' + \frac{1}{2}c_3' &= \frac{1}{2}c_1, \\ c_2' &= c_2, \\ c_3' &= c_3. \end{aligned}$$

Hence

$$\begin{aligned} c_0' &= \frac{1}{2}(c_0 - c_2), \\ c_1' &= \frac{1}{2}(c_1 - c_3), \\ c_2' &= c_2, \\ c_3' &= c_3 \end{aligned}$$

are rational integers. Therefore,

$$\left\{1, \sqrt{a}, \frac{1}{2}(1+\sqrt{b}), \frac{1}{2}(\sqrt{a}+\sqrt{a'b'})\right\}$$

is an integral basis of $\mathbb{K}$ in Case (b).

We have in Case (c) the relation

$$\gamma = c_0' + c_1'\sqrt{a} + c_2'\sqrt{b} + \frac{1}{2}c_3'(\sqrt{a}+\sqrt{a'b'}) \quad (c_i' \in \mathbb{Q}).$$

From (B.14) we know this time that

$$\begin{aligned} c_0' &= \frac{1}{2}c_0, \\ c_1' + \frac{1}{2}c_3' &= \frac{1}{2}c_1, \\ c_2' &= \frac{1}{2}c_2, \\ c_3' &= c_3. \end{aligned}$$

Again we conclude that

$$\begin{aligned} c_0' &= \frac{1}{2}c_0, \\ c_1' &= \frac{1}{2}(c_1 - c_3) \\ c_2' &= \frac{1}{2}c_2, \\ c_3' &= c_3 \end{aligned}$$

are rational integers. Note that $c_0 \equiv c_2 \equiv 0 \pmod 2$ and $c_1 \equiv c_3 \pmod 2$ in Case (c). Hence, in Case (c)

$$\left\{1, \sqrt{a}, \sqrt{b}, \frac{1}{2}\left(\sqrt{a} + \sqrt{a'b'}\right)\right\}$$

is an integral basis of $\mathbb{K}$. □

Schmal [186], [187] obtains a similar integral basis for the multiquadratic field $\mathbb{K}$ over $\mathbb{Q}$, which for biquadratic fields $\mathbb{K}$ boils down to

(a')

$$\left\{1, \frac{1}{2}(\sqrt{a} - a), \frac{1}{2}(\sqrt{b} - b), \frac{1}{4t}(\sqrt{a} - a)(\sqrt{b} - b)\right\}$$

for $(a, b) \equiv (1, 1) \pmod 4$,

(b')

$$\left\{1, (\sqrt{a} - a), \frac{1}{2}(\sqrt{b} - b), \frac{1}{2t}(\sqrt{a} - a)(\sqrt{b} - b)\right\}$$

for $(a, b) \equiv (2, 1)$ or $(3, 1) \pmod 4$,

(c')

$$\left\{1, (\sqrt{a} - a), (\sqrt{b} - b), \frac{1}{2t}(\sqrt{a} - a)(\sqrt{b} - b)\right\}$$

for $(a, b) \equiv (2, 3) \pmod 4$,

where as before $t = \gcd(a, b)$, $a' = \dfrac{a}{t}$, $b' = \dfrac{b}{t}$.

The transition matrix from (a') to (a) is

$$\begin{pmatrix} 1 & 0 & 0 & 0 \\ -\frac{1}{2}(a+1) & 1 & 0 & 0 \\ -\frac{1}{2}(b+1) & 0 & 1 & 0 \\ \frac{1}{4}\left(\pm 1 + a' + b' + \frac{ab}{t}\right) & -\frac{1}{2}(b' \pm 1) & -\frac{1}{2}(a' + 1) & 1 \end{pmatrix}$$

with entries

$$\frac{1}{2}(a+1),\ \frac{1}{2}(b+1),\ \frac{1}{2}(a'+1),\ \frac{1}{2}(b'\pm 1),\ \frac{1}{4}\left(\pm 1+a'+b'+\frac{ab}{t}\right)\in\mathbb{Z}$$

for $(a, b) \equiv (1, 1) \pmod 4$ with $(a', b') \equiv (1, 1)$ resp. with $(a', b') \equiv (3, 3) \pmod 4$.
(The minus sign in front of the two ones occurs in case $(a', b') \equiv (3, 3) \pmod 4$.)

The transition matrix from (b') to (b) is

$$\begin{pmatrix} 1 & 0 & 0 & 0 \\ -a & 1 & 0 & 0 \\ -\frac{1}{2}(b+1) & 0 & 1 & 0 \\ \frac{1}{2}a'(b+1) & -\frac{1}{2}(b'+1) & -a' & 1 \end{pmatrix}$$

Again, the entries are integers:

$$\frac{1}{2}(b+1),\ \frac{1}{2}(b'+1),\ \frac{1}{2}a'(b+1)\in\mathbb{Z}$$

because $(a, b) \equiv (2, 1)$ or $(3, 1) \pmod 4$.

Finally, the transition matrix from (c') to (c) is

$$\begin{pmatrix} 1 & 0 & 0 & 0 \\ -a & 1 & 0 & 0 \\ -b & 0 & 1 & 0 \\ \frac{1}{2}\frac{ab}{t} & -\frac{1}{2}(b'+1) & -\frac{1}{2}a' & 1 \end{pmatrix}$$

In this case, the entries are rational integers too:

$$\frac{1}{2}\frac{ab}{t},\ \frac{1}{2}(b'+1),\ \frac{1}{2}a'\in\mathbb{Z}$$

because $(a, b) \equiv (2, 3) \pmod 4$.

These matrices are each *unimodular*, so that Schmal indeed exhibited another integral basis of $\mathbb{K}|\mathbb{Q}$.

An integral basis $\{\mu_1, \mu_2, \mu_3, \mu_4\}$ of the biquadratic field $\mathbb{K} = \mathbb{Q}(\sqrt{a}, \sqrt{b})$ being known, it is no problem to compute its discriminant $d_{\mathbb{K}}$. For $d_{\mathbb{K}}$ is defined in terms of any integral basis $\{\mu_1, \mu_2, \mu_3, \mu_4\}$ of $\mathbb{K}$ as the expression (mentioned already)

$$d_{\mathbb{K}} = \det\left(\mathrm{Tr}_{\mathbb{K}|\mathbb{Q}}(\mu_i\mu_j)_{i,j=1,2,3,4}\right)$$

where $\mathrm{Tr}_{\mathbb{K}|\mathbb{Q}}$ is the *trace* of $\mathbb{K}$ over $\mathbb{Q}$ (see Section B.2). In this way, a somewhat tedious computation (for which we again used SIMATH) leads to (see Williams [238])

Theorem B.8. *The discriminant $d_{\mathbb{K}}$ of the biquadratic field $\mathbb{K} = \mathbb{Q}(\sqrt{a}, \sqrt{b})$ is*

$$d_{\mathbb{K}} = t^2 a'^2 b'^2$$

if $(a, b) \equiv (1, 1) \pmod 4$,

$$d_{\mathbb{K}} = 2^4 t^2 a'^2 b'^2$$

if $(a, b) \equiv (2, 1)$ *or* $(3, 1) \pmod 4$,

$$d_{\mathbb{K}} = 2^6 t^2 a'^2 b'^2$$

if $(a, b) \equiv (2, 3) \pmod 4$.

Here we have chosen a different approach compared to Section B.2, because we first determined an integral basis of $\mathbb{K}$ and then applied it to calculate the discriminant $d_{\mathbb{K}}$ of $\mathbb{K}$ by the above formula.

We mention that the field $\mathbb{K}$ need not have a *relative* integral basis over one of its quadratic subfields. Schmal (see [187], Theorem 8.1) has given a criterion for a biquadratic field $\mathbb{K}$ over $\mathbb{Q}$ to possess a relative integral basis over a quadratic subfield. According to this criterion, for $\mathbb{K} = \mathbb{Q}(\sqrt{a}, \sqrt{b})$ to possess an integral basis over say $\Bbbk_a = \mathbb{Q}(\sqrt{a})$ it is necessary that

$$\sqrt{t} \in \Bbbk_a \quad \text{resp.} \quad \sqrt{2t} \in \Bbbk_a$$

in the first two cases resp. in the third case of Theorem B.8. However, aside from trivial cases this is not satisfied.

The composition formula considering that the quadratic fields $\Bbbk_a = \mathbb{Q}(\sqrt{a})$, $\Bbbk_b = \mathbb{Q}(\sqrt{b})$, $\Bbbk_{ab} = \mathbb{Q}(\sqrt{a} \cdot \sqrt{b}) = \mathbb{Q}(\sqrt{a'b'})$ are each subfields of $\mathbb{K} = \mathbb{Q}(\sqrt{a}, \sqrt{b})$ can be applied to derive Theorem B.8 in the biquadratic situation also. For the composition formula reads in general (see Hasse [92])

$$d_{\mathbb{K}} = \mathcal{N}_{\mathbb{K}_0|\mathbb{Q}} \left(\vartheta_{\mathbb{K}|\mathbb{K}_0}\right) d_{\mathbb{K}_0}^2,$$

where $\mathbb{K}_0 \leq \mathbb{K}$ is here one of the quadratic subfields of $\mathbb{K}$, $\vartheta_{\mathbb{K}|\mathbb{K}_0}$ denotes the relative discriminant of $\mathbb{K}$ with respect to $\mathbb{K}_0$ and $\mathcal{N}_{\mathbb{K}_0|\mathbb{Q}}$ is the *norm* of $\mathbb{K}_0$ over $\mathbb{Q}$. The relative discriminant $\vartheta_{\mathbb{K}|\mathbb{K}_0}$ is a divisor of $\mathbb{K}_0$, but if it is associated to an element of $\mathbb{Q}$, the norm $\mathcal{N}_{\mathbb{K}_0|\mathbb{Q}}$ amounts to the squaring of $\vartheta_{\mathbb{K}|\mathbb{K}_0}$ if $\mathbb{K}_0|\mathbb{Q}$ is quadratic.

(1) If $(a, b) \equiv (1, 1) \pmod 4$, we choose $\mathbb{K}_0 = \mathbb{Q}\left(\sqrt{a}\right) = \Bbbk_a$ and get

$$d_{\Bbbk} = \mathcal{N}_{\Bbbk_a|\mathbb{Q}} \left(\vartheta_{\mathbb{K}|\Bbbk_a}\right) d_{\Bbbk_a}^2 = b'^2 a^2 = t^2 a'^2 b'^2.$$

The relative discriminant is $\vartheta_{\mathbb{K}|\mathbb{K}_a} \cong b'$ because the ramified primes in $\mathbb{K}$ not yet ramified in $\Bbbk_a$ are exactly those contained in b'.

(2) If $(a, b) \equiv (2, 1)$ or $(3, 1) \pmod 4$, we choose $\mathbb{K}_0 = \mathbb{Q}(\sqrt{b}) = \Bbbk_b$ and get

$$d_{\mathbb{K}} = \mathcal{N}_{\Bbbk_b|\mathbb{Q}}\left(\vartheta_{\mathbb{K}|\Bbbk_b}\right) d^2_{\Bbbk_b} = (4a')^2 b^2 = 4^2 t^2 a'^2 b'^2.$$

The relative discriminant is $\vartheta_{\mathbb{K}|\Bbbk_b} \cong 4a'$ because by Theorem B.9 the ramified primes in $\mathbb{K}$ not yet ramified in $\Bbbk_b$ are exactly 2 and the primes dividing a'.

(3) If $(a, b) \equiv (2, 3) \pmod 4$, we choose again $\mathbb{K}_0 = \mathbb{Q}\left(\sqrt{a}\right) = \Bbbk_a$ and get

$$d_{\mathbb{K}} = \mathcal{N}_{\Bbbk_a|\mathbb{Q}}\left(\vartheta_{\mathbb{K}|\Bbbk_a}\right) d^2_{\Bbbk_a} = (2b')^2 (4a)^2 = 2^2 \cdot 4^2 t^2 a'^2 b'^2.$$

The relative discriminant is this time $\vartheta_{\mathbb{K}|\Bbbk_a} \cong 2b'$ because by Theorem B.9 , 2 not only ramifies in $\Bbbk_a|\mathbb{Q}$, but also in $\mathbb{K}|\Bbbk_a$, and the other primes not yet ramified in $\Bbbk_a$ but in $\mathbb{K}$ are those contained in b'.

The discriminant $d_{\mathbb{K}}$ of the biquadratic field $\mathbb{K} = \mathbb{Q}(\sqrt{a}, \sqrt{b})$ can be related to the *inessential discriminant divisor* $\mathfrak{m}(\alpha)$ for a suitable element $\alpha \in \mathbb{K}$ (see Hasse [92]). We choose $\alpha := \sqrt{a} + \sqrt{b}$, remembering that $\mathbb{K} = \mathbb{Q}(\sqrt{a} + \sqrt{b})$. We obtain for the *different* of α

$$D(\alpha) = f'(\alpha) \in \mathbb{K},$$

where $f'(X)$ is the first derivative of the polynomial $f(X) = \mathrm{Irr}(\alpha, \mathbb{Q})(X)$ according to formula (B.16) (which will appear below)

$$f(X) = X^4 - 2(a+b)X^2 + (a-b)^2 \in \mathbb{Z}[X].$$

The discriminant of the element $\alpha \in \mathbb{K}$ is the norm of the different of α:

$$d(\alpha) = \mathcal{N}_{\mathbb{K}|\mathbb{Q}}(D(\alpha)).$$

We get

$$d\left(\sqrt{a} + \sqrt{b}\right) = 2^{12} t^6 a'^2 b'^2 (a' - b')^2,$$

where, as before,

$$t = \gcd(a, b), \quad a' = \frac{a}{t}, \quad b' = \frac{b}{t}.$$

Expressed through a and b, we have

$$d\left(\sqrt{a} + \sqrt{b}\right) = 2^{12} a^2 b^2 (a - b)^2.$$

By Theorem B.8 this yields

$$d\left(\sqrt{a} + \sqrt{b}\right) = \left\{ \begin{array}{c} 2^{12} t^4 (a' - b')^2 \cdot d_{\mathbb{K}} \\ 2^8 t^4 (a' - b')^2 \cdot d_{\mathbb{K}} \\ 2^6 t^4 (a' - b')^2 \cdot d_{\mathbb{K}} \end{array} \right\}$$

for

$$(a,b) \equiv \left\{ \begin{array}{l} (1,1) \pmod 4 \\ (2,1) \text{ or } (3,1) \pmod 4 \\ (2,3) \pmod 4 \end{array} \right\}.$$

Hence the inessential discriminant divisor in these cases is

$$\mathfrak{m}(\sqrt{a}+\sqrt{b}) = \left\{ \begin{array}{l} 2^6 t^2 (a'-b') \\ 2^4 t^2 (a'-b') \\ 2^3 t^2 (a'-b') \end{array} \right\}.$$

The decomposition law in $\mathbb{K}$ reads (see Cohn [36] or Hollinger [1], [100] and Theorem B.5) as follows.

Theorem B.9. (a) *Let p be a (non-archimedean) prime number.*

(1) $p \in \mathbb{P}$, $p \neq 2$:

If $\left(\frac{a}{p}\right) = 1$ *and* $\left(\frac{b}{p}\right) = 1$, $p \cong \mathfrak{p}_1\mathfrak{p}_2\mathfrak{p}_3\mathfrak{p}_4$.

If $\left(\frac{a}{p}\right) = -1$ *and* $\left(\frac{b}{p}\right) = 1$, $p \cong \mathfrak{p}_1\mathfrak{p}_2$.

If $\left(\frac{a}{p}\right) = 0$ *and* $\left(\frac{b}{p}\right) = 1$, $p \cong \mathfrak{p}_1^2\mathfrak{p}_2^2$.

If $\left(\frac{a}{p}\right) = 0$ *and* $\left(\frac{b}{p}\right) = -1$, $p \cong \mathfrak{p}_1^2$.

(2) $p = 2$:

If $a \equiv 1 \pmod 8$ *and* $b \equiv 1 \pmod 8$, $p \cong \mathfrak{p}_1\mathfrak{p}_2\mathfrak{p}_3\mathfrak{p}_4$.
If $a \equiv 5 \pmod 8$ *and* $b \equiv 1 \pmod 8$, $p \cong \mathfrak{p}_1\mathfrak{p}_2$.
If $a \equiv 2, 3 \pmod 4$ *and* $b \equiv 1 \pmod 8$, $p \cong \mathfrak{p}_1^2\mathfrak{p}_2^2$.
If $a \equiv 2, 3 \pmod 4$ *and* $b \equiv 5 \pmod 8$, $p \cong \mathfrak{p}_1^2$.
If $a \equiv 2 \pmod 4$ *and* $b \equiv 3 \pmod 4$, $p \cong \mathfrak{p}_1^4$.

(b) $p = \infty$ *(the archimedean prime):*

If $a > 0$ *and* $b > 0$, $p \cong \mathfrak{p}_1\mathfrak{p}_2\mathfrak{p}_3\mathfrak{p}_4$.
If $a < 0$ *and* $b > 0$, $p \cong \mathfrak{p}_1^2\mathfrak{p}_2^2$.

Here, utilizing again the symmetry in a and b, we have occasionally defined a different $b \in \mathbb{Z}$ (cf. Hollinger [1],[100]). (For instance, we replaced ab by a new b.) Of course, the case $p \cong \mathfrak{p}_1^4 \dots \mathfrak{p}_{2^r-3}^4$ cannot occur if $r = 2$.

We remark that the case where simultaneously

$$a \equiv 2 \pmod 4 \quad \text{and} \quad b \equiv 2 \pmod 4$$

is avoided here (see the justification of Condition (B.2)). If $2|a$ and $2|b$ we replace b by $\tilde{b} := 2^{-2}ab$ and obtain

$$2 \nmid 2^{-2}ab = \tilde{b}.$$

We have then with $\tilde{a} := a$

$$\mathbb{K} = \mathbb{Q}(\sqrt{a}, \sqrt{b}) = \mathbb{Q}(\sqrt{a}, \sqrt{2^{-2}ab}) = \mathbb{Q}\left(\sqrt{\tilde{a}}, \sqrt{\tilde{b}}\,\right)$$

with $\tilde{a} \equiv 2 \pmod 4$ and $\tilde{b} \not\equiv 2 \pmod 4$ (see also Schmal [186], [187]).

We consider now the Galois groups of biquadratic fields.

$$\mathbb{K} = \mathbb{Q}(\sqrt{a}, \sqrt{b}), \quad a, b \in \mathbb{Z}, \text{ both } \neq 1, \text{ and } a \neq b.$$

The quadratic subfields

$$\Bbbk_a := \mathbb{Q}(\sqrt{a}) \quad \text{and} \quad \Bbbk_b := \mathbb{Q}(\sqrt{b})$$

are linearly disjoint over $\mathbb{Q}$, and $\mathbb{K} = \Bbbk_a\Bbbk_b$ is their compositum. Of course, these fields are Galois extensions of $\mathbb{Q}$, and we may also write $\mathbb{K} = \Bbbk_a\left(\sqrt{b}\right) = \Bbbk_b\left(\sqrt{a}\right)$.

If $\sigma_a : \mathbb{K} \to \mathbb{K}$ is given by $\sqrt{b} \mapsto -\sqrt{b}$, $\sigma_a|_{\Bbbk_a} = \mathrm{id}_{\Bbbk_a}$, and $\sigma_b : \mathbb{K} \to \mathbb{K}$ is the automorphism analogously defined, then $\sigma_a = \sigma_2$ and $\sigma_b = \sigma_1$ (see Section B.1). The non-trivial automorphism of $\mathbb{K}_a|\mathbb{Q}$ is σ_b and the non-trivial automorphism of $\mathbb{K}_b|\mathbb{Q}$ is σ_a (by abuse of the notation), and the Galois groups of $\Bbbk_a|\mathbb{Q}$ resp. of $\Bbbk_b|\mathbb{Q}$ are

$$G_b := \langle\sigma_b\rangle \quad \text{resp.} \quad G_a := \langle\sigma_a\rangle.$$

We obtain the Hasse diagrams (see also Section B.1):

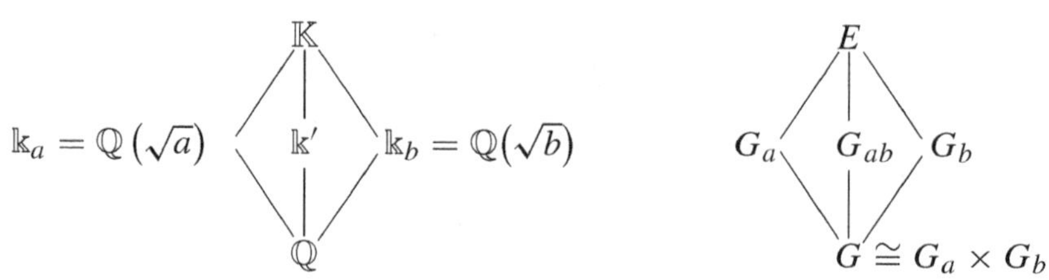

(this time with the notation

$$\Bbbk' := \mathbb{Q}(\sqrt{a'b'}), \quad G_{ab} := \langle\sigma_a \circ \sigma_b\rangle.$$

The unit group is E and G is the Galois group (up to isomorphism)

$$G \cong \langle\sigma_a, \sigma_b\rangle = G_a \times G_b,$$

of the biquadratic field $\mathbb{K}|\mathbb{Q}$. From these diagrams it is immediately clear that $\mathbb{K}$ contains exactly three quadratic subfields of which either all three are real or one is real and the two other are complex. We call $\mathbb{K}$ *totally real* in the first case and *totally complex* in the second. In the first case, we may assume that $a, b > 0$, in the second that $a, b < 0$.

We have also

$$\mathbb{K} = \mathbb{Q}\big(\sqrt{a} + \sqrt{b}\big).$$

For, by the theorem on the primitive element, the biquadratic field has the form

$$\mathbb{K} = \mathbb{Q}(\alpha)$$

with an element $\alpha \in \mathbb{C}, \alpha \notin \mathbb{Q}$. The degree of $\mathbb{K}|\mathbb{Q}$ is 4 so that $\mathbb{K}$ can be regarded as a quartic field. In general, $\mathbb{K}$ has the minimal polynomial over $\mathbb{Q}$

$$\operatorname{Irr}(\alpha, \mathbb{Q})(X) = X^4 - s_1X^3 + s_2X^2 - s_3X + s_4 \in \mathbb{Z}[X]. \tag{B.15}$$

If $\alpha = \sqrt{a} + \sqrt{b}$, this amounts to

$$\operatorname{Irr}\big(\sqrt{a} + \sqrt{b}, \mathbb{Q}\big)(X) = X^4 - 2(a+b)X^2 + (a-b)^2 \tag{B.16}$$

with the roots

$$\sqrt{a} + \sqrt{b}, \sqrt{a} - \sqrt{b}, -\sqrt{a} + \sqrt{b}, -\sqrt{a} - \sqrt{b}.$$

In particular,

$$\begin{aligned}
\mathrm{id} &: \sqrt{a} + \sqrt{b} \mapsto \sqrt{a} + \sqrt{b},\\
\sigma_a &: \sqrt{a} + \sqrt{b} \mapsto \sqrt{a} - \sqrt{b},\\
\sigma_b &: \sqrt{a} + \sqrt{b} \mapsto -\sqrt{a} + \sqrt{b},\\
\sigma_a \circ \sigma_b &: \sqrt{a} + \sqrt{b} \mapsto -\sqrt{a} - \sqrt{b}.
\end{aligned}$$

B.6 Totally real and totally complex biquadratic fields

The biquadratic field $\mathbb{K} = \mathbb{Q}\big(\sqrt{a}, \sqrt{b}\big)$ has degree

$$[\mathbb{K} : \mathbb{Q}] = 2^2$$

and hence is generated by one root α of the quartic polynomial (B.15) , i.e. of

$$\operatorname{Irr}(\alpha, \mathbb{Q})(X) := f(X) = X^4 - s_1X^3 + s_2X^2 - s_3X + s_4 \quad (s_i \in \mathbb{Q}).$$

Let $\alpha_1 = \alpha, \alpha_2, \alpha_3, \alpha_4 \in \mathbb{C}$ be the distinct roots of the polynomial $f(X) = \operatorname{Irr}(\alpha, \mathbb{Q})(X)$. Put

$$\mathbb{F} := \mathbb{Q}\,(\alpha_1, \alpha_2, \alpha_3, \alpha_4)\,.$$

The splitting field $\mathbb{F}$ of f is Galois over $\mathbb{Q}$ with Galois group $G := G_{\mathbb{F}}$, a subgroup of the symmetric group S_4 on four letters:

$$G \leq S_4.$$

In fact, this holds for an arbitrary quartic field $\mathbb{K}$. If $\mathbb{K}$ is biquadratic, it is, of course, Galois over $\mathbb{Q}$ and its Galois group is the *Kleinian* 4-*group*:

$$G := V_4 = \langle \sigma_a, \sigma_b \rangle = \{(1), (12)(34), (13)(24), (14)(23)\} \cong \mathbb{Z}/2\mathbb{Z} \times \mathbb{Z}/2\mathbb{Z},$$

where $(i_1 ... i_n)$ is the cycle notation.

V_4 is a normal subgroup of S_4:

$$V_4 \trianglelefteq S_4.$$

Put

$$\begin{aligned}\beta_1 &:= \alpha_1\alpha_2 + \alpha_3\alpha_4, \\ \beta_2 &:= \alpha_1\alpha_3 + \alpha_2\alpha_4, \\ \beta_3 &:= \alpha_1\alpha_4 + \alpha_2\alpha_3.\end{aligned}$$

Then the polynomial

$$\begin{aligned}g(X) &= (X - \beta_1)(X - \beta_2)(X - \beta_3) \\ &= X^3 - s_2 X^2 + (s_1 s_3 - 4s_4)X - s_1^2 s_4 + 4 s_2 s_4 - s_3^2 \in \mathbb{Z}[X] \quad \text{(B.17)}\end{aligned}$$

is the *cubic resolvent* of $f(X)$ (see e.g. Stein [215] or Hungerford [101], v. d. Waerden [229]).

According to (B.16) the cubic resolvent of $\mathrm{Irr}(\sqrt{a} + \sqrt{b}, \mathbb{Q})(X) = f(X)$ is thus

$$g(X) = X^3 + 2(a + b)X^2 - 4(a - b)^2 X - 8(a^2 - b^2)(a - b) \in \mathbb{Z}[X]. \quad \text{(B.18)}$$

Let

$$\mathbb{L} := \mathbb{Q}(\beta_1, \beta_2, \beta_3).$$

The splitting field $\mathbb{L}$ of the cubic resolvent $g(X)$ of $f(X)$ is Galois over $\mathbb{Q}$ with Galois group

$$\bar{G} = S_4/V_4 \cong S_3.$$

More generally, if we replace $\mathbb{Q}$ by any ground field $\mathbb{K}_0$ of characteristic $\neq 2, 3$ and consider a general irreducible separable quartic polynomial

$$f(X) \in \mathbb{K}_0[X]$$

with distinct roots $\alpha_1, \alpha_2, \alpha_3, \alpha_4$ and splitting field

$$\mathbb{F} := \mathbb{K}_0(\alpha_1, \alpha_2, \alpha_3, \alpha_4)$$

and with cubic resolvent $g(X) = (X - \beta_1)(X - \beta_2)(X - \beta_3)$ as well as splitting field

$$\mathbb{L} = \mathbb{K}_0(\beta_1, \beta_2, \beta_3),$$

we obtain the following result (see Stein [215] or Hungerford [101]).

Proposition B.10. *Let $G = Gal(\mathbb{F}|\mathbb{K}_0)$ be the Galois group of $\mathbb{F}|\mathbb{K}_0$ with $f(X) \in \mathbb{K}_0[X]$, a separable quartic polynomial which is irreducible over $\mathbb{K}_0$ (except in the case $m = 2$ below). We denote by*

$$m := [\mathbb{L} : \mathbb{K}_0]$$

the field degree of $\mathbb{L}|\mathbb{K}_0$. Then

$$\begin{aligned} m = 6 &\Leftrightarrow G = S_4, \\ m = 4 &\Leftrightarrow G = A_4, \\ m = 2 &\Leftrightarrow G = D_4 \text{ or } Z_4, \\ m = 1 &\Leftrightarrow G = V_4. \end{aligned}$$

Here, A_4 is the alternating group on four letters so that $A_4 \trianglelefteq S_4$, D_4 is the dihedral group of order 8 (occurring if f is irreducible over $\mathbb{K}_0$) and Z_4 is the cyclic group of order 4 (occurring otherwise).

For a proof, we refer to Stein [215] or Hungerford [101].

The corresponding Hasse diagram in the general situation $G = S_4$ (where $(i_1 \dots i_n)$ is again the cycle notation) reads for $\mathbb{K}_0 = \mathbb{Q}$:

$$\begin{array}{lcl}
\mathbb{F} = \mathbb{Q}(\alpha_1, \alpha_2, \alpha_3, \alpha_4) & ------ & E = \langle (1) \rangle \\
\quad \Big| \, 2 & & \quad \Big| \, 2 \\
\mathbb{F}_0 = \mathbb{Q}(\beta_1, \beta_2, \beta_3, \alpha_1\alpha_2, \alpha_1 + \alpha_2) & --- & Z_2 = \langle (12)(34) \rangle \\
\quad \Big| \, 2 & & \quad \Big| \, 2 \\
\mathbb{L} = \mathbb{Q}(\beta_1, \beta_2, \beta_3) & ------- & V_4 = \langle (12)(34), (13)(24) \rangle \\
\quad \Big| \, 3 & & \quad \Big| \, 3 \\
\mathbb{L}_0 = \mathbb{Q}\left(\sqrt{\bar{d}_{\mathbb{F}}}\right) & ------- & A_4 = \langle (123), (124) \rangle \\
\quad \Big| \, 2 & & \quad \Big| \, 2 \\
\mathbb{Q} & ---------- & S_4 = \langle (12), (13), (14) \rangle
\end{array}$$

The quantity $\bar{d}_{\mathbb{F}} = d_f$ is the *discriminant* of the 4-th degree polynomial $f(X) = (X - \alpha_1)(X - \alpha_2)(X - \alpha_3)(X - \alpha_4)$ (see v. d. Waerden [229]). Since

$$\begin{aligned} \bar{d}_{\mathbb{F}} &= (\alpha_1 - \alpha_2)^2(\alpha_1 - \alpha_3)^2(\alpha_1 - \alpha_4)^2(\alpha_2 - \alpha_3)^2(\alpha_2 - \alpha_4)^2(\alpha_3 - \alpha_4)^2 \\ &= (\beta_1 - \beta_2)^2(\beta_1 - \beta_3)^2(\beta_2 - \beta_3)^2 \\ &= \bar{d}_{\mathbb{L}}, \end{aligned}$$

$\bar{d}_{\mathbb{F}} = d_f$ is hence the discriminant of the polynomial $f(X)$ defined by (B.15), and $\bar{d}_{\mathbb{L}} = d_g$ is the discriminant of its cubic resolvent $g(X)$ exhibited in (B.17). The identity $\bar{d}_{\mathbb{F}} = \bar{d}_{\mathbb{L}}$ means simply that $f(X)$ and $g(X)$ have the same discriminant:

$$d_f = d_g.$$

In this connection, E is again the unit group, Z_2 is a cyclic group of order 2, $V_4 \cong \mathbb{Z}/2\mathbb{Z} \times \mathbb{Z}/2\mathbb{Z}$ the Kleinian 4-group, A_4 the alternating and S_4 the symmetric group on 4 letters.

The chain of groups

$$S_4 \supset A_4 \supset V_4 \supset Z_2 \supset E$$

corresponding to the chain of fields

$$\mathbb{Q} \subset \mathbb{L}_0 \subset \mathbb{L} \subset \mathbb{F}_0 \subset \mathbb{F}$$

is a *composition series* because the indices of the factor groups

$$S_4|A_4, \;\; A_4|V_4, \;\; V_4|Z_2, \;\; Z_2|E$$

are the primes 2 or 3 (see v. d. Waerden [229]).

The only non-trivial normal subgroups of S_4 are V_4 and A_4. The group A_4 can be shown to possess no subgroup of order 6. (We remark that the field $\mathbb{Q}$ can again be replaced by any field $\mathbb{K}_0$ of characteristic $\neq 2, 3$.)

In our case, we have the biquadratic field of degree 4

$$\mathbb{K} = \mathbb{Q}(\sqrt{a}, \sqrt{b})$$

with $a, b \in \mathbb{Z}$ square-free and such that say $\sqrt{b} \notin \mathbb{Q}\left(\sqrt{a}\right)$. Then $\mathbb{K}$ contains the quadratic fields

$$\Bbbk_a = \mathbb{Q}\left(\sqrt{a}\right), \quad \Bbbk_b = \mathbb{Q}(\sqrt{b}) \quad \text{and} \quad \Bbbk_{ab} = \mathbb{Q}(\sqrt{a'b'}).$$

Hence its Galois group over $\mathbb{Q}$ is

$$G \cong V_4$$

(see also Proposition B.10, Equation (B.16) and Exercise 9 in Hungerford [101], Section 4 of Chapter V).

We have the identities

$$\mathbb{F} = \mathbb{K}, \quad \mathbb{L} = \mathbb{L}_0 = \mathbb{K}_0 = \mathbb{Q}.$$

These identities can also be obtained from the fact that

$$\mathbb{K} = \mathbb{Q}(\sqrt{a} + \sqrt{b})$$

with generating polynomial (see (B.16))

$$f(X) = X^4 - 2(a+b)X^2 + (a-b)^2$$

and cubic resolvent (see (B.18))

$$g(X) = X^3 + 2(a+b)X^2 - 4(a-b)^2 X - 8(a+b)(a-b)^2.$$

In any case in the notation

$$\begin{aligned} \alpha = \alpha_1 &= \sqrt{a} + \sqrt{b} = \alpha^{\mathrm{id}}, \\ \alpha_2 &= \sqrt{a} - \sqrt{b} = \alpha^{\sigma_a}, \\ \alpha_3 &= -\sqrt{a} + \sqrt{b} = \alpha^{\sigma_b}, \\ \alpha_4 &= -\sqrt{a} - \sqrt{b} = \alpha^{\sigma_a \circ \sigma_b}, \end{aligned}$$

we get

$$\begin{aligned} \beta_1 &= 2(a-b), \\ \beta_2 &= -2(a-b), \\ \beta_3 &= -2(a+b). \end{aligned}$$

We consider in more detail two cases of Proposition B.10. Let again

$$\mathbb{F} = \mathbb{Q}(\alpha_1, \alpha_2, \alpha_3, \alpha_4), \quad \mathbb{L} = \mathbb{Q}(\beta_1, \beta_2, \beta_3), \quad \mathbb{K}_0 = \mathbb{Q},$$

$$m = [\mathbb{L} : \mathbb{Q}]$$

and

$$G = \mathrm{Gal}(\mathbb{F}|\mathbb{Q}).$$

Then Proposition B.10 yields as one case

(1) $m = 6 \Leftrightarrow G = S_4$.

We obtain the simple Hasse diagrams for fields and groups:

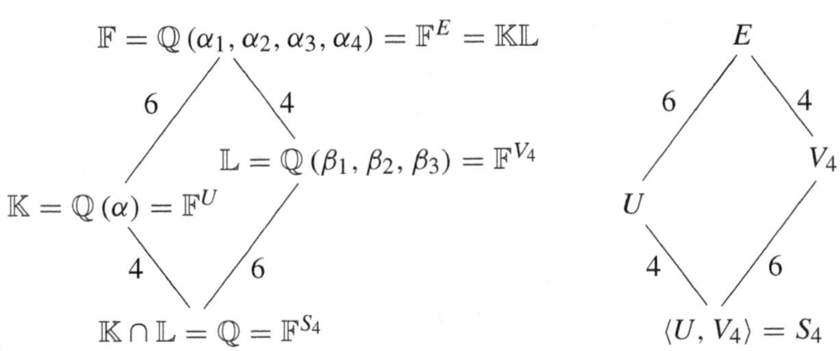

Here, as usual, $\mathbb{F}^H$ stands for the fixed field of the group H. Furthermore, as can be shown, the subgroup

$$U = \langle (23), (34) \rangle \leq S_4$$

is generated by the transpositions (23) and (34). We have added to the diagrams the field degrees resp. the group indices.

Observe that the alternating group A_4 cannot be inserted on the left of the group diagram because A_4 does not contain a subgroup of order 6.

The second case of Proposition B.10 we consider is

(2) $m = 1 \Leftrightarrow G = V_4$.

This is the case of special interest to us. The biquadratic field is

$$\mathbb{K} = \mathbb{Q}(\sqrt{a}, \sqrt{b}) = \mathbb{Q}(\sqrt{a} + \sqrt{b})$$

such that $a, b \in \mathbb{Z}$ with

$$[\mathbb{K} : \mathbb{Q}] = 4.$$

Here $\mathbb{K} = \mathbb{F}$ is its own splitting field and $\mathbb{L} = \mathbb{Q}$ is the basic field of rationals. This way we have the trivial diagrams:

$$\begin{array}{l} \mathbb{F} = \mathbb{K} = \mathbb{Q}(\sqrt{a} + \sqrt{b}) = \mathbb{F}^E \\ \quad \Big| \; 4 \\ \mathbb{K} \cap \mathbb{L} = \mathbb{F}^{V_4} = \mathbb{Q} \end{array} \qquad\qquad \begin{array}{c} E \\ \Big| \; 4 \\ V_4 \end{array}$$

We recall that the group S_4 has only the alternating group A_4 and the Kleinian 4-group V_4 as non-trivial normal subgroups, and that the factor groups are (up to isomorphisms)

$$S_4/A_4 \cong Z_2, \quad S_4/V_4 \cong S_3.$$

Hence, in the former case, $\mathbb{L}|\mathbb{Q}$ is always a Galois extension with Galois group $\mathrm{Gal}(\mathbb{L}|\mathbb{Q}) \cong S_3$.

We have for the various groups the generators

$$\begin{aligned} S_4 &= \langle (12), (13), (14) \rangle, \\ A_4 &= \langle (12)(13), (12)(14), (13)(14) \rangle, \\ V_4 &= \langle (12)(34), (13)(24) \rangle. \end{aligned}$$

If we put

$$\begin{aligned} Z_2 &= \langle (12)(34) \rangle, \\ Z_2' &= \langle (13)(24) \rangle, \\ Z_2'' &= \langle (14)(23) \rangle, \end{aligned}$$

these latter groups are all isomorphic of order

$$|Z_2| = |Z_2'| = |Z_2''| = 2.$$

Since $|U| = 6$ and

$$|A_4| = 12, \quad |V_4| = 4,$$

U is a non-normal subgroup of S_4 so that, in the former case (1), $\mathbb{K}$ is not normal over $\mathbb{Q}$.

Of course (see the last case),

$$Z_2, Z_2', Z_2'' \trianglelefteq V_4;$$

these groups are normal subgroups of V_4. We could insert these groups and their fields in the diagrams of the last case (2) noticing that

$$\begin{aligned} 2\sqrt{a} &= \alpha_1 + \alpha_2 = -(\alpha_3 + \alpha_4), \\ 2\sqrt{b} &= \alpha_1 - \alpha_2 = (\alpha_3 - \alpha_4), \\ 4\sqrt{ab} &= \alpha_1^2 - \alpha_2^2 = -(\alpha_3^2 - \alpha_4^2), \end{aligned}$$

so that

$$\begin{aligned} \Bbbk_a &= \mathbb{Q}\left(\sqrt{a}\right) = \mathbb{F}^{Z_2}, \\ \Bbbk_b &= \mathbb{Q}(\sqrt{b}) = \mathbb{F}^{Z_2'}, \\ \Bbbk_{ab} &= \mathbb{Q}(\sqrt{ab}) = \mathbb{Q}(\sqrt{a'b'}) = \mathbb{F}^{Z_2''}. \end{aligned}$$

When is now a biquadratic field $\mathbb{K} = \mathbb{Q}(\sqrt{a}, \sqrt{b})$ totally real or totally complex?

If $\mathbb{K}$ is totally real, all the roots of a generating 4-th degree polynomial $f(X)$ (see (B.15)) must be real:

$$\alpha_1, \alpha_2, \alpha_3, \alpha_4 \in \mathbb{R}.$$

When is this condition satisfied? The answer is given by Delone and Faddeev [46] (see also Stein [215]):

Theorem B.11. *The polynomial $f(X)$ (according to (B.15)) has real roots*

$$\alpha_1, \alpha_2, \alpha_3, \alpha_4 \in \mathbb{R}$$

if it satisfies the relations

(a) $3s_1^2 - 8s_2 > 0$,

(b) $s_1^2 s_2 - s_2^2 - \dfrac{3}{16} s_1^4 - s_1 s_3 + 4s_4 > 0$,

(c) $4(s_2^2 - 3s_1 s_3 + 12 s_4)^3 - (2s_2^3 - 72 s_2 s_4 + 27 s_1^2 s_4 - 9 s_1 s_2 s_3 + 27 s_3^2)^2 > 0.$

Condition (c) can be simplified in the following way. The discriminant of the polynomial $f(x)$ is

$$\begin{aligned} d_f = & -3^3 s_1^4 s_4^2 + 2 \cdot 3^2 s_1^3 s_2 s_3 s_4 - 2^2 s_1^3 s_3^3 - 2^2 s_1^2 s_2^3 s_4 + s_1^2 s_2^2 s_3^2 + 2^4 \cdot 3^2 s_1^2 s_2 s_4^2 \\ & -2 \cdot 3 s_1^2 s_3^2 s_4 - 2^4 \cdot 5 s_1 s_2^2 s_3 s_4 + 2 \cdot 3^2 s_1 s_2 s_3^3 - 2^6 \cdot 3 s_1 s_3 s_4^2 + 2^4 s_2^4 s_4 \\ & -2^2 s_2^3 s_3^2 - 2^7 s_2^2 s_4^2 + 2^4 3^2 s_2 s_3^2 s_4 - 3^3 s_3^4 + 2^8 s_4^3. \end{aligned}$$

We have the relation

$$3^3 d_f = 2^2 (s_2^2 - 3 s_1 s_3 + 2^2 3 s_4)^3 - (2 s_2^3 - 2^3 3^2 s_2 s_4 + 3^3 s_1^2 s_4 - 3^2 s_1 s_2 s_3 + 3^3 s_3^2)^2.$$

Thus Condition (c) means simply that

(c′) $d_f > 0$.

If $f(X)$ has the special form (B.16) and its cubic resolvent $g(X)$ is thus given by (B.18), we have

$$s_1 = s_3 = 0, \quad s_2 = -2(a+b), \quad s_4 = (a-b)^2.$$

Then the Conditions (a), (b), (c) amount to

(a) $a + b > 0$ (note that a and b can be interchanged)

(b) $|a + b| > |a - b|$

(c′) $a^2 b^2 > 0$

Of course, (c′) is always satisfied, whereas (a) and (b) are tantamount to $a > 0$ and $b > 0$.

The elliptic curve

$$E_f : Y^2 = X^3 + (s_2^2 - 3 s_1 s_3 + 12 s_4) X + \sqrt{d_f}$$

has discriminant

$$\Delta_f = -16(4(s_2^2 - 3 s_1 s_3 + 12 s_4) + 27 d_f).$$

Hence d_f is a square in $\mathbb{Q}$ if and only if E_f is defined over $\mathbb{Q}$. This is of course the case if $f(X)$ has the special form (B.16), that is, if $\mathbb{K}$ is biquadratic. For we have then

$$d_f = 2^{12} a^2 b^2 (a-b)^2 = d_g$$

which was mentioned already above.

So when does an irreducible quartic polynomial $f(X)$ define a totally real biquadratic field $\mathbb{K}$? Is it sufficient that its roots $\alpha_1, \alpha_2, \alpha_3, \alpha_4$ are real?

Corollary B.12. *A quartic polynomial* $f(X) \in \mathbb{Q}[X]$*, given by* (B.15)*, defines a totally real biquadratic number field*

$$\mathbb{K} = \mathbb{Q}(\alpha), \quad (\alpha = \alpha_1 \in \mathbb{R} \text{ a root of } f(X), \text{ that is, } \mathrm{Irr}(\alpha, \mathbb{Q})(X) = f(X))$$

if the above Conditions (a), (b), (c′) *are satisfied.*

Moreover it can be shown (see Stein [215]) that the following assertion is true:

Theorem B.13. *If (and only if) the discriminant* d_f *of the polynomial* $f(X)$ *(see* (B.15)) *is a square and if the cubic resolvent* $g(X)$ *(see* (B.17)) *of* $f(X)$ *has all its roots in* $\mathbb{Q}$*, then the Galois group of the splitting field* $\mathbb{F}$ *of* $\mathbb{K}$ *is*

$$G = \mathrm{Gal}(\mathbb{F}|\mathbb{Q}) = V_4.$$

Of course, for a biquadratic field $\mathbb{K} = \mathbb{Q}(\sqrt{a}, \sqrt{b})$, $a, b \in \mathbb{Z}$, $b \notin \mathbb{Q}(\sqrt{a})$, the discriminant d_f of $f(X)$ is always a square in $\mathbb{Q}$ and the cubic resolvent $g(X)$ of $f(X)$ has three rational roots.

This theorem is especially interesting if $\mathbb{K}$ is its own splitting field, i.e. if $\mathbb{F} = \mathbb{K}$, which holds if $\mathbb{K}$ is biquadratic.

Examples of totally real quartic fields $\mathbb{K}$ defined by a polynomial (B.15) whose Galois group is the Kleinian 4-group V_4 can be taken from a table in Delone and Faddeev [46]:

$$\begin{aligned}(s_1, s_2, s_3, s_4) = \; & (0, 6, 0, 4),\ (0, 4, 0, 1),\ (2, 7, -8, 1), \\ & (0, 9, 0, 4),\ (0, 5, 0, 1),\ (0, 11, 0, 9).\end{aligned}$$

B.7 Exercises

1) Let $\mathbb{K} \subseteq \mathbb{K}' \subseteq \mathbb{K}''$ be extensions of Galois number fields of ramification orders

$$e' = e(\mathbb{K}'|\mathbb{K}), \quad e'' = e(\mathbb{K}''|\mathbb{K}'), \quad e = e(\mathbb{K}''|\mathbb{K})$$

and residue degrees

$$f' = f(\mathbb{K}'|\mathbb{K}), \quad f'' = f(\mathbb{K}''|\mathbb{K}'), \quad f = f(\mathbb{K}''|\mathbb{K})$$

so that the field degrees are

$$n' = [\mathbb{K}' : \mathbb{K}] = g'e'f', \quad n'' = [\mathbb{K}'' : \mathbb{K}'] = g''e''f'', \quad n = [\mathbb{K}'' : \mathbb{K}] = gef,$$

where g', g'', g are the numbers of primes of $\mathbb{K}'|\mathbb{K}$, $\mathbb{K}''|\mathbb{K}'$, $\mathbb{K}''|\mathbb{K}$ lying over a prime $\mathfrak{p}$ of $\mathbb{K}$ resp. $\mathfrak{p}'$ of $\mathbb{K}'$.

a) Prove that

$$e = e'e'' \quad \text{and} \quad f = f'f''.$$

b) Corresponding relations hold if the extensions are non-Galois. Generalize your proof to this case.

2) Carry out the proof of Theorem B.2 in the case

$$(a, b) \equiv (1, 1) \pmod 4, (a', b') \equiv (3, 3) \pmod 4.$$

3) How can one say when a general quartic (separable) polynomial $f(X)$ according to (B.15) generates a totally real or totally complex field

$$\mathbb{K} = \mathbb{Q}(\alpha),$$

where $\alpha = \alpha_1 \in \mathbb{C}$ is a root of $f(X)$?

Bibliography

[1] Abel-Hollinger, C. S., Zimmer, H. G.: Torsion groups of elliptic curves with integral j-invariant over multiquadratic fields. In: *Proceedings of the Int. Conf. Number Theoretic and Algebraic Methods in Computer Science, Moscow, 1993*, A. J. van der Poorten, I. Shparlinski, H. G. Zimmer (ed.), 69–87, World Sci. Publishing, 1993.

[2] Agrawal, M., Kayal, N., Saxena, N.: PRIMES is in P. Preprint, 2003.

[3] Ahlfors, L. V.: *Complex analysis.* Intern. Series in Pure and Applied Math., McGraw-Hill Book Company, 1966.

[4] Arthaud, N.: On Birch and Swinnerton-Dyer's conjecture for elliptic curves with complex multiplication. *Comp. Math.* **37**, 209–232, 1978.

[5] Baker, A.: Linear forms in the logarithms of algebraic numbers I, II, II, IV. *Mathematica* **13**, 204–216, 1966; **14**, 102–107, 220–228, 1967; **15**, 204–216, 1968.

[6] Baker, A.: The diophantine equation $y^2 = ax^3 + bx^2 + cx + d$. *J. London Math. Soc.* **43**, 1–9, 1968.

[7] Baker, A.: Contribution to the theory of Diophantine equations. *Philos. Trans. Roy. Soc. London Ser.* A **263**, 173–208, 1968.

[8] Baker, A.: Bounds for the solutions of the hyperelliptic equation. *Math. Proc. Cambridge Philos. Soc.* **65**, 439–444, 1969.

[9] Baker, A., Davenport, H.: The equations $3x^2 - 2 = y^2$ and $8x^2 - 7 = z^2$. *Quart. J. Math. Oxford Ser.* (2) **20**, 129–137, 1969.

[10] Baker, A., Coates, J.: Integer points on curves of genus 1. *Math. Proc. Cambridge Philos. Soc.* **67**, 595–602, 1970.

[11] Baker, A.: *Transcendental Number Theory.* Cambridge University Press, 1975.

[12] Baker, A., Wüstholz, G.: Linear forms and group varieties. *J. Reine Angew Math.* **442**, 19–62, 1993.

[13] Becker, T., Weispfennig, V.: *Gröbner Bases. A computational approach to commutative algebra.* Springer-Verlag 1993.

[14] Birch, B. J., Swinnerton-Dyer, H. P. F.: Notes on elliptic curves I. *J. Reine Angew. Math.* **212**, 7–25, 1963.

[15] Birch, B. J., Swinnerton-Dyer, H. P. F.: Notes on elliptic curves II. *J. Reine Angew. Math.* **218**, 79–108, 1965.

[16] Birch, B. J.: Heegner points on elliptic curves. *Sympos. Math. INDAM Rome* **15**, 441–445, 1975.

[17] Birch B. J., Stephens, N. M.: Heegner's construction of points on the curve $y^2 = x^3 - 1728e^3$. In: *Seminar on number theory (Paris, 1981/1982)*, 1–19, Progr. Math. 38, Birkhäuser Verlag, 1983.

[18] Blake, I. F., Seroussi, G., Smart, N. P.: *Elliptic curves in cryptography.* London Math. Soc. Lecture Notes Ser. 265, Cambridge University Press, 1999.

[19] Bloch, S.: A note on height pairings, Tamagawa numbers, and the Birch and Swinnerton-Dyer conjecture. *Invent. Math.* **58**, 65–76, 1980.

[20] Bremner, A., Stroeker, R. J., Tzanakis, N.: On sums of consecutive squares. *J. Number Theory* **62**, 39–70, 1997.

[21] Bronstein, I. N., Semendjajew, K. A.: *Taschenbuch der Mathematik*. Teubner Verlag, 1996.

[22] Brumer, A., McGuinnes, O.: The behavior of the Mordell–Weil group of elliptic curves. *Bull. Amer. Math. Soc.* **23**, 375–382, 1990.

[23] Bugeaud, Y., Győry, K.: Bounds for the solutions of Thue–Mahler equations and norm form equations. *Acta Arith.* **74**, 273–292, 1996.

[24] Bugeaud, Y.: Bounds for the solutions of superelliptic equations. *Compositio Math.* **107**, 187–219, 1997.

[25] Cassels, J. W. S.: A note on the division values of $\wp(u)$. *Math. Proc. Cambridge Philos. Soc.* **45**, 167–172, 1949.

[26] Cassels, J. W. S.: *An introduction to the geometry of numbers*. Springer-Verlag, 1959.

[27] Cassels, J. W. S.: Diophantine equations with special reference to elliptic curves. *J. London Math. Soc.* **41**, 193–291, 1966.

[28] Cassels, J. W. S.: *Rational quadratic forms*. London Math. Soc. Monogr. Ser. 13, Academic Press, 1978.

[29] Cassels, J. W. S.: The Mordell–Weil group of curves of genus 2. In: *Arithmetic and Geometry, Papers Dedicated to I. R. Shafarevich on the Occasion of His Sixtieth Birthday, Vol. I Arithmetic*, 27–60, Progr. Math. 35, Birkhäuser Verlag, 1983.

[30] Cassels, J. W. S.: *Lectures on elliptic curves*. London Math. Soc. Stud. Texts 24, Cambridge University Press, 1991.

[31] Chahal, J. S.: *Topics in number theory*. Univ. Ser. Math., Plenum Press, 1988.

[32] Coates, J.: An effective p-adic analogue of a theorem of Thue III; the diophantine equation $y^2 = x^3 + k$. *Acta. Arith.* **74**, 425–435, 1970.

[33] Coates, J., Wiles, A.: On the conjecture of Birch and Swinnerton-Dyer. *Invent. Math.* **39**, 223–251, 1977.

[34] Cohen, H.: *A Course in Computational Algebraic Number Theory*. Grad. Texts in Math. 138, Springer-Verlag, 1993.

[35] Cohen, H., Miyaji, A., Ono, T.: Efficient Elliptic Curve Exponentiation Using Mixed Coordinates. In: *Advances in Cryptology - Asiacrypt'98*, Lecture Notes in Comput. Sci. 1514, 51–65, Springer-Verlag, 1998.

[36] Cohn, H.: *A Classical Invitation to Algebraic Numbers and Class Fields*. Universitext, Springer-Verlag, 1978.

[37] Connell, J.: Addendum to a paper of Harada and Lang. *J. Algebra* **145**, 463–467, 1992.

[38] Conrad, B., Diamond, F., Taylor, R.: Modularity of certain potentially Barsotti-Tate Galois representations. *J. Amer. Math. Soc.* **12**, 521–567, 1999.

[39] Couveignes, J.-M., Morain, F.: Schoof's algorithm and isogeny cycles. In: *ANTS-1: Algorithmic Number Theory'94,* Lecture Notes in Comput. Sci. 877, 43–58, Springer-Verlag, 1994.

[40] Cox, D. A.: *Primes of the form* $x^2 + ny^2$. J. Wiley & Sons, 1989.

[41] Cremona, J. E., Whitley, E.: Periods of cusp forms and elliptic curves over imaginary quadratic fields. *Math. Comp.* **62**, 407–429, 1994.

[42] Cremona, J. E.: *Algorithms for modular elliptic curves. Second edition.* Cambridge University Press, 1997.

[43] Cremona, J. E., Serf, P.: Computing the rank of elliptic curves over real quadratic number field of class number 1. *Math. Comp.* **68**, 1187–1200, 1999.

[44] Darmon, H.: A proof of the full Shimura-Taniyama-Weil conjecture is announced. *Notices Amer. Math. Soc.* **46**, 1397–1401, 1999.

[45] David, S.: Minorations de formes linéaires de logarithmes elliptiques. *Mém. Soc. Math. France* **62**, 1995.

[46] Delone, B. N., Faddeev, D. K.: *The Theory of Irrationalities of the Third Degree.* Amer. Math. Soc. Transl. Ser. 10, 1964.

[47] Deuring, M.: Invarianten und Normalformen elliptischer Funktionenkörper. *Math. Z.* **47**, 47–56, 1940.

[48] Deuring, M.: Die Typen der Multiplikatorenringe elliptischer Funktionenkörper. *Abh. Math. Sem. Hamburg* **14**, 197–272, 1941.

[49] Deuring, M.: Die Zetafunktion einer algebraischen Kurve vom Geschlecht Eins i. *Nachr. Akad. Wiss. Göttingen*, 85–94, 1953.

[50] Deuring, M.: Die Zetafunktion einer algebraischen Kurve vom Geschlecht Eins ii. *Nachr. Akad. Wiss. Göttingen*, 13–42, 1955.

[51] Deuring, M.: On the zeta-function of an elliptic function field with complex multiplication. In: *Proc. Internat. Sympos. Alg. Number Theory, Tokyo, Nikko*, 47–50, 1955.

[52] Deuring, M.: Die Zetafunktion einer algebraischen Kurve vom Geschlecht Eins iii. *Nachr. Akad. Wiss. Göttingen*, 37–76, 1956.

[53] Deuring, M.: Die Zetafunktion einer algebraischen Kurve vom Geschlecht Eins iv. *Nachr. Akad. Wiss. Göttingen*, 55–80, 1957.

[54] Deuring, M.: Die Klassenkörper der komplexen Multiplikation. In: *Enz. der Math. Wiss. mit Einschl. ihrer Anw.* **1, 10**, Teil 2, 1958.

[55] Doud, D.: A procedure to calculate torsion of elliptic curves over $\mathbb{Q}$. *Manuscripta Math.* **95**, 463–469, 1998.

[56] Edixhoven, B.: Rational torsion points on elliptic curves over number fields (after Kamienny and Mazur) *Astérisque* **227**, 209–227, 1994.

[57] ElGamal, T.: A public key cryptosystem and signature scheme based on discrete logarithms. *IEEE Trans. Inform. Theory* **31** (4), 469–472, 1985.

[58] Elistratov, I. V.: Elementary proof of Hasse's theorem. In: *Studies in Number Theory* **I** (in Russian), 21–26, 1966.

[59] Ellison, W. J.: Recipes for solving diophantine problems by Baker's method. *Publ. Math. Univ. Bordeaux Année I*, no. 1, Exp. no. 3, 10 pp., (1972/73)

[60] Elkies, N. D.: The existence of infinitely many supersingular primes for every elliptic curve over $\mathbb{Q}$. *Invent. Math.* **89**, 561–567, 1987.

[61] Evertse, J.-H., Győry, K., Stewart, C. L., Tijdeman, R.: S-unit equations and their applications. In: *New Advances in Transcendence Theory* (A. Baker ed.), Cambridge University Press., 110–174, 1988.

[62] Fermigier, S.: Une courbe elliptique définie sur $\mathbb{Q}$ de rang ≥ 22. *Acta Arith.* **82**, 359–363, 1997.

[63] Fischer, W., Lieb, I.: *Funktionentheorie* Vieweg Verlag, 1980.

[64] Folz, H. G.: *Ein Beschränktheitssatz für die Torsion von 2–defizienten elliptischen Kurven über algebraischen Zahlkörpern*. Dissertation, Universität des Saarlandes, Saarbrücken, 1985.

[65] Folz, H. G., Zimmer, H. G.: A boundedness theorem for the torsion of elliptic curves over algebraic number fields. In: *Number theory* **II** (Budapest, 1987), 697–721, Colloq. Math. Soc. János Bolyai, 51, North-Holland, 1990.

[66] Frey, G.: Elliptische Kurven über bewerteten Körpern. Manuscript, Heidelberg.

[67] Frey, G.: Some remarks concerning points of finite order on elliptic curves over global fields. *Ark. Math.* **15**, 1–19, 1977.

[68] Frey, G.: Der Rang der Lösungen von $Y^2 = X^3 \pm p^3$ über $\mathbb{Q}$. *Manuscripta Math.* **48**, 71–101, 1984.

[69] Frey, G.: Some aspects of the theory of elliptic curves over number fields. *Exposition. Math.* **4**, 35–66, 1986.

[70] Frey, G., Rueck, H.-G.: A remark concerning m-divisibility and the discrete logarithm in the divisor class group of curves. *Math. Comp.* **62**, 865–874, 1994.

[71] Fricke, R.: *Die elliptischen Funktionen und ihre Anwendungen. Zweiter Teil: Die algebraischen Ausführungen*. Teubner Verlag, 1922.

[72] Fueter, R.: Über kubische diophantische Gleichungen. *Comment. Math. Helv.* **2**, 69–89, 1930.

[73] Fulton, W.: *Algebraic curves. An introduction to algebraic geometry.* Notes written with collab. of R. Weiss. New ed. Addison-Wesley, 1989.

[74] Fung, G. W., Ströher, H., Williams, H. C., Zimmer, H. G.: Torsion groups of elliptic curves with integral j-invariant over pure cubic fields. *J. Number Theory* **36**, 12–45, 1990.

[75] Gaál, I.: *Diophantine Equations and Power Integral Bases. – New Computational Methods*, Birkhäuser Verlag, 2002.

[76] Gantmacher, F. R.: *The theory of matrices I.* AMS Chelsea Publishing, 1998.

[77] Gebel, J., Pethő, A., Zimmer, H. G.: Computing integral points on elliptic curves. *Acta Arith.* **68**, 171–192, 1994.

[78] Gebel, J., Zimmer, H. G.: Computing the Mordell–Weil group of an elliptic curve over $\mathbb{Q}$. In: *Elliptic curves and related topics*. CRM Proc. Lecture Notes **4**, 61–83, 1994.

[79] Gebel, J.: *Bestimmung aller ganzen und S-ganzen Punkte auf elliptischen Kurven über den rationalen Zahlen mit Anwendung auf die Mordellschen Kurven.* Dissertation, Universität des Saarlandes, Saarbrücken, 1996.

[80] Gebel, J., Pethő, A., Zimmer, H. G.: On Mordell's equation. *Compositio Math.* **110**, 335–367, 1998.

[81] Gebel, J., Pethő, A., Zimmer, H. G.: Computing all S-integral points on elliptic curves. *Math. Proc. Cambridge Philos. Soc.* **127**, 383–402, 1999.

[82] Gibson, C. G.: *Elementary geometry of algebraic curves: an undergraduate introduction.* Cambridge University Press, 1998.

[83] Goldwasser, S., Kilian, J.: Almost all primes can be quickly certified. In: *Proc. 18th STOC,* 316–329, 1986.

[84] Gonzalez-Avilés, C. D.: On the conjecture of Birch and Swinnerton-Dyer. *Trans. Amer. Math. Soc.* **349**, 4181–4200, 1997.

[85] Greenberg, R.: On the Birch and Swinnerton-Dyer conjecture. *Invent. Math.* **72**, 241–265, 1983.

[86] Gross, B. H., Zagier, D. B.: Heegner points and derivatives of L-series. *Invent. Math.* **84**, 225–320, 1986.

[87] Győry, K.: Sur les polynômes à coefficients entiers et de discriminant donné III., *Publ. Mat. Debrecen* **23**, 141–165, 1976.

[88] Győry, K.: On the number of solutions of linear equations in units of an algebraic number field. *Comment. Math. Helv.* **54**, 585–600, 1979.

[89] Győry, K.: Bounds for the solutions of decomposable form equations, *Publ. Math. Debrecen* **52**, 1–31, 1998.

[90] Hajdu, L., Herendi, T.: Explicit bounds for the solutions of elliptic equations with rational coefficients. *J. Symbolic Comput.* **25**, 361–366, 1998.

[91] Hasse, H.: Zur Theorie der abstrakten elliptischen Funktionenkörper. III. Die Struktur des Meromorphismenringes. Die Riemannsche Vermutung. *J. Reine Angew. Math.* **175**, 193–208, 1936.

[92] Hasse, H.: *Number Theory.* Springer-Verlag, 1980.

[93] Heiser, A. J.: *Berechnung der Néron-Tate-Höhe auf elliptischen Kurven über quadratischen Zahlkörpern mit Anwendungen auf die Regulatorberechnung.* Diplomarbeit, Universität des Saarlandes, Saarbrücken, 1997.

[94] Herrmann, E.: *Bestimmung aller ganzzahligen Lösungen quartischer elliptischer diophantischer Gleichungen unter Verwendung von Linearformen in elliptischen Logarithmen.* Diplomarbeit, Universität des Saarlandes, Saarbrücken, 1998.

[95] Herrmann, E., Pethő, A.: S-integral points on elliptic curves - Notes on a paper of B. M. M. de Weger. *J. Théor. Nombres Bordeaux* **13**, 443–451, 2001.

[96] Herrmann, E.: *Bestimmung aller S-ganzen Punkte auf elliptischen Kurven.* Dissertation, Universität des Saarlandes, Saarbrücken, 2002.

[97] Heuberger, C.: On general families of parametrized Thue equations. In: *Algebraic Number Theory and Diophantine Analysis, Proc. Conf. Graz, 1998,*, 215–238, Eds.: F. Halter-Koch and R. F. Tichy, Walter de Gruyter, 2000.

[98] Heuß, J.: *Zum schwachen Endlichkeitssatz von Mordell.* Diplomarbeit, Mathematisches Institut II, Universität (TH) Karlsruhe, 1975.

[99] *Mathematical developments arising from Hilbert problems.* Proceedings of the Symposium in Pure Mathematics of the American Mathematical Society held at Northern Illinois University, De Kalb, Ill., May, 1974. Edited by Felix E. Browder. Proceedings of Symposia in Pure Mathematics, Vol. XXVIII. American Mathematical Society, 1976.

[100] Hollinger, C. S.: *Die Torsionsgruppe elliptischer Kurven mit ganzer j-Invariante über totalkomplexen biquadratischen Zahlkörpern.* Diplomarbeit, Universität des Saarlandes, Saarbrücken, 1990.

[101] Hungerford, T. W.: *Algebra.* Springer-Verlag, 1974.

[102] Hurwitz, A., Courant, R.: *Funktionentheorie*, 4. Aufl. Grundlehren Math. Wiss. 3, Springer-Verlag, 1964.

[103] Husemöller, D.: *Elliptic curves.* Grad. Texts in Math. 111, Springer-Verlag, 1987.

[104] Jacobson, M., Menezes, A., Stein, A.: Solving elliptic curve discrete logarithm problems using Weil descent. *J. Ramanujan Math. Soc.* **16**, 231–260, 2001.

[105] Kamienny, S.: Torsion points on elliptic curves. *Bull. Amer. Math. Soc.* **23**, 371–373, 1990.

[106] Kenku, M. A., Momose, F.: Torsion points on elliptic curves defined over quadratic fields. *Nagoya Math. J.* **109**, 125–149, 1988.

[107] Kida, M.: Galois descent and twists of an abelian variety. *Acta Arith.* **73**, 51–57, 1995.

[108] Kida, M.: Good reduction of elliptic curves over imaginary quadratic fields. *J. Théor. Nombres Bordeaux* **13**, 201–209, 2001.

[109] Knapp, A. W.: *Elliptic curves.* Math. Notes 40, Princeton University Press, 1992.

[110] Kolyvagin, V. A.: Euler systems. In: *Birkhäuser volume in honor of Grothendieck*, 435–493. Progr. Math. 87, Birkhäuser Verlag, 1990.

[111] Kubert, D. S.: Universal bounds on the torsion of elliptic curves. *Proc. London Math. Soc.* **33**, 193–237, 1976.

[112] Lang, S.: Diophantine approximation on toruses. *Amer. J. Math.* **86**, 521–533, 1964.

[113] Lang, S.: *Algebra.* Addison-Wesley, 1965.

[114] Lang, S.: *Algebraic Number Theory.* Addison-Wesley Publishing Co., Inc., Reading, MA, 1970.

[115] Lang, S.: *Elliptic Curves; Diophantine Analysis.* Grundlehren Math. Wiss. 231, Springer-Verlag, 1978.

[116] Lang, S.: *Complex Multiplication.* Springer-Verlag, 1983.

[117] Lang, S.: *Fundamentals of Diophantine Geometry.* Springer-Verlag, 1983.

[118] Lang, S.: *Elliptic Functions. With an appendix by J. Tate.* Second edition, Springer-Verlag, 1987.

[119] Laska, M.: An algorithm for finding a minimal Weierstrass equation for an elliptic curve. *Math. Comp.* **38**, 257–260, 1982.

[120] Laska, M.: *Elliptic curves over number fields with prescribed reduction type*. Vieweg Verlag, Aspects of Mathematics, 1983.

[121] Laska, M., Lorenz, M.: Rational points on elliptic curves over $\mathbb{Q}$ in elementary abelian 2-extensions of $\mathbb{Q}$. *J. Reine Angew. Math.* **355**, 163–172, 1985.

[122] Laurent, M., Mignotte, M., Nesterenko, Y.: Formes linéaires en deux logarithmes et déterminants d'interpolation. *J. Number Theory* **55**, 101–111, 1995.

[123] Lay, G.-J.: *Konstruktion elliptischer Kurve mit gegebener Gruppenordnung über endlichen Primkörpern*. Diplomarbeit, Universität des Saarlandes, Saarbrücken, 1994.

[124] Lay, G.-J., Zimmer, H. G.: Constructing elliptic curves with given group order over large finite fields. In: *ANTS-1: Algorithmic Number Theory '94*, 250–263. Lecture Notes in Comput. Sci. 877, Springer-Verlag, 1994.

[125] Lehmann, F., Maurer, M., Müller, F., Shoup, V.: Counting the number of points on elliptic curves over finite fields of characteristic greater than three. In: *ANTS-1: Algorithmic Number Theory'94*, 60–70. Lecture Notes in Comput. Sci. 877, Springer-Verlag, 1994.

[126] Lemmermeyer, F.: On Tate-Shafarevich groups of some elliptic curves. In: *Algebraic number theory and Diophantine analysis (Graz, 1998)*, 277–291. Walter de Gruyter, 2000

[127] Lemmermeyer, F.: *Reciprocity laws. From Euler to Eisenstein*. Springer Monogr. in Math., Springer-Verlag, 2000.

[128] Lenstra, A. K., Lenstra Jr., H. W., Lovász, L.: Factoring polynonials with rational coefficients. *Math. Ann.* **261**, 515–534, 1982.

[129] Lenstra Jr., H. W.: Factoring integers with elliptic curves. *Ann. of Math.* **126**, 649–673, 1987.

[130] Leutbecher, A.: *Zahlentheorie. Eine Einführung in die Algebra*. Springer-Verlag, 1996.

[131] Lind, C.-E.: *Untersuchungen über die rationalen Punkte der ebenen kubischen Kurven vom Geschlecht Eins*. Thesis, University of Uppsala, 1940.

[132] Lutz E.: Sur l'equation $y^2 = x^3 - Ax - B$ dans les corps p-adic. *J. Reine Angew. Math.* **177**, 237–247, 1937.

[133] Mahler, K.: Über die rationalen Punkte auf Kurven vom Geschlecht Eins. *J. Reine Angew. Math.* **170**, 168–178, 1934.

[134] Manin, Ju. I.: On cubic congruences to a prime modulus. (Russian) *Izv. Akad. Nauk SSSR. Ser. Mat.* **20**, 673–678, 1956.

[135] Manin, Ju. I.: The p-torsion of elliptic curves is uniformly bounded. *Izv. Akad. Nauk SSSR.* **33**, 459–465, 1969.

[136] Manin, Ju. I.: Cyclotomic fields and modular curves. *Russ. Math. Surveys* **26**, 7–78, 1971.

[137] Matiyasevič, Yu. V.: The Diophantineness of enumerable sets. (Russian) *Dokl. Akad. Nauk SSSR* **191**, 279–282, 1970.

[138] Matiyasevič, Yu. V., Robinson, J.: Reduction of an arbitrary Diophantine equation to one in 13 unknowns. *Acta Arith.* **27**, 521–553, 1975.

[139] Matiyasevič, Yu. V.: *Hilbert's tenth problem*. Translated from the 1993 Russian original by the author. With a foreword by Martin Davis. Found. Comput. Ser., MIT Press, 1993.

[140] Mazur, B.: Modular curves and the Eisenstein ideal. *Inst. Hautes Études Sci. Publ. Math.* **47**, 33–186, 1977.

[141] Mazur, B.: Rational isogenies of prime degree (with an appendix by D. Goldfeld). *Invent. Math.* **44**, 129–162, 1978.

[142] Menezes, A. J., Okamoto, T., Vanstone, S. A.: Reducing elliptic curve logarithms to logarithms in a finite field. *IEEE Trans. Inform. Theory* **39**, 1639–1646, 1993.

[143] Merel, L.: Bornes pour la torsion des courbes elliptiques sur les corps de nombres. *Invent. Math.* **124**, 437–449, 1996.

[144] Merel, L.: Le problème de la torsion des courbes elliptiques. In: *Doc. Math. Extra Vol. ICM 1998* **II**, 183–186, 1998.

[145] Merriman, J. R., Siksek, S., Smart, N. P.: Explicit 4-descents on an elliptic curve. *Acta Arith.* **77**, 385–404, 1996.

[146] Meyer, C.: Zur Theorie und Praxis der elliptischen Einheiten. *Ann. Univ. Sarav., Ser. Math.* **6**, 215–572, 1995.

[147] Mignotte, M.: *Mathematics for Computer Algebra*. Springer-Verlag 1992.

[148] Mordell, L. J.: On the rational solutions of the indeterminate equations of the third and fourth degrees. *Math. Proc. Cambridge Philos. Soc.* **21**, 179–192, 1922.

[149] Müller, H. H., Ströher, H., Zimmer, H. G.: Torsion groups of elliptic curves with integral j-invariant over quadratic fields. *J. Reine Angew. Math.* **397**, 100–161, 1989.

[150] Mumford, D.: *Abelian Varieties*. Oxford University Press, Oxford, 1974.

[151] Mumford, D.: *Algebraic geometry I: Complex projective varieties*. Reprint of the corr. 2nd print. Springer-Verlag, 1995.

[152] Nagell, T.: Solution de quelques problèmes dans la théorie arithmétique des cubiques planes du premier genre. *Wid. Akad. Skrifter Oslo I.* **1**, 1935.

[153] Nagell, T.: Les points exceptionnels sur les cubiques planes du premier genre. *Nova Acta Soc. Scient. Upsaliensis, Ser. IV,* **14**, 3-40, 1946.

[154] Nathanson, M. B.: *Additive Number Theory. The classical bases*. Grad. Texts in Math. 164, Springer-Verlag, 1996.

[155] Neukirch, J.: *Class field theory*. Grundlehren Math. Wiss. 280, Springer-Verlag, 1986.

[156] Ogg, A. P.: Elliptic curves and wild ramification. *Amer. J. Math.* **89**, 1–21, 1967.

[157] Neukirch, J.: *Algebraische Zahlentheorie*. Springer-Verlag, 1992.

[158] Neukirch, J., Schmidt, A., Wingberg, K.: *Cohomology of number fields*. Grundlehren Math. Wiss. 323, Springer-Verlag, 2000.

[159] Papadopoulos, I.: Sur la classification de Néron des courbes elliptiques en caractéristique résiduelle 2 et 3. *J. Number Theory* **44**, 119–152, 1993.

[160] Parent, P.: Bornes effectives pour la torsion des courbes elliptiques sur les corps de nombres. *J. Reine Angew. Math.* **506**, 85–116, 1999.

[161] Pethő A. und Schulenberg R.: Effektives Lösen von Thue Gleichungen. *Publ. Math. Debrecen* **34** 189–196, 1987.

[162] Pethő, A., Weis, T., Zimmer, H. G.: Torsion groups of elliptic curves with integral j-invariant over general cubic number fields. *Internat. J. Algebra Comput.* **7**, 353–413, 1997.

[163] Pethő, A., Zimmer, H. G.: On a system of norm-equations over cyclic cubic number fields. *Publ. Math. Debrecen* **53**, 317–332, 1998.

[164] Pethő, A., Zimmer, H. G., Gebel, J., Herrmann, E.: Computing all S-integral points on elliptic curves. *Math. Proc. Cambridge Philos. Soc.* **127**, 383–402, 1999.

[165] Pethő, A., Zimmer, H. G.: S-integer points on elliptic curves, theory and practice. In: *Algebraic Number Theory and Diophantine Analysis, Proc. Conf. Graz, 1998,* Eds.: F. Halter-Koch and R. F. Tichy, Walter de Gruyter, 351–363, 2000.

[166] Pethő, A., Schmitt, S.: Elements with bounded height in number fields. *Period. Math. Hungar.* **43**, 31–41, 2001.

[167] Pfeifer, M.: A boundedness theorem for the torsion of a class of elliptic curves over algebraic number fields. *Arch. Math.* **62**, 519–527, 1994.

[168] Pohlig, G. C., Hellmann, M. E.: An improved algorithm for computing logarithms over $GF(p)$ and its cryptographic significance. *IEEE Trans. Inform. Theory* **24**, 106–110, 1978.

[169] Pohst, M.: On the computation of number fields of small discriminants including the minimum discriminants of sixth degree fields. *J. Number Theory* **14**, 99–117, 1982.

[170] Pollard, J. M.: Monte Carlo methods for index computation $(\bmod\ p)$. *Math. Comp.* **32**, 918–924, 1978.

[171] Poonen, B.: An explicit algebraic family of genus-one curves violating the Hasse principle. *J. Théor. Nombres Bordeaux* **13**, 263–274, 2001.

[172] Quer, J.: *Sobra el* 3*-rang del cossos quadraàtics i la corba elliptica* $Y^2 = X^3 + M$. PhD thesis, Univ. Autònoma de Barcelona, 1987.

[173] Reichardt, H.: Einige im Kleinen überall lösbare, im Grossen unlösbare diophantische Gleichungen. *J. Reine Angew. Math.* **184**, 12–18, 1942.

[174] Reichert, M. A.: Explicit determination of nontrivial torsion structures of elliptic curves over quadratic number fields. *Math. Comp.* **46**, 637–658, 1986.

[175] Rémond, G., Urfels, F.: Approximation diophantienne de logarithmes elliptiques p-adiques. *J. Number Theory* **57**, 133–169, 1996.

[176] Rónyai, L.: Galois groups and factoring polynomials over finite fields. *SIAM J. Discrete Math.* **5**, 345–365, 1992.

[177] Roquette, P.: *Analytic theory of elliptic functions over local fields.* Vandenhoeck & Ruprecht, 1970.

[178] Rubin, K.: Elliptic curves with complex multiplication and the conjecture of Birch and Swinnerton-Dyer. *Invent. Math.* **64**, 455–470, 1981.

[179] Rubin, K.: Congruences for special values of L-functions of elliptic curves with complex multiplication. *Invent. Math.* **71**, 339–364, 1983.

[180] Rubin, K.: Tate-Shafarevich groups and L-functions of elliptic curves with complex multiplication. *Invent. Math.* **89**, 527–560, 1987.

[181] Rubin, K.: Elliptic curves with complex multiplication and the conjecture of Birch and Swinnerton-Dyer. In: *Arithmetic Theory of Elliptic Curves, Cetraro, Italy*, 1997. Lecture Notes in Math. 1716, Springer-Verlag, 1997.

[182] Rück, H.-G.: A note on elliptic curves over finite fields. *Math. Comp.* **49**, 301–304, 1987.

[183] Sato, A.: The Behavior of Mordell–Weil Groups under Field Extensions. Preprint.

[184] Satoh, T., Araki, K.: Fermat quotients and the polynomial time discrete log algorithm for anomalous elliptic curves. *Comment. Math. Univ. St. Paul.* **47**, 81–92, 1998.

[185] Schertz, R.: Die singulären Werte der Weberschen Funktionen $\mathfrak{f}$, $\mathfrak{f}_1$, $\mathfrak{f}_2$, γ_2, γ_3. *J. Reine Angew. Math.* **286/287**, 46–74, 1976.

[186] Schmal, B.: Diskriminanten, $\mathbb{Z}$-Ganzheitsbasen und relative Ganzheitsbasen bei multiquadratischen Zahlkörpern. *Arch. Math.* **52**, 245–257, 1989.

[187] Schmal, B.: Diskriminanten, $\mathbb{Z}$-Ganzheitsbasen und relative Ganzheitsbasen bei Komposita von Radikalerweiterungen vom Grad p über $\mathbb{Q}$. Dissertation, Universität des Saarlandes, Saarbrücken, 1991.

[188] Schmidt, F. K.: Analytische Zahlentheorie in Körpern der Charakteristik p. *Math. Z.* **33**, 1–32, 1931.

[189] Schmitt, S.: *Bestimmung der Mordell–Weil Gruppe elliptischer Kurven über algebraischen Zahlkörpern.* Dissertation, Universität des Saarlandes, Saarbrücken, 1999.

[190] Schmitt, S.: *A conditional algorithm for the computation of the rank of elliptic curves over quadratic number fields.* Preprint.

[191] Schneiders, U.: *Rangverhalten elliptischer Kurven beim Übergang vom rationalen Zahlkörper zu quadratischen Erweiterungen.* Diplomarbeit, Universität des Saarlandes, Saarbrücken, 1990.

[192] Schneiders, U., Zimmer, H. G.: The rank of elliptic curves upon quadratic extension. In: *Computational number theory (Debrecen, 1989)*, 239–260, Walter de Gruyter, 1991.

[193] Schoof, R.: Elliptic curves over finite fields and the computation of square roots mod p. *Math. Comp.* **44**, 483–494, 1985.

[194] Schoof, R.: Counting points on elliptic curves over finite fields. *J. Théor. Nombres Bordeaux* **7**, 219–254, 1995.

[195] Selmer, E. S.: The Diophantine equation $ax^3 + by^3 + cz^3 = 0$. *Acta Math.* **85**, 203–362, 1951.

[196] Selmer, E. S.: The Diophantine equation $ax^3 + by^3 + cz^3 = 0$. Completion of the tables. *Acta Math.* **92**, 191–197, 1954.

[197] Serf, P.: *The rank of elliptic curves over real quadratic number fields of class number* 1. Dissertation, Universität des Saarlandes, Saarbrücken, 1995.

[198] Serre, J.-P.: *Lectures on the Mordell-Weil Theorem.* Translated from the French and edited by Martin Brown from notes by Michel Waldschmidt. Aspects of Mathematics, E15. Friedr. Vieweg & Sohn, Braunschweig/Wiesbaden, 1989.

[199] Shafarevich, I. R.: *Basic algebraic geometry 1: Varieties in projective space.* Transl. from the Russian by Miles Reid. 2nd, rev. and exp. ed., Springer-Verlag, 1994.

[200] Shorey, T. N., Tijdeman, R.: *Exponential Diophantine Equations*. Cambridge University Press, 1986.

[201] Siegel, C. L.: Über einige Anwendungen diophantischer Approximationen. *Abh. Preuss. Akad. Wiss.*, 1–41, 1929.

[202] Siegel, C. L.: *Lectures on the geometry of numbers*. Springer-Verlag, 1988.

[203] Siksek, S.: Infinite descent on elliptic curves. *Rocky Mountain J. Math.* **25**, 1501–1538, 1995.

[204] Silverman, J. H.: *The arithmetic of elliptic curves*. Graduate Texts in Math. 106, Springer-Verlag, 1985.

[205] Silverman, J. H.: Computing heights on elliptic curves. *Math. Comp.* **51**, 339–358, 1988.

[206] Silverman, J. H.: The difference between the Weil height and the canonical height on elliptic curves. *Math. Comp.* **55**, 723–743, 1990.

[207] Silverman, J. H.: *Advanced topics in the arithmetic of elliptic curves*. Grad. Texts in Math. 151, Springer-Verlag, 1994.

[208] Silverman, J. H.: Computing canonical heights with little (or no) factorization. *Math. Comp.* **66,** 787–805, 1997.

[209] Silverman, J. H.: Computing rational points on rank 1 elliptic curves via L-series and canonical heights. *Math. Comp.* **68**, 835–858, 1999.

[210] Simon, D.: Computing the rank of elliptic curves over number fields. Preprint.

[211] Smart, N. P.: S-integral points on elliptic curves. *Math. Proc. Cambridge Philos. Soc.* **116**, 391–399, 1994.

[212] Smart, N. P., Stephens, N. M.: Integral points on elliptic curves over number fields. *Math. Proc. Cambridge Philos. Soc.* **122**, 9–16, 1997.

[213] Smart, N. P.: *The algorithmic resolution of diophantine equations*. London _math. Soc. Student Texts 41, Cambridge University Press, 1998.

[214] Smart, N. P.: The discrete logarithm problem on elliptic curves of trace one. *J. Cryptology* **12**, 193–196, 1999.

[215] Stein, J.: *Die Torsionsgruppe elliptischer Kurven mit ganzer j-Invariante über total-reellen biquadratischen Zahlkörpern*. Diplomarbeit, Universität des Saarlandes, Saarbrücken, 1994.

[216] Stroeker, R. J.: Aspects of elliptic curves; an introduction. *Nieuw Arch. Wisk.* (3) **26,** 371–412, 1978.

[217] Stroeker, R. J., Tzanakis, N.: Solving elliptic diophantine equations by estimating linear forms in elliptic logarithms. *Acta Arith.* **67**, 177–196, 1994.

[218] Swinnerton-Dyer, H. P. F.: The conjectures of Birch, Swinnerton-Dyer, and of Tate. In: *Proceedings of a Conference on Local Fields*, 132–157, 1966.

[219] Tate, J.: On the conjectures of Birch and Swinnerton-Dyer and a geometric analog. In: *Séminaire Bourbaki* **306**, 1966.

[220] Tate, J.. Algorithm for determining the type of a singular fiber in an elliptic pencil. In: *Modular functions of one variable* **IV**, 33–52. Lecture Notes in Math. 476, Springer-Verlag, 1975.

[221] Taylor, R., Wiles, A.: Ring-theoretic properties of certain Hecke algebras. *Ann. Math.* **141**, 553–572, 1995.

[222] Thome, A.: Die Existenz von Ganzheitsbasen bei endlichen separablen Erweiterungen von Dedekindringen. Dissertation, Universität des Saarlandes, Saarbrücken, 1986.

[223] Thue, A.: Über Annäherungswerte algebraischer Zahlen. *J. Reine Angew. Math.* **135**, 284–305, 1909.

[224] Tijdeman, R.: Applications of the Gel'fond-Baker method to rational number theory. *Topics in Number Theory, Proc. Conf. Debrecen 1974, Collog. Math. Soc. János Bolyai 13*, 399–416, North-Holland, 1976.

[225] Tschöpe, H. M., Zimmer, H. J.: Computation of the Néron-Tate height on elliptic curves. *Math. Comp.* **48,** 351–370, 1987

[226] Tzanakis, N.: Effective solution of two simultaneous Pell equations by the elliptic logarithm method. *Acta Arith.* **103**, 119–135, 2002.

[227] Vessis, T.: *Points of finite order on elliptic curves*. Thesis, Iraklio 1995.

[228] Villegas, F. R., Zagier, D.: Which primes are sums of two cubes? In: *Number theory*, 295–306, CMS Conf. Proc. 15, Amer. Math. Soc., 1985.

[229] van der Waerden, B. L.: *Algebra*, Vol. I. Springer-Verlag, 1971.

[230] Waterhouse, W. C.. Abelian varieties over finite fields. *Ann. Sci. École Norm. Sup.* **2**, 521–560, 1969.

[231] Weber, H. (1908). *Algebra III*. Vieweg, 1908.

[232] de Weger, B. M. M.: *Algorithms for diophantine equations*. Centrum Wisk. Inform., CWI Tract 65, 1989.

[233] de Weger, B. M. M.: S-integral solutions to a Weierstraß equation. *J. Théor. Nombres Bordeaux* **9**, 281–301, 1997.

[234] Weil, A.: L'arithmétique sur les courbes algébriques. *Acta Math.* **52**, 281–315, 1928.

[235] Weil, A.: Dirichtlet Series and Automorphic Forms. Springer-Verlag, Lecture Notes in Math. 189, 1971.

[236] Weis, T.: *Die Torsionsgruppe elliptischer Kurven mit ganzer j-Invariante über kubischen Zahlkörpern*. Diplomarbeit, Universität des Saarlandes, Saarbrücken, 1992.

[237] Wiles, A.: Modular elliptic curves and Fermat's last theorem. *Ann. of Math.* **141**, 443–551, 1995.

[238] Williams, K. S.: Integers of biquadratic fields. *Canad. Math. Bull.* (4) **13**, 519–526, 1970.

[239] Yoshida, S.: On the equation $y^2 = x^3 + pqx$. *Comment. Math. Univ. St. Paul.* **49**, 23–42, 2000.

[240] Yu, K. R.: Linear forms in p-adic logarithms. *Acta Arith.* **53**, 107–186, 1989.

[241] Yui, N., Zagier, D.: On the singular values of Weber modular forms. *Math. Comp.* **66**, 1645–1662, 1997.

[242] Zagier, D.: *Zetafunktionen und quadratische Körper*. Springer-Verlag, 1981.

[243] Zagier, D.: Large integral points on elliptic curves. *Math. Comp.* **48**, 425–436, 1987.

[244] Zagier, D.: The Birch-Swinnerton-Dyer conjecture from a naive point of view. In: Arithmetic algebraic geometry (Texel, 1989), 377–389, Progr. Math. 89, Birkhäuser Verlag, 1989.

[245] Zagier, D.: Elliptische Kurve. *Jber. Deutsch. Math. Verein.* **92,** 58–76, 1990.

[246] Zassenhaus, H.: On Hensel Factorization, I. *J. Number Theory* **1**, 291–311, 1969.

[247] Zimmer, H. G.: An elementary proof of the Riemann hypothesis for an elliptic curve over a finite field. *Pacific J. Math.* **36**, 267–278, 1971.

[248] Zimmer, H. G.: Ein Analogon des Satzes von Nagell-Lutz über die Torsion einer elliptischen Kurve. *J. Reine Angew. Math.* **268/269,** 360–378, 1974.

[249] Zimmer, H. G.: On the difference of the Weil height and the Néron-Tate height. *Math. Z.* **147**, 35–51, 1976.

[250] Zimmer, H. G.. Generalization of Manin's Conditional Algorithm. In: *Proc. of the 1976 ACM Symp. on Symbolic and Algebraic Computation*, 285–299, 1976.

[251] Zimmer, H. G.: Torsion points on elliptic curves over a global field. *Manuscripta Math.* **29**, 119–145, 1979.

[252] Zimmer, H. G.: Quasifunctions on elliptic curves over local fields. *J. Reine Angew. Math.* **307/308**, 221–246, 1979.

[253] Zimmer, H. G.: Correction and remarks concerning: "Quasifunctions on elliptic curves over local fields" *J. Reine Angew. Math.* **343**, 203–211, 1982.

[254] Zimmer, H. G.: *Zur Arithmetik der Elliptischen Kurven.* Bericht der Math.-Stat. Sektion der Forschungsges. Joanneum, 271, 1986.

[255] Zimmer, H. G.: Computational aspects of the theory of elliptic curves. In: *Number Theory and Applications, Banff Proceedings*, NATO ASI Ser., Ser. C 265, R. A. Mollin (ed.), 279–324, 1988.

[256] Zimmer, H. G.: A limit formula for the canonical height of an elliptic curve and its application to height computations. In: *Number theory (Banff)*, 641–659, 1990.

[257] Zimmer, H. G.: Torsion groups of elliptic curves over cubic and certain biquadratic number fields. *Contemp. Math.* **174**, 203–220, 1994.

[258] Zimmer, H. G.: Height functions on elliptic curves. In: *Public-key Cryptography and Computational Number Theory*, 303–322. Walter de Gruyter, 2001.

Index

CPSIA information can be obtained
at www.ICGtesting.com
Printed in the USA
BVHW060744100519
547884BV00011B/259/P